Our Genes

Situated at the intersection of natural scienc   philosophy, *Our Genes* explores historical practices, investigates current trends, and imagines future work in genetic research to answer persistent, political questions about human diversity. Readers are guided through fascinating thought experiments, complex measures and metrics, fundamental evolutionary patterns, and in-depth treatment of exciting case studies. The work culminates in a philosophical rationale, based on scientific evidence, for a moderate position about the explanatory power of genes that is often left unarticulated. Simply put, human evolutionary genomics – our genes – can tell us much about who we are as individuals and as collectives. However, while they convey scientific certainty in the popular imagination, genes cannot answer some of our most important questions. Alternating between an up-close and a zoomed-out focus on genes and genomes, individuals and collectives, species and populations, *Our Genes* argues that the answers we seek point to rich, necessary work ahead.

RASMUS GRØNFELDT WINTHER is a philosopher of science, researcher, writer, educator, diver, and explorer. He is Professor of Humanities at University of California, Santa Cruz and Affiliate Professor of Transformative Science at the GLOBE Institute at University of Copenhagen.

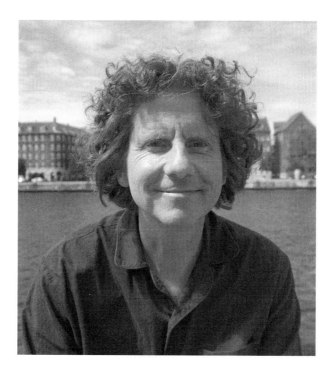

Photo credit: Marie Raffn

Our Genes

A Philosophical Perspective on Human Evolutionary Genomics

RASMUS GRØNFELDT WINTHER

Professor of Humanities at University of California, Santa Cruz

Affiliate Professor of Transformative Science at the GLOBE Institute at University of Copenhagen

CAMBRIDGE
UNIVERSITY PRESS

CAMBRIDGE
UNIVERSITY PRESS

University Printing House, Cambridge CB2 8BS, United Kingdom

One Liberty Plaza, 20th Floor, New York, NY 10006, USA

477 Williamstown Road, Port Melbourne, VIC 3207, Australia

314–321, 3rd Floor, Plot 3, Splendor Forum, Jasola District Centre, New Delhi – 110025, India

103 Penang Road, #05–06/07, Visioncrest Commercial, Singapore 238467

Cambridge University Press is part of the University of Cambridge.

It furthers the University's mission by disseminating knowledge in the pursuit of education, learning, and research at the highest international levels of excellence.

www.cambridge.org
Information on this title: www.cambridge.org/9781107170407
DOI: 10.1017/9781316756324

First published 2022

Printed in the United Kingdom by TJ Books Limited, Padstow Cornwall

A catalogue record for this publication is available from the British Library.

ISBN 978-1-107-17040-7 Hardback
ISBN 978-1-316-62150-9 Paperback

For Us.

And with extreme gratitude to
Aage Bisgaard Winther, Helen Longino, and Amir Najmi.

Contents

Color plates can be found between pages 210 and 211.

Preface

Our Genes would not exist without Helen Longino's push to collect and revise some of my papers, and to write new material. Nor would it exist without constant dialogue with statistician and polymath Amir Najmi, nor without Aage Bisgaard Winther's gentle paternal nudging and multifarious support.

The origins of my thinking about these matters can be discerned in the acknowledgments of the published papers that served as springboards for various chapters. Here I wish to focus on interlocutors who have helped with this book, with *Our Genes*, in particular. The evolutionary genomics material has been made as strong as possible by dialogue with various geneticists and statisticians: Michael "Doc" Edge, A.W.F. Edwards, Marcus Feldman, Ryan Giordano, Melissa Ilardo, Richard C. Lewontin, Søren Mørk, Amir Najmi, Rasmus Nielsen, John Novembre, Noah Rosenberg, Mark Thompson, Michael J. Wade, and Eske Willerslev. Philosophers of science with whom I've discussed the relevant technical and general topics over the years include John Dupré, Sébastien Dutreuil, Claus Emmeche, Peter Godfrey-Smith, Christopher D. Green, James Griesemer, Ian Hacking, Philippe Huneman, Jonathan M. Kaplan, Elisabeth A. Lloyd, Helen Longino, Carlos López Beltrán, Roberta Millstein, Richard Otte, Hans-Jörg Rheinberger, Alex Rosenberg, Elliott Sober, and Denis Walsh. No book gets written without "invisible support" – friends with whom one can discuss the ebbs and flows, the chaos and order, involved in building a book. For their ears and for their words of wisdom, I am especially grateful to Marie Raffn, Ann Lipson, Céline Malraux, Marco del Seta, Mette Bannergaard Johansen, Finn Bannergaard Johansen, Heidi Svenningsen Kajita, Annika Döring, Lars Friis Mikkelsen, Jácome "Jay" Armas, Annette Spicker Bruhn, Edward Lipson, April Snøfrid Kleppe, Jasmine Alinder, Brian Cantwell Smith, Marit Liv Miners, Andrew Miners, and Tom Ryckman. Laura Laine and Mats Wedin assisted with the figures; Lucas McGranahan and Gloria Sturzenacker expertly edited various chapters;

Molly Gage gave the whole project a magnificent, deep edit; Amy Marks and Gary Smith skillfully copy edited the final manuscript; and Helmut Filacchione helped compile the index. Also wonderful was the team at Cambridge University Press: Katrina Halliday, Jenny van der Meijden, Aleksandra Serocka, and Susan Francis. Artists who produced images and served as sounding boards for the visual aspect of the book were Marie Raffn, Pablo Carlos Budassi, and Larisa DePalma. Finally, University of California, Santa Cruz, Universidad Nacional Autónoma de México, Stanford University, and University of Copenhagen, especially the Niels Bohr Institute, are the institutions where I have been graciously afforded ample opportunity to develop many of the ideas in this book.

In completing this book, now my fourth, I have come to appreciate an analogy. When you have multiple children, I am told that you love each of them intensely, in unique ways. I have learned that this certainly holds for your books – you love each in its own way. Writing books and raising children present similar challenges and gifts. Near-infinite amounts of attention, patience, and unconditional love are required. Ultimately, just as you are forever changed in raising a child, so are you in writing a book. Both processes lead to the interactive emergence of self-knowledge and self-awareness. I can only hope that those of us lucky enough to receive these gifts can make use of them in a mindful way, putting our acquired consciousness to good use far beyond our engagement with our families and with our books.

Original Sources

This book was initially based on a number of articles that were substantively revised and pared down, and significant amounts of new material was added. All reprint permissions have been secured from the publishers, and all coauthors have given their permission.

Chapters 2 and 8: Rasmus Grønfeldt Winther. 2019. A Beginner's Guide to the New Population Genomics of *Homo sapiens*: Origins, Race, and Medicine. *The Harvard Review of Philosophy* 26: 135–151.

Chapter 3: Rasmus Grønfeldt Winther, Ryan Giordano, Michael D. Edge, and Rasmus Nielsen. 2015. The Mind, the Lab, and the Field: Three Kinds of Populations in Scientific Practice. *Studies in History and Philosophy of Science Part C: Studies in History and Philosophy of Biological and Biomedical Sciences* 52: 12–21.

Chapter 4: Jonathan M. Kaplan and Rasmus Grønfeldt Winther. 2013. Prisoners of Abstraction? The Theory and Measure of Genetic

Variation, and the Very Concept of "Race." *Biological Theory* 7(4): 401–412 (reprinted and adapted, with permission, from *Biological Theory*, volume 7 special issue, The Meaning of "Theory" in Biology. © Springer Nature 2012); and Rasmus Grønfeldt Winther and Jonathan M. Kaplan. 2013. Ontologies and Politics of Biogenomic "Race." *Theoria. A Journal of Social and Political Theory* 60(136): 54–80.

Chapter 5: Rasmus Grønfeldt Winther. 2014. The Genetic Reification of "Race"? A Story of Two Mathematical Methods. *Critical Philosophy of Race* 2(2): 204–223 (reprinted and adapted, with permission from The Pennsylvania State University Press).

Chapter 6: Rasmus Grønfeldt Winther. 2018. "Race and Biology." In *The Routledge Companion to the Philosophy of Race*, ed. P. Taylor, L. Alcoff, and L. Anderson. London: Routledge, pp. 305–320.

Chapter 9: Jonathan M. Kaplan and Rasmus Grønfeldt Winther. 2014. Realism, Antirealism, and Conventionalism about Race. *Philosophy of Science*, 81(5): 1039–1052 (copyright 2014 by the Philosophy of Science Association).

Chapter 10: Rasmus Grønfeldt Winther. 2014. Determinism and Total Explanation in the Biological and Behavioral Sciences. *eLS*. Available: https://onlinelibrary.wiley.com/doi/10.1002/9780470015902.a0024143

A Note on the Graphics

Although color graphics are often desirable, we have aimed to simplify production and keep costs down for readers. Therefore a selection of figures has been included in color, either in a color plate section (with black and white versions of all figures in the running text) or in situ dependent on format.

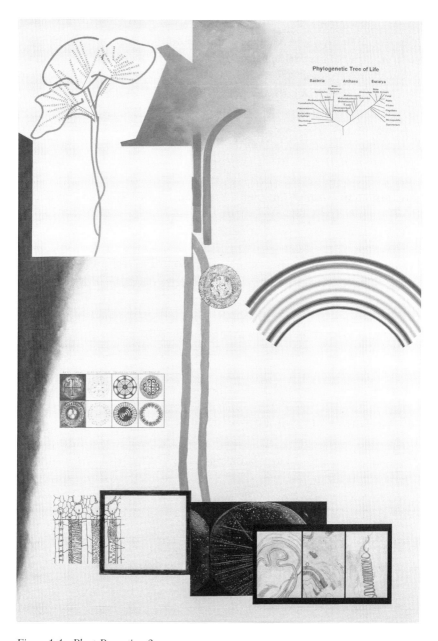

Figure 1.1. Plant Perception 2
A map of the perception of plants, including a cellular-level intersection of a plant stem (center) and a phylogenetic tree (based on work by Carl Woese, who used 16S ribosomal RNA to discover a new kingdom, or domain, of microorganisms – the Archaea). Shared ancestry and genetic hereditary systems connect the entire tree

1 · *Introduction*

As humans, we have a graspable identity that has been shaped by our individual and collective attempts to seek answers to fundamental questions: *Who am I? Where did I come from? Where am I going? Where and to whom do I belong, and how can I help myself and others?* The persistence of such questions may signal the insatiability of our human curiosity, but they also offer evidence of our possibly endless search for substantive, finite meaning. We yearn to identify who we are and to be part of something greater than our own limited individuality. This desire leads people to draw strong, even vicious, us-versus-them boundaries in political and social life. But it is also a spiritual wish to connect to all of humanity, indeed to all of life and the cosmos, and to take benevolent action accordingly (Figure 1.1, chapter opener). Even when answers to our questions about identity prove inconclusive, changeable, or otherwise unsatisfying, our search continues apace.

In our quest for identity, we often discover that in addition to being an active and intentional *subject*, we are perhaps as much a passive *object* dancing to the tune of greater forces of many kinds – familial, social, political, ecological, and even spiritual. In fact, the search for identity and

---

Caption for Figure 1.1. (*cont.*) of life, not just eukaryotes. The rainbow represents light, photosensitivity, and how plants can orient themselves by identifying colors. On the left is a schematic image of electromagnetic patterns in the development of a vegetable; in *Clavis Medicinae Duplex*, Carl Linnaeus concludes that all life consists of bark, marrow, and electricity. The depictions of xylem and phloem, osmosis, and cell morphology at the bottom reflect Nikola Tesla's attempts to decipher electrical brain waves. Ödlund keeps a black frame empty, opening our imagination for speculation on connection, intelligence, and the possibility of communication with other life forms – the privilege of the artist. (Co-written with Ödlund.) *Plant Perception 2* by Christine Ödlund (2015), painting. Copyright by the artist. Photo by Christian Saltas. Reprinted with permission. (A black and white version of this figure will appear in some formats. For the color version, please refer to the plate section.)

belonging can become especially urgent when these greater forces for maintaining identity are weakened. We see this in the worker who has lost her job, the refugee who has lost his state, or in the individuals who are discriminated against due to basic sexuality or ethnicity. Threats to our identities force us to a radical outside, like a fish out of water.

Who am I? This fundamental question is so central to religious and philosophical traditions across the globe because it suggests discoverable origins: *Where did my talents, skills, and weaknesses come from? Where did my ancestors, and my family, come from? Which group(s), not to mention ethnicities or tribes, might have a claim on me? Why do I have the body I have?* It also suggests fungibility and futurity: *Can I change? Can I freely alter any of the talents, skills, and weakness that originated within me? Now? Later?* So many of us, across varied cultural contexts, have sought answers both in and outside ourselves, in appeals to the soul, supernatural creation, or human nature. We look to religion, to philosophy, and to other sources for the possible causal powers and explanations we seek.[1]

Of course, we also look to science. Indeed, some of the most fecund and provocative answers to our questions of origins have been pursued within the natural sciences. In the West, the field of natural science dates back to the Greeks and the very origin of reason and philosophy, but it was rebooted in the seventeenth century, when the founding of the British and French academies of sciences in the 1660s helped usher in the Scientific Revolution. Physics, chemistry, and physiology were reborn and became newly applicable to humanity's age-old questions. Discoveries and inventions furthered our search. In England, Isaac Newton (1643–1727) articulated the universal law of gravitation and was a co-developer of calculus; Robert Boyle (1627–1691) and Robert Hooke (1635–1703) invented the air pump to create a partial vacuum and thereby started articulating the physical laws of heat and work; and William Harvey (1578–1657) discovered closed blood circulation in the body of many animals, including humans. Scientific theory and experiment took up new meanings. Mathematics became the theoretical gold standard for developing explanations, predictions, and understanding. Controlled and idealized experiments, with the air pump as a paradigm, came to be seen as central to acquiring scientific knowledge.[2]

---

[1] A useful text is Partridge (2018).
[2] See Shapin and Schaffer (1985), Shapin (1996), and Dear (2019).

Natural science, in both content and method,[3] has proven effective in providing the platform and tools for understanding how the world works. By measuring objects and processes of many kinds, organizing and managing data, and postulating and abstracting out hypotheses, models, and theories, science builds integrated best guesses of why and how the causal swirls within and outside us happen. Background theory and empirical information feed on each other, and novelty and discovery emerge. Science is partial, dynamic, and a communal effort of inquiry. We use it to answer questions of all kinds, including those that concern us about ourselves and each other.

However, philosophy also supplies an avenue for pursuing answers to our questions of identity. Viewed as the practice of asking big questions about the nature of knowledge, reality, and human life, and as a set of methodologies for answering such questions, philosophy may not provide answers, but it certainly helps clarify our questions. As a historical intellectual tradition, it, too, can be traced in the West back to ancient Greece, where it was inseparable from science. Among the important questions posed by ancient and modern philosophers alike, philosophy helps us to ask and answer these questions:

- Why is there something rather than nothing?
- Are ideas real? And if so, how so?
- What is human nature? Is it anything?
- How is knowledge justified and tested?
- What is the relation between theory and experiment, and theory and data?
- What is it to be rational?
- How might thoughts and feelings relate?
- What is happiness? What is wisdom? What is suffering?
- What do we owe to each other? To nonhuman animals? To nature?
- What is the role of science in a just and fair society?
- Are we free? What should we do with our freedom?
- Does God or gods or any kind of spiritual dimension of reality exist? How might spirituality influence us?

These are not questions easily solved: Nearly all of us think about some of them occasionally – or we almost certainly did as children or teenagers, even if we now keep them in a glass jar in the back of our heart's closet.

---

[3] See, e.g., Hacking (2002), Longino (2002), and Winther (2012a, 2020a).

Some of us study these questions professionally. We see philosophy as a medium by which to ask questions, and by which to trouble or nuance our answers. Why? Because, in the words of John Dewey, "philosophy is criticism":

criticism of the influential beliefs that underlie culture; a criticism which traces the beliefs to their generating conditions as far as may be, which tracks them to their results, which considers the mutual compatibility of the elements of the total structure of beliefs.[4]

When asking big questions, we return to the fundamental issues of our existence and our identity. When we ask big questions critically, we refuse to automatically accept the answers handed to us by tradition, society, or family. As British philosopher Bertrand Russell said:

Philosophy, though unable to tell us with certainty what is the true answer to the doubts which it raises, is able to suggest many possibilities which enlarge our thoughts and free them from the tyranny of custom. Thus, while diminishing our feeling of certainty as to what things are, it greatly increases our knowledge as to what they may be; it removes the somewhat arrogant dogmatism of those who have never travelled into the region of liberating doubt, and it keeps alive our sense of wonder by showing familiar things in an unfamiliar aspect.[5]

Far from existing cleaved from natural science, philosophical questions help drive the development of science. Albert Einstein (1879–1955) asked penetrating questions about space and time, energy, and matter as he developed his two relativity theories.[6] Isaac Newton's conceptual questions about what makes for a proper and effective theory, and what confirms a theory experimentally, were central to his own mechanistic project and to his great influence during the Scientific Revolution and beyond. In fact, the philosophical cadence of *Who am I?* and related questions about identity, origins, futurity, and fungibility have a special relation to the sciences of life and mind. In addition to the variety of creation stories about how humans came to be, and to the religious narratives that seek to explain human nature, such questions have explicitly driven work on evolution and genetics. Questions about our features, our personalities, and our predilections, and about the nature and dynamics of the groups and collectives to which we each belong in

---

[4] Dewey (1985 [1931], p. 19).   [5] Russell (1997 [1912], p. 157).
[6] See Galison (2003) and Ryckman (2017).

multiple ways, are central to the birth and progress of evolutionary theory from the nineteenth century onward.[7] Indeed, human evolutionary genomics offers a particularly rich, if also necessarily limited, way to explore philosophical questions about why we – and not just *we*, but other life forms, as well – are as we are, and why we desire what we desire.

★★★

This book sits at the crossroads of natural science and philosophy. In it, I zoom in on crucial causal objects that help co-make who we are – genes. I argue that the field of human evolutionary genomics in general, and the study of genes in particular, permits researchers, critical thinkers, and the lay public to integrate and substantiate philosophical questions about identity and collectivity within the natural sciences. After all, genetics and genomics ultimately concern processes of the emergence, as well as the potentials and limits, of our bodies, our minds, and our selves. Through the study of genes, we learn more about who and what we are, and who and what we are not.

The scientists who brought forth genetics and statistics in the nineteenth and twentieth centuries were, like us, motivated by their curiosity about questions of origins and evolution. They were also, like us, driven by a political desire to intervene in society. Early men of genetics seem to have assumed the superiority of a specific class of educated, moneyed, white male Anglo, as well as the ubiquity of strongly heritable differences in physical features and cognitive capacities among individuals. Their work reflects that. Some, especially Francis Galton (1822–1911) and his protégé (and a committed socialist) Karl Pearson (1857–1936), promoted eugenic discoveries that proved profoundly problematic in ideation and execution.[8] Others, like strongly left-wing J.B.S. Haldane (1892–1964),

---

[7] Although *evolutionary genetics* and *evolutionary genomics* are used somewhat interchangeably throughout the book, they mean different things. Whereas evolutionary genetics (or population genetics) is more the mathematical evolutionary theory first developed by, especially, R.A. Fisher, Sewall Wright, and J.B.S. Haldane and integrated into the "modern synthesis" of Darwinian evolutionary theory and chromosomal, material genetics by their less mathematically inclined colleagues (e.g., Theodosius Dobzhansky, Julian Huxley, Ernst Mayr, and George Gaylord Simpson), evolutionary genomics is more about the data-driven knowledge surrounding demographic, genealogical, forensic, and medical applications of contemporary genomics (see the glossary at the end of Chapter 2).

[8] As just one peek into their contributions to the history of statistics, Galton discovered, not to say invented, the very concept of *correlation* between two statistical variables, and Pearson developed the chi-square test of statistical significance.

worried about the medical promises and perils of genetics. It cannot be forgotten that eugenic views, so prevalent among geneticists until roughly World War II, cut across the political spectrum, from left to right.[9] In our age of hopeful diversity and equality, we justly find many of these pursuits morally suspect, if not reprehensible. Yet our contemporary investigations into the discoveries these researchers made offer insight into the questions we continue to ask of science and of philosophy about the origin and nature of body and mind, over evolutionary time and over our individual and collective lifetimes.

Biological scientists and practitioners like those mentioned above were as obsessed with questions of identity, of ancestry, and of futurity as we are today. They sought answers not in religion or philosophy, but in measurable analyses, especially as applied to questions about nature versus nurture. Statistics offered a stable, operational tool with which to locate continuity and ascertain the future. In the hands of R.A. Fisher (1890–1962), in particular, efforts to hone statistics through interpretation and application resulted in the development of evolutionary genetics. Of course, we have learned much since this time. In particular, we have learned (and continue to learn) that statistics cannot fix the future, in the sense of either prediction or repair. It is a tool, and as such it has and will always be used to serve explicitly political, and therefore limited, ends.

Nonetheless, statistics helps us negotiate the space between what is measurable and what is unmeasurable. This is to say, statistics constructs one bridge between the meaningful and knowable on one side *and* the meaningless or unknowable on the other. Its earliest inventors and practitioners likely grasped this, applying statistics to the space and links between similarities and differences, individuals and populations, and past, present, and future, to find answers to their most persistent questions. The answers yielded by statistics are necessarily partial and therefore fraught. In fact, the essential limitations of statistics informed the field of evolutionary genetics: Even today, evolutionary genetics is frequently understood in terms of a history theorizing and emphasizing eugenics, race, and IQ. The so-called IQ wars, which made evolutionary genetics culturally important, stand as a case in point: In 1969, on the political right, Arthur Jensen (1923–2012) argued for *intelligence* as a legitimate scientific concept and trait, of which genetic variance both across and

---

[9] See Paul (1984).

within populations is highly explanatory and predictive. In the early 1970s, Richard Charles Lewontin (1929–2021), and later, Stephen Jay Gould (1941–2002), responded from the left, arguing for the effective irrelevance of genetic variance to intelligence. Meanwhile, *intelligence* itself came to be seen as a problematic and ambiguous property and process to define. The IQ wars raged half a century ago, but the debates, which continue to this day, as we shall see in Chapters 8 and 9, help illustrate the limits of the field.

Evolutionary genetics is not just a limiting lens, however. We can also use genetics and genomics – fields dedicated to human variance and difference – to *expand* rather than contract our knowledge of human identity and origins, particularly in terms of understanding community, connectivity, and collective and individual potentials. When released from the political agendas in which they are often trapped, genetics and genomics, informed by statistics, can tell us incredible things about ourselves and other species. We can learn about the past through the study of our ancestral populations; we can learn about the present through the study of our individuated and yet common experience with our bodies and minds, their presentations, and their degradations; we can learn about the future through the study of adaptation to extreme environments, the genetic basis of diseases, sexualities, and cognitive capacities, and the way genes and environment richly interact during the history and development of the individual.

Because evolutionary genomics provides a varied candidate set of answers – ranging from contemporary "genetic reductionists" and "genetic determinists" emphasizing selfish genes and human nature (Richard Dawkins and Steven Pinker) to "group selectionists" and "developmental interactionists" defending a complex human reality of genomes, brains, and bodies in social and cultural environments at many emergent levels of analysis (David Sloan Wilson and Richard Lewontin) – to our deepest questions about our place in the world. Because of this, the field has, can, and will continue to address the questions growing out of that kernel query *Who am I?* in political and politicized ways. Further, reasonable questions remain about the best interpretation of genomic data, statistical analyses, and theoretical results; the role of assumptions and other background context in theoretical development and mathematical modeling; and the nature of self, causation, and equality, given genomic results.

Ultimately, human evolutionary genomics can tell us much about who we are as individuals and as a collective or collectives. But also, and as I make clear throughout *Our Genes*, it has limited power in fully answering some of our deepest questions. The *population* is that entity which, according to biologists, *evolves*. Over the past few decades, experimental and diagnostic technologies across the life sciences and biomedicine have grown, and computational machinery has greatly increased in power and capacity. The study of the genomics of populations – especially human populations – has therefore also evolved in many new theoretical and conceptual directions, including:

- developing fundamental Darwinian evolutionary theory;
- reconstructing the evolutionary history within and among species, both extinct and extant, and even the entire tree or network of life;
- suggesting consequences of various conservation actions for biodiversity (and other ecological measures and metrics);
- enabling inferences about human population history and demography, including the history and demography of other species in the genus *Homo* and beyond;
- studying the structure of human genomic variation for biomedical or neuroscientific purposes (e.g., disease etiology identified through genome-wide association studies, or GWAS);
- assisting an understanding of the role of genes in development;
- identifying genes that are targets of natural selection, for example, genes increasing survival and reproduction in epidemics and pandemics, at high latitudes or elevations, or underwater;
- inferring the ancestral populations for a given individual (e.g., 23andMe or AncestryDNA); and
- assessing candidate suspects in forensic criminology.

In the pages that follow, and in an effort to produce yet more answers to questions about identity, origin, and community, we shall try to make sense of the accelerating work emerging out of theoretical population genetics and human genomics, including biomedical genomics. Geneticist Kärt Tomberg brought to life genetic methods and concepts in her painting *Acrylic Genetics* (Figure 1.2). My main perspective is *population-level phenomena*: I want to know how much individuals differ genetically, both within their own groups and populations, and with respect to very different groups and populations, such as those on other continents, and how this difference is measured, and what it means. How can we use comparative studies of individuals – twin studies or genome-wide association studies (GWAS), for instance – to draw inferences about the relative causal role of genes and environment (or their interaction) in

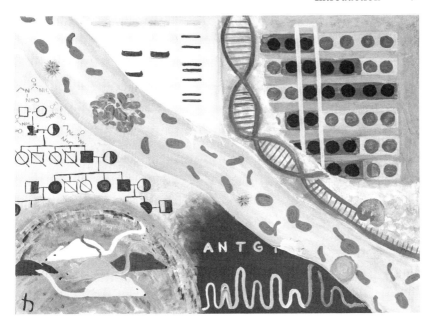

Figure 1.2. Acrylic Genetics
The blood vessel in the center of the painting contains a thrombus, the phenotypic target for Tomberg and her collaborators' genetic screen. The painting's left side illustrates a pedigree generated from a male mouse following treatment with the chemical mutagen ENU (red or gray formula) as well as a DNA sequence tracer sequence (the bottom curves, with corresponding nucleotide sequence, where "N" indicates unknown or error), with a representative ENU-induced DNA variant sequence. The right side of the painting depicts (left) electrophoretic genotyping and (right) genetic region mapping across experimental mice to identify the causal gene variant. *Acrylic Genetics* by Kart Tomberg (2018), painting. Copyright by the artist. Reprinted with permission. (A black and white version of this figure will appear in some formats. For the color version, please refer to the plate section.)

building the individual's observable characteristics, or about the relatively recent human history of migrations and invasions over the past 12,000 years or so, or about the deepest origins of *Homo sapiens* and our nearest kin species and subspecies? How can we design studies and experiments using model systems such as mice or fruit flies, informed by necessarily limited interpretive frames, and assay different human populations for genetic susceptibility to disease, to try to develop medical diagnoses and treatments? By pursuing answers to these questions, we stand to learn much about our ancestry and population structure. We also stand to gain a view into the future of human evolutionary genomics, including its potentials (and its perils) for biomedicine or neuroscience, and its relevance for conservation biology.

My work assumes the robustness of population genetic theory, with excellent and time-proven credentials rooted in the work of R.A. Fisher, Sewall Wright (1889–1988), and J.B.S. Haldane. I strive to illustrate the coherence, depth, and evidentiary status of the subtle, rich, and explanatory theoretical *organism* – not to say machinery – of human evolutionary genomics. I call evolutionary genomic theory an organism because it is robust and stable, complex and ever-developing, and made up of functionally integrated parts (with clear models, methods, and assumptions). This may not be a popular argument. Indeed, there has been recent concern that the theoretical organism is moribund, and pleas for an "extended evolutionary synthesis" have lately filled the halls.[10] Proponents clamor for a broader and more integrated evolutionary theory that includes more evolutionary forces and that respects processes of development and complex ecological organization. These skeptics suggest that modern evolutionary genomics cannot take into account variable rates of evolutionary change and speciation, phenotypic plasticity and developmental bias, niche construction and the organism's role in influencing its environment, or nonrandom genetic mutation.

Proponents of the extended evolutionary synthesis make reasonable points: It is true that developmental dynamics and ecological conditions are often left out of standard evolutionary genomic theory. Leigh Van Valen presciently observed as much in 1973: "A plausible argument could be made that evolution is the control of development by ecology. Oddly, neither area has figured importantly in evolutionary theory since Darwin, who contributed much to each."[11] However, it certainly does seem imperialist to call for a single theory covering all biological properties across the great web of life. Evolutionary genomics works well in its defined corner, investigating the change of allele frequencies over generations (that is, describing and explaining the changes in the relative percentages, for any given gene, of all alleles of that gene) and the distribution and composition of genetic variation across populations. And it can play exceedingly well with fields and theories close to it, such as molecular and developmental biology, and biogeography and behavioral ecology. But, while the field draws on such neighboring fields and

---

[10] See, e.g., Pigliucci and Müller (2010) and Laland et al. (2015).

[11] Van Valen (1973, p. 488). In this review of a festschrift dedicated to G.G. Simpson, Van Valen also provided an adaptationist gloss on Lewontin (1972).

theories, it cannot act as a substitute. Organismic homeostasis and cell differentiation, or food web networks and species–area relationships, cannot be explained through population or evolutionary genetics. Instead, human evolutionary genomics offers one very useful but very limited tool with which to answer some of our most pressing questions about ourselves, and about each other. No single theory – inside or outside of science – can or even should explain everything.

Instead, evolutionary genomics fares well, to my mind, *for what it was designed to do* – track allele frequency changes in and across populations due to standard evolutionary forces such as natural selection, random genetic drift, migration, and meiotic drive, and thereby explain patterns in genetic variation across populations.[12] The findings it yields can usefully inform, and also complicate, other questions and other answers. I am therefore a localist: One field of biology may work locally on its defined problems of interest. However, I am also a pluralist: Scientists, doctors, philosophers, and others should use different definitions of evolution, with their differently related models, and contemplate which predictions and understandings might follow. Indeed, population genetics is a kind of flexible theoretical organism. Theoretical evolutionary genomics works, but it is not – and never was – a "master theory." It therefore should be "disunified" and pluralized – both within itself and with respect to other sister theories – not "extended" in some kind of monolithic, grand evolutionary synthesis.[13] This book follows this implicit injunction.

Because science and philosophy can further our understanding of ourselves, our politics, and our societies, I hope *Our Genes* reaches at least three audiences: First, I hope to reach philosophers and other humanities scholars, as well as social scientists, who may not be familiar with evolutionary genomic science, but who recognize its relevance to one of their central concerns – the human condition. The second audience comprises scientists who may have the technical knowledge but might be interested in and benefit from learning more about philosophical investigation both in their domain and in general. I strive to make the science accessible to the former and the philosophy relevant

---

[12] See Futuyma (2017).

[13] See Stoltzfus (2017). Danchin et al. (2019) also adopt a pluralist position, combining standard population genetics with "early in life effects" that some would label neo-Lamarckian (or, roughly, the belief in the inheritance of acquired characteristics). Such phenomena are indeed important to development and evolution, as proponents of the extended evolutionary synthesis also correctly insist.

and interesting to the latter. Finally, I hope to reach the general reader who has an active interest in science or philosophy and who is preoccupied, as we all are to some degree, with topics of identity, origins, mutability, and futurity.

This book addresses and explores a set of specific questions within these topics. I ask about our genetic endowment, our evolutionary ancestry and future, and our heritable limits and potential. More specifically, I ask how our genes can help us understand the origin and nature of our body and mind, if not our soul. Who are the *we* here on Earth, in terms of tribes, nations, social classes, populations, and even local ecologies and larger ecosystems, or entire branches of the tree of life? Does our genomic dowry preordain you and me to certain behavioral or professional destinies? If so, to what extent does freedom remain possible? Outside of genes, but still at the intersection of natural science, philosophy, and human genomics, how do culture, ecology, and "randomness" or "chance" shape us? *Our Genes* engages with these features and functions of evolutionary genomics. I admit that I am pooling features and research questions of related fields, such as medical genetics and behavioral genetics, with evolutionary genetics. But these areas of genetics are rarely sharply distinguishable. And a panoramic view of the appropriate breadth and depth of human genomics in the context of mathematical evolutionary theory is my goal.[14]

★★★

Our Genes started, in thought, as a reprint collection, but developed into a much more coherent and integrated book, with much reorganization, new research, and ample rewriting in light of novel research and critical reflection. Chapters have been heavily edited in multiple ways. Furthermore, this chapter and Chapters 2, 7, and 8 are new. In Chapter 2, we situate ourselves within the fields of genomics and statistics, which emerged out of the late-nineteenth-century preoccupation with questions of individual and collective identities. We also begin to familiarize ourselves with the various stages of humanity's deep, historical genomic journeys. To enable our study, I also supply a genomics glossary at the end of Chapter 2. Chapter 3 discusses the contextual frames that

---

[14] Standard textbooks on evolutionary genetics include *Principles of Population Genetics*, 2nd ed. (Hartl and Clark, 1989), *Introduction to Quantitative Genetics*, 4th ed. (Falconer and Mackay, 1996), *The Genetics of Populations*, 3rd ed. (Hedrick, 2005), and *An Introduction to Population Genetics* (Nielsen and Slatkin, 2013).

shape the interpretation and meaning of populations and collectivities, in evolutionary genetic theory and beyond. Chapters 4 and 5 address methodological matters about how human population genetics is statistically modeled and, in some cases, philosophically conceptualized. Metrics and measures (Chapter 4) and models and methodologies (Chapter 5) in human evolutionary genomics come in many shapes and sizes – there are no one-size-fits-all questions or themes. We must be careful about using the contextually objective, proper tool for a given question.[15] Chapter 6 presents six basic empirical patterns of human evolutionary genomics and their representation of the evolutionary history and demographic structure of global populations of *Homo sapiens*. These patterns suggest that our genome, and the multileveled populations and individual bodies housing it, are more similar than different (although our differences are equally suggestive). Chapter 7 presents an in-depth discussion of selection, including a case study investigation of one of the clearest depictions of selection as a shaping influence in human populations. In so doing it also explores, philosophically and conceptually, the power that distinctions can have and the way we can integrate different explanatory paradigms.

Chapters 8 and 9 explore some consequences of evolutionary genomics for our ontological, personal, and political understandings of intelligence, sexuality, disease, and race. Chapter 8 discusses two case studies relevant to the past development of human evolutionary genomics as a field but also to its future: intelligence and the evolution of female orgasm. By focusing on two ambiguous topics, topics that are still the subjects of ongoing investigation and analysis, I illuminate what human evolutionary genomics can help us see, but I also show what human evolutionary genomics can help to hide. Moreover, I indicate how the distinction between gene and environment, and the associated dichotomy between nature and nurture, which drove the development of both genomics and statistics, remain vigorous drivers in and of the field today.

Chapter 9 applies information from the preceding chapters to a topic that has long preoccupied geneticists, in both incredibly limiting and somewhat expansive ways, to plead emphatically for the separation of genomics from political and social deliberation. Chapter 9 takes as a principal philosophical concern the potential dangers of conflating or

---

[15] See Winther (2020a).

perniciously reifying model with world.[16] This occurs when we fall in love with a model or theory and take it to be absolutely and irrevocably true. I return to this motif throughout the pages of this book, since one of my central worries in how we "do" science is the all-too-frequent systematic and communal ignoring or silencing of the multiple, reasonable alternatives to the few, dominant, orthodox representations. All theories or models are partial and satisfy only a particular subset of specific purposes. The concern with the dangers of reification is best synopsized as the question at the core of Chapter 9: In what sense, if any, can and should we say that race is real?

Chapter 10 takes a step back, considering the relation between statistics and evolutionary genomics in light of a thought experiment of different kinds of universes, and with the exploration of certain foundational matters surrounding statistics, including its possible futures.

★★★

While very different tasks, reading and writing books share at least one requirement: *empathy*. Writing means trying to place oneself in the mind and heart of many different readers – imagining what would make them giggly, fascinated, or engaged; trying to avoid boring them with irritating pedantry; and working hard to engage their sense of curiosity and wonder. Of course, there is some hit-and-miss, and the best an author can do is try. Similarly, you, as reader, have many choices about what to do with your time, including which books to hold in your hands or stare at on a screen. If your experience is anything like mine, once a book catches your attention, you try to give it an honest chance by bending your mind to the book's perspective or paradigm. You do this because you hope to gain something from it. Why else spend your precious resources on it?

A number of exciting books about human evolutionary genomics are available. Spencer Wells's *The Journey of Man* (2003), David Reich's *Who We Are and How We Got Here* (2018), and Adam Rutherford's *How to Argue with a Racist* (2020) are each written by geneticists who know their core material exceedingly well. Other books present the scientific

---

[16] Winther (2020a) develops the concept of *pernicious reification* in detail: "When we fall in love with our [map] projection or theory, we treat the theory or projection as if it were a real, concrete thing that also *is* and *describes* the entire world. We overestimate our representation's capacities and promises" (p. 89). In my book on maps and mapping, I carefully define and deploy this concept, which is ubiquitous to modeling and theorizing across the natural and social sciences.

material in a more ideological manner, whether from the left or the right, including Jonathan Marks's *Human Biodiversity* (1995) and Charles Murray's *Human Diversity* (2020). What I bring to the subject is a philosophical investigation of both our subject and the results of our investigations. This includes addressing ethical and political issues, and considering deeply rooted psychological questions of identity and belonging, outside of any ideology (to the extent that is possible). It also means exploring the late-nineteenth-century historical and conceptual interweaving of statistics, genetics, and evolutionary theory. The philosophical and statistical-mathematical level of detail in *Our Genes* will perhaps require more theoretical stamina on the reader's part than will engaging with the other books on the market. But I have worked to make this approach appealing and comprehensible (by focusing on assumptions, definitions, and the stakes of the questions, rather than on mathematical equations as such). With that said, readers who might be put off by even a hint of mathematics or statistics could jump from Chapter 3 to Chapter 6. I believe our shared exploration, however challenging, will be worthwhile for those who wish to dig deeper into the political promises and limits, as well as the technical apparatus, surrounding the coevolving fields of genomics, evolutionary theory, and statistics.

This book is inspired by two overarching theses: one political and one philosophical. The contemporary political landscape, somewhat independently of national context, tends to deploy our best genomic science in one of two ways. The political left generally interprets human evolutionary genomics as unequivocal: Contemporary social and economic inequalities cannot be justified – or explained in a morally salient sense – on the basis of human genetics. After all, genomics has shown or proven that we are all "the same," biologically. The political right almost invariably reads a moral hierarchy into the human genetic story, emphasizing genomic differences at both the individual and population levels. The former finds a genomic basis for its desire for a flat and equal society; the latter reads off genes a vertical and hierarchical society.[17] Who is right? Does evolutionary genetics prove that humans are essentially the same or that we are fundamentally different, and what does or would this answer tell us about how we ought to behave socially?

[17] This classification is a useful idealization – see, e.g., Frase (2016).

As I have indicated, the radically understated political thesis of this book is that *genomics is not a good basis for arguing about human rights, regardless of political inclinations, left or right.* This is meant in at least two ways. First, genes tell a *changing* story. As scientific findings change, any attempt to use them as a moral basis is precarious. *Our Genes* points out problematic examples (e.g., sickle-cell anemia, "Out of Africa" and "Out of Europe" migrations, Neanderthal genes in African populations). Second, genes tell an *ambiguous* story. Even when what we know stays constant, interpretation can be taken in different directions. There is a plurality of views regarding how to interpret genomic data and model results, as this book demonstrates. Assuming that some scientific fact was found regarding, for example, the robustness of five continental genomic clusters, what broader social and ethical inferences would then be possible?

These fundamental contingencies should inspire a humility and cautiousness toward reading strong political lessons from genomics: The genome neither proves human equality nor demonstrates human hierarchy. The contingencies should also underscore the importance of learning some of the details of human genomics. We must learn so as to begin to understand the simultaneous power and limits of science, especially with respect to complex social and ecological systems. In fact, justifying moral positions on the basis of genes, such as an embrace or grounding of human rights, is *strategically, conceptually, and ethically flawed*. Genetics has been used to justify all kinds of political positions and actions, especially oppressive, violent, and discriminatory ones. But even if one human population were clearly and explicitly shown to have an intrinsically high cognitive capacity, whether genetically based or not, no such biological facts would actually justify a particular political or ethical norm.

The philosophical thesis of this book is that *we should be vigilant and wary of using genomics (and biology more generally) to perniciously reify population-level or category-level differences.* In particular, such differences should not be viewed as absolute or strongly explanatory of individual-level differences – or as painting a full picture of individuals in the absence of complex information about personality or culture.

Bluntly put, even if evolutionary genomics is sometimes useful in the doctor's office, in forensic criminology, or for identity-seekers using DNA tests such as 23andMe, it does not and cannot provide stable ground for our social visions in Parliament or Congress, or in our public discourses. We are each much more than members of particular populations or categories – even if we also stand in deep relationality to other life – and we must be wary of conflating our properties with those of our groups.

We simply cannot chart a path toward equality and freedom using only scientific information. We must move forward according to morality and law. Political and moral philosophy provide arguments premised on intrinsic equality and fairness, not genes. Philosophy may also help us expand the moral universe of our empathy and of our concern for others, including other *kinds* of living beings. Accordingly, *Our Genes* invites you to continue contemplating *Who am I?* as a philosophical question. We will explore themes of individuality and connectivity, of accurate representation and deceptive misrepresentation, of similarity and identity, and of difference and multiplicity in the context of critical thinking, the history of genetics, and the targeted use of statistics. We will find only incomplete, inconclusive answers. These will nonetheless shape our search.

Figure 2.1. From Galápagos-Writ-Large to Planet Unity 1

The extreme ends of the thought experiment on human population structure: (left) Galápagos-Writ-Large and (right) Planet Unity. The Galápagos map includes equidistant contour lines at 250 meters on land and 500 meters in the ocean. (Galápagos map, using data sources from GEBCO Compilation Group 2020, Natural Earth, earthexplorer.usgs.gov and www.ncei.noaa.gov.) Concept by Rasmus Grønfeldt Winther, illustrated by Larisa DePalma (aliens) and Mats Wedin (Galápagos map).

© 2021 Rasmus Grønfeldt Winther.

2 · *Origins and Histories*

Modern human evolutionary genomics applies natural scientific knowledge to address ancient and fundamental *Why?* questions, enabling us to deepen our philosophical reflection on identity and futurity. But to understand contemporary human evolutionary genomics, we must comprehend the field's explicit and implicit connections to its origins and early development. By delving into the basic theoretical outline and historical sources of human evolutionary genomics, we become more familiar with the layered and fractal history of *Homo sapiens*.

Humans have long been asking questions about identity, origins, and belonging, but only in the nineteenth century did scientists begin to look for answers in a systematic, coordinated way. During the nineteenth century, Charles Robert Darwin (1809–1882) developed an integrated theory of evolution, and findings in cell theory and comparative morphology and embryology permitted significant progress on questions of development and heredity. These advances are at the heart of the *genetic paradigm*, an approach to understanding the structure, physiology, and behavior of organisms based on the transgenerational inheritance of genes, including attention to the interaction of genes and environment during organismic development. In agreement with many historians, critics, and others, I locate the emergence of the genetic paradigm in the nineteenth century because it marks the extended moment during which progress on the puzzle of why children resemble their parents began to be made in earnest.

The Emergence of the Genetic Paradigm

The genetic paradigm began to coalesce in the nineteenth century, and was stabilized as a result of the rediscovery of Gregor Mendel's work in the early twentieth century. Although Mendel published his work in 1866 and died in relative obscurity in 1884, scientists in 1900 saw his investigations as newly relevant to the nascent genetic paradigm. Many readers have probably heard of Mendel – perhaps you have even tried your luck at drawing a few Punnett squares. While there is another book to be written about the history from which Mendel drew, what exactly he sought to accomplish in his argument, and why his work was ignored or diminished for 35 years, here I focus more narrowly on his formative contributions to genetics.[1]

Scientists in the early 1900s rediscovered and recognized two principles in Mendel's work that proved central to genetics: the law of segregation and the law of independent assortment. For each kind of heritable factor (or what we today call *gene*), an individual that is a product of sexual reproduction has received one of its two heritable factors from each parent. According to the *law of segregation*, the two, paired heritable factors in each parent had separated – segregated – in the process of forming gametes, namely egg or sperm, such that each gamete received only one of the two factors.[2] Again, this is the case for each kind of heritable factor. Furthermore, heritable factors are *difference-makers* on the phenotype. That is, changing a gene from two identical copies of one of its versions – or what we today call *alleles* – to two identical copies of another of its versions changes certain features or parts of the individual's structure, physiology, or behavior – its phenotype. You may recall that Mendel developed the law of segregation by breeding pea plants and observing the results of seven heritable factors including those affecting plant height (tall or short) and flower color (purple or white). Mendel also discovered that, in many cases, for a factor affecting a given character, one heritable version of the factor is dominant, meaning it is expressed and shows its causal power even if there is only one copy of the allele (i.e., even when the individual is heterozygous), while the other is

---

[1] Mendel presented his paper in 1865, but it was not published until 1866 (Mendel, 1996). Olby (1979), Henig (2001), and Franklin et al. (2008) provide focused analyses of Mendel's work. Robinson (1979) and Bowler (1989) advance simplified and synoptic views of heredity theory during the second half of the nineteenth century; see also Winther (2000, 2001) and Churchill (2015). Provine (1971) and MacKenzie (1981) illuminate the obstacles and nonlinear (social) history, in the early twentieth century, associated with the rediscovery of Mendel's work. Mukherjee (2016) deploys Mendel as part of his autobiographical engagement with genetics.

[2] Multicellular eukaryotes (roughly and loosely, plants, fungi, and animals) have both cells that are *diploid* (here, paired chromosomes), as well as *haploid* (e.g., egg and sperm, with a single copy of each chromosome). In flowering plants, including Mendel's peas, the haploid gametes are found within a haploid individual phase – the so-called gametophyte. In a process known as *alternation of generations*, these plants cycle between haploid gametophytes (embryo sacs or pollen) and diploid sporophytes via fertilization (sporophytes from gametophytes) and meiosis (gametophytes from sporophytes). See Figure 20.4 in Gilbert (2000).

recessive, expressing its causal power only when there are two copies of that allele (i.e., the individual is homozygous).[3] Mendel found in his peas that, for instance, tall is dominant over short and purple is dominant over white. Mendel then tracked the ratios of plant appearance over generations, inferring which factors caused which phenotypic effects. Mendel used an early, unique, and correct combinatorial notation, but a Punnett square is the tool used ubiquitously today to depict such ratios.

In addition to identifying the law of segregation, Mendel also detected the *law of independent assortment*. According to this latter law, different kinds of heritable factors – that is, factors associated with different features or parts of the phenotype – segregate and are passed on independently of one another. For instance, in his pea plants, Mendel discovered that a 3:1 ratio of purple to white offspring flowers when each parent was heterozygous for purple flowers occurred independently of the 3:1 ratio of tall to short offspring plants (with analogously tall but heterozygous parents). In other words, Mendel observed that three-quarters of short plants would have purple flowers and one-quarter would have white flowers, and likewise for tall plants.

Mendel's observations provided relevant background for the development of the genetic paradigm and also illustrate how, from its very inception, this paradigm necessarily involved individual difference and variation, as well as family-level and population-level comparisons. But today we know that Mendel was very lucky in his choice of the seven factor pairs and associated features. Had Mendel chosen factors on the same chromosome that were much closer together, the ratios of different factors would *not* have been independent. This is because when chromosomes pair up before each one is segregated to subsequently find itself in a different egg or sperm, the chromosomal pairs can join, twist, and break, with different chromosomal segments subsequently pasted together. To put this another way: Chromosomal pairs *recombine*. The closer that genes are to each other along the chromosome, the less they tend to recombine with respect to each other. Neighboring kinds of factors tend to be inherited together. As a result, the closer two genes are, the higher the relative number of egg or sperm of the two parental types (i.e., the two

---

[3] Dominance of one allele over the other is not the only possibility for heterozygous individuals. There can also be *incomplete dominance*, whereby heterozygous individuals have a phenotype somewhere in between what is typical of two homozygous individuals, or even *co-dominance*, where both alleles are expressed (e.g., for many primates, including humans, when individuals are AB for the ABO blood group; Ségurel et al., 2012).

combinations of two alleles sitting adjacent to each other on each of the two chromosomes of each of two parents) as opposed to egg or sperm of the two recombinant types (i.e., the two sets of mixed-and-matched alleles, each sitting on two different chromosomes of each of two parents). If, for instance, the genes for flower color and plant height of Mendel's peas had been quite close to each other on the same chromosome, the ratios would not be independent.[4]

This scenario, which Mendel did not discuss, given his choice of focus, illustrates the *law of linkage*, the third law of early-twentieth-century genetics. Its emergence was crucial to furthering the materialist grounding in chromosomes of the inheritance of genes. In fact, deviations in counting ratios from what would be expected with independent assortment of many pairs of genes permitted Thomas Hunt Morgan's "Fly Group" at Columbia University to start mapping genes along the four chromosomal pairs of fruit flies around 1910. Mendel's rediscovered experiments and laws helped early-twentieth-century scientists grasp the principles of inheritance. Later work in the field led to substantive discoveries in the material relationship between chromosomes and genes. Through Mendel, the genetics paradigm became a clear contender for explaining that which is passed on across generations and thereby contributes to the emergence of organismal form and function.[5] When applied to humans, this paradigm concretizes as human evolutionary genomics.

Mathematical, Material, and Computational Genetics

Subsequent to the rediscovery of Mendel's 1866 paper in 1900, which helped cement the nascent genetic paradigm as foundational, three important, broad categories of genetic work emerged: mathematical evolutionary genetics in the 1920s; material genetics in the 1940s; and "big data," or computational genomics, which exploded in the twenty-first century.

---

[4] That two of Mendel's seven chosen characters are likely quite close on chromosome 4 of *Pisum sativum*, and therefore linked – i.e., plant height and pod form – is discussed in Reid and Ross (2011). See also Offner (2011), which includes a linear genetic map of the seven chromosomes of peas, three of which alone carry the seven genes for Mendel's seven characters.

[5] For a fuller explication of Mendel's logic and contribution of the laws of segregation and independent assortment, and of Morgan's Fly Group and their use of the law of linkage, see chapter 8, "Mapping Genetics," of Winther (2020a).

Because of their role in heredity and development, and because their different versions could, in principle at least, be counted in populations in which evolutionary forces operated, genes were useful theoretical units to the mathematical theory of evolutionary genetics. The central currency of the mathematical models of evolution first developed by R.A. Fisher, J.B.S. Haldane, and Sewall Wright in the 1920s was *allele frequencies*. To grasp this concept, think of a *population* of individuals, whether they be beavers or oak trees or some other group of a given species: They live in the same area, have the potential to mate, and are subject to roughly the same environmental conditions. Within that population, there will be at least one version – or *allele* – of each gene (the difference-makers).[6] When there is more than one version of a given gene, at least one of those alleles is likely to make a unique difference to the phenotype of the individuals who possess it, compared to the other alleles, such that individuals with that allele will, on average, do better than other individuals *without* that allele – in what Darwin called the "struggle for life" and "struggle for existence." All else equal, individuals with the relatively beneficial allele will tend to leave more offspring than individuals with other alleles that are not so beneficial. The allele will therefore iteratively increase its proportional representation in the population gene pool of downstream generation upon generation. This is, of course, the well-known principle of *natural selection*, discussed at length in Chapter 7. What matters for mathematical evolutionary genetics is that we can model changes in allele frequencies over many generations. Such capability gives us explanatory and even predictive power for understanding which evolutionary forces are at play, in which populations, and subject to which environments.

Although the material basis of genetics and heredity was first established *molecularly* as DNA in the 1940s, the history is, as ever, complex. In fact, the physical and material basis of genetics and heredity was first located on the chromosomes when Morgan's Fly Group managed to map the four chromosomal pairs of *Drosophila melanogaster* by abstractly calculating the relative distances of gene pairs across the chromosome according to the logic of the law of linkage outlined above.[7] In the

---

[6] If there is only one version, though, its frequency will be 100%, and evolution will be boring, with no change, barring mutation. Variation is not only the spice of life; it is the engine of evolution.

[7] A pair of homologous chromosomes is also sometimes called a *linkage group*. Although two genes sufficiently far from each other on the same chromosome assort independently – that is, they recombine as if they were on different chromosomes, as was even the case for some of Mendel's factors – the far-away genes are respectively linked with two other loci between them that

1930s, it became clear that these abstract series of genes arranged in linkage group maps could be made to correspond to – in other words, they could be themselves mapped onto – microscopic features or landmarks of physical chromosomes, such as those on the four chromosomal pairs of *Drosophila*. Although this signaled major progress for understanding the material basis of genetics, it was not until the 1940s that the molecular underpinning of heredity took off. Up until World War II, the standard view was that hereditary information was carried by proteins in the egg and sperm. Through work on the bacterium causing pneumonia in 1944, however, Canadian American biologist Oswald Avery and his collaborators were able to show that DNA, rather than proteins, carried hereditary information, and that DNA was the relevant molecular basis of genetics. Meanwhile, in the late 1940s, Rosalind Franklin learned X-ray crystallography in Paris, and, in the early 1950s, her X-rays of DNA laid the foundation for James Watson and Francis Crick's discovery of the double helix. In the 1960s, geneticists were able to sort out the genetic code – the mapping of three-letter nucleotide words to proteins (see Figure 7.5). In the 1970s and 1980s, DNA amplification and sequencing technologies were developed.[8]

The study of genetics and genomics hit a major milestone in 2001, with the publication of a first draft of the human genome. But, as in the prehistory of the genetic paradigm, genomic computational technologies started being created many years before this. Global human genetic diversity projects, such as Luigi Luca Cavalli-Sforza's (1922–2018) Human Genome Diversity Project (now stored in cell lines located in Paris), were first developed in the 1990s as a form of human natural history – a complex set of field studies of human populations. Around the same time, nucleotide sequencing technologies were being perfected, further establishing and refining laboratory and experimental protocols. Computational algorithms, programs, and databases were also articulated to harness, store, and explore an increasingly massive amount of genomic data. This was the theoretical mind in action. In the twenty-first century, the integration of global genomic diversity data, experimental protocols,

---

themselves either do not assort independently (i.e., are linked) or do assort independently but are respectively linked with two other loci between them that themselves either do not assort independently (i.e., are linked), and so on. The recursion here ends when the two in-between genes are linked, and always ends after just a small number of recursions, for any of the 23 human linkage groups.

[8] Olby (1994), Judson (1996), and Keller (2002) tell the rich history of the rise of material genetics from various perspectives.

and computational approaches gave rise to genomic "big data," on which many of the arguments in *Our Genes* rest. The theoretical mind offers a provocative but partial perspective: Genes and their alleles are small, almost fundamental entities that help explain the large entities – the bodies and minds – within which they are housed.[9]

The Power and Pervasiveness of Statistics

Today, the genetic paradigm can tell us much about identity, but also, as I point out in Chapter 1, very little. The paradigm's power, limited though it may be, depends on the fundamental role played by statistics. The genetic paradigm helps answer questions about origins, causes, and the past, in part because it came to depend on statistics, which helps answer those and other questions about ends, outcomes, and the future. Statistics is an omnipresent part of modern life, so much so that it is hard to overestimate its centrality and ubiquity. Government agencies, doctors, environmental conservationists, physicists, and engineers use a diverse toolbox of data collection techniques, statistical computational packages, and mathematical models to summarize, classify, and draw meaning out of complex data; to build abstract and idealized models with a goal of explaining the data; and to test the models and, if confirmed, use them to make tentative predictions about – and interventions in – the future or other, hitherto uninvestigated contexts of application. Statistics is so tremendously powerful that it is considered by many to be key to genuinely new knowledge. Indeed, to an ever-increasing extent, we learn about the world today by measuring and modeling it according to statistical methods. These methods sate our human desire to ameliorate uncertainty about the present, and the present's relationship to the future or to other new and unknown domains, by making model-based inferences bridging from the known to the unknown.

Somewhat similarly to the development of the genetic paradigm, the study and practice of statistics in the eighteenth and nineteenth centuries were shaped by flurries of activity and discovery. The rise of probability theory, or the mathematical study of random events, stands as an important example.[10] A crucial source for probabilistic thinking was the formal

---

[9] On "smallism," see note 29, p. 188 of Winther (2020a).

[10] As philosopher Ian Hacking (1990) states: "our idea of probability is a Janus-faced mid-seventeenth-century mutation in the Renaissance idea of signs ... [which] came into being with a frequency aspect and a degree-of-belief aspect" (p. 96).

investigation of so-called games of chance, such as die-rolling or cards.[11] This was accompanied by a strong drive to find patterns in social data, such as statistics associated with governmental tax income, crimes and criminals, and disease or suicide rates. Even so, key probability theory work by great minds, such as Jacob Bernoulli (1655–1705), Leonhard Euler (1707–1783), Pierre-Simon Laplace (1749–1827), and Carl Friedrich Gauss (1777–1855), was less about making sense of data and more about investigating the mathematical laws, and even philosophical foundations, of chance.

Such work included diverse yet careful definitions and conceptualizations of probability, frequency, and patterns, and delineated different probability distribution functions, such as the famous Gaussian "normal" or "bell curve" distribution, or the perhaps less well-known Poisson and binomial distributions, which are superbly useful (because of their applicability to many events in the world). The two great interpretations of probability, the frequentist "probability-is-what-happens-in-the-long-run" and the Bayesian "probability-is-the-degree-of-belief-or-confidence-I-can-have-in-a-hypothesis-or-model," crystallized and were formalized in the eighteenth and nineteenth centuries, even if some of the concepts on which both are based emerged with the Scientific Revolution of the 1660s, or even earlier. Such basic and general probability theory concepts, methods, and frameworks were a gateway to applied mathematics and, eventually, to deploying the genetic paradigm as a creative smithy for statistical thinking.

If probability theory is a set of fundamental mathematical investigations, then statistics is a kind of applied mathematics, or applied probability theory. But the rise of statistics as data interpretation and as model-building testing is, to a large extent, a direct outcome of late-nineteenth- and early-twentieth-century concerns with heredity and development. Bluntly put, statistics is dependent on the rise of biometry, heredity theory, and, in particular, genetics. Why? Because biometricians Francis Galton and his protégé and eminent statistician Karl Pearson, among others, were obsessed with biological measurement, including height, weight, and countless other biological features of individuals, and with which patterns and lessons we could draw from such measurements.

---

[11] A probability distribution function maps out the probability or frequency of the particular values or *events* of a process, such as that of the number of dots on a fair die, which can take on one of six values. The events here are evenly distributed – that is, the probabilities or frequencies of each of the six values are equal. A fair die thus has a discrete uniform distribution.

Biological phenomena are fascinating, and they were especially enthralling for many early statisticians, not only because they offer a straightforward suite of continuous and discontinuous (discrete) characters that can be measured, but also because there is a presumption that we understand – or can at least postulate a theoretical mechanism for – the *causes* or *development* of such characters. That is, the hereditary material explains the morphology, physiology, and behavior of individuals. Galton and Pearson, who focused on continuous traits rather than on discrete Mendelian factors with large effects on the phenotype, assumed that phenotypic variation was caused by, or at least associated with, differences in the hereditary material. Environmental variation also made a difference to the phenotype, but differences in physical and cognitive features of individuals were correlated, according to Galton and Pearson, primarily with cross-individual and cross-family (to not say cross-race or cross-social class) variation in hereditary endowments. The hereditary system thus became an excellent idealized system for statistical investigation because there were, in modern statistical parlance, clear independent (explanatory) and dependent (response) variables that could be correlated or *regressed* on one another. (See Chapter 10 for further discussion of the power and pervasiveness of statistics.)

R.A. Fisher: Statistics, Genetics, and Eugenics

To the extent it is possible to attribute any one discovery to any one individual, it was R.A. Fisher who integrated the streams of statistical thought flowing from the nineteenth-century fin de siècle into the early twentieth century. Fisher domesticated statistical vocabulary and articulated statistical tools for experimental design (such as randomization), causal analysis (such as analysis of variance), and model–world evaluation (such as maximum likelihood estimation). It is almost a cliché to state that Fisher invented modern statistics, and it is no exaggeration to claim his 1922 paper "On the Mathematical Foundations of Theoretical Statistics" as the single most influential paper in twentieth-century statistics.

What is more interesting to argue (and perhaps less predictable in its suggestion) is that we should not underestimate the synergistic effects of Fisher's simultaneous interests in statistics, genetics, and eugenics. For example, in 1918, Fisher showed the futility of the debate between the Mendelians (such as William Bateson, 1861–1926), who argued that many, if not most, hereditary factors had large effects (think back to the purple versus white flowers in peas), and the biometricians (such as

Pearson and his protégé, Raphael Weldon, 1860–1906), who focused on the statistics of continuity of hereditary effects (think height). Fisher handily resolved this debate by suggesting that there was a large ensemble of genes, each with a small effect on the overall phenotypic variation.

Fisher also recognized he could build mathematical models of evolution. That is, he could sketch out in formal terms the evolutionary pressures on populations, including the forces of selection, mutation, and migration. Again, Mendelian genetics proved a particularly powerful machine for Fisher's models because it could be clearly conceptualized and formalized. According to Fisher:

- Every individual was postulated to have hundreds if not thousands of hereditary factors or genes or loci.
- Each of these genes had two "slots," each of which could take on one of all possible versions or alleles – think of each gene as a dial or switch that can be set (heritable factors with only one version are not per se interesting in explaining variation).
- Genes and their allele versions had small yet clearly correlated effects on the operationalizable, measurable phenotype (i.e., the morphology, physiology, and behavior of individuals – namely, the output of development).
- Changes of allele frequencies at any number of loci in any given population(s) could be modeled and calculated as a consequence of core evolutionary forces, including natural selection and mutation.

Fisher's evolutionary thinking was a probabilistic and statistical dream come true: Clear explanatory and predictive models of frequency changes of hypothetical genetic factors and of changes in correlated, measurable phenotypic traits could be built, tested, and hacked based on ample biological data. In building out his models, Fisher developed his statistical toolbox. He articulated randomization protocols in his studies of experimental design – for him, a well-designed experiment was analogous to providing nature with "a logical and carefully thought out questionnaire."[12] Indeed, his experimental methods, developed via his breeding experiments at Rothamsted Experimental Station, and documented especially in *Statistical Methods for Research Workers* (1925) and *The Design of Experiments* (1935), were instrumental in the subsequent rise of modern randomized controlled trials (RCTs) in biomedical

---

[12] Fisher (1926, p. 511); cf. G.E.P. Box (1976) and J. Box (1980).

contexts. Second, he also developed an early causal analysis in terms of abstracting out different independent sources of variation (and, importantly, their potential interactions) correlated to an overarching dependent variable. This analysis – his famous analysis of variance or ANOVA – appears throughout the pages of this book. Such analysis of variance provides a statistical method for correlating changes in explanatory or independent variables (e.g., hereditary factors or genes) with changes in the measured dependent variables (e.g., phenotypes or appearances).[13] Finally, Fisher also found a robust way to estimate, from given data, the best values of the ideal parameters associated with mathematical models – maximum likelihood estimation (MLE), which remains a powerful statistical estimation method to this day.[14]

Although *Our Genes* takes one cue from Fisher's synergistic efforts in genetics and statistics, we cannot fully understand the emergence of statistics through the "Fisherian needle-eye," without also considering Fisher's energetic commitment to differential propagation, controlled breeding, and eugenics. Of particular relevance here is Fisher's model-building and statistical data analysis in investigating data on yields of crops at Rothamsted Experimental Station in the 1920s, and his historical and sociological investigations of the role of fertility differences in families of humans, particularly of different social classes, in powerful civilizations. Furthermore, in 1934 Fisher replaced Karl Pearson as the editor of *Annals of Eugenics* (today, and since 1954, called *Annals of Human Genetics*). In his editorial foreword, Fisher identified "Genetics and Mathematical Statistics" as "the two primary disciplines" "contributing to the further study and elucidation of the genetic situation in man."[15] We absolutely cannot support Fisher's eugenic efforts, which are also on display in *The Genetical Theory of Natural Selection*.[16] However, we can see that he, like many scientists of his time, was obsessed with the relevance of biological features to the ongoing evolution or dynamics of society. In fact, we must recognize and critique the, broadly speaking, eugenic concerns expressed by Fisher, Galton, Pearson, and Haldane – the latter another architect of modern evolutionary genetics, and one who directly

---

[13] Fisher (1918, 1925); cf. Rao (1992).

[14] Fisher (1912, 1922); cf. Edwards (1972), Aldrich (1997), and Thompson (2018). Jerzy Neyman (1894–1981), who, like Fisher, was also associated with agricultural studies, developed hypothesis testing (with Karl Pearson's son, Egon Pearson) in two papers (Neyman and Pearson, 1928, 1933), whereby rival hypotheses are tested against each other in terms of ratios of their respective MLEs. Amir Najmi provided constructive feedback here.

[15] Fisher (1934).   [16] Fisher (1958 [1930]); cf. Esposito (2018).

influenced Aldous Huxley's *Brave New World* – as neither incidental nor accidental to the rise of modern statistics.

The emergences of modern evolutionary genetics and of modern statistics are fundamentally and interestingly – which is not to say inevitably – intertwined. Fisher's discoveries were contingent and conditional to several, related domains – including agriculture, evolution, and genetics – that were remarkably adapted to statistical thinking. Why? As we have already begun to see, these domains feature manipulatable components, which require empirical measurements based not solely on human judgment. In addition, the manipulations of the components have potential sizable and replicable collective effects that can be controlled and randomized. Indeed, manipulations of the components in each domain are multidimensional, both in terms of those very manipulations, otherwise recognizable as independent variables (x), and in terms of responses, recognizable as dependent variables (y). In agriculture, as in evolution, as in genetics, multiple combinations of explanatory factors can be run over moderate periods of time in order to assess the measurable effects of the factors: Even if an exhaustive search is impossible, well-designed experiments (namely, Fisher's "questionnaires" to nature) are feasible.

Naturally, statistics evolved, and it continues to evolve today. Methods and applications of statistical inference keep on developing and progressing. As statistics grows, nonlinear dynamics, complex causal interactions, and the "curse of dimensionality" become increasingly central phenomena to understand and domesticate – Fisherian statistics, as it were, cannot really do justice to them. What fascinates me here, however, is the strong historical interrelation and, I believe, almost necessary intertwinement of the simultaneous emergence of statistics and evolutionary genetics. This is a key theme for *Our Genes*. Unlike many critics, who find in evolutionary genetics and statistics a propped-up rationale for identifying population-based inferiorities and superiorities, I see beyond such fractious conclusions drawn by some twentieth-century geneticists and genomicists. I argue that modern statistics has helped show our genomic history as marked by similarity, not difference. Our genes are a mirror through which we can see, and reflect on, our ultimate connection to one another and to all of life.

A Thought Experiment

Now, with this background on the development of the genetic paradigm and statistics, we can begin to dream of the many ways we might use their findings. For instance, let's imagine that we have landed in the largest city of an alien planet. Let's say we are in the planet's capital. We quickly notice that the inhabitants are all within a few centimeters of the same height, all have a nearly identical muscular body, and all have exquisitely similar facial features. Perhaps most surprisingly, every inhabitant has purple skin. We learn from the ambassador who greets us that every person on the planet looks the same. She is no exception. Why the purple skin? The ambassador says it is a developmental effect of the planetary system's double sun interacting with the inhabitants' skin pigment proteins. These proteins, she continues, are coded for by a few genes, almost all of which have only one allele shared by all individuals. She explains the uniformity by acknowledging that, "exactly 619 years ago, we experienced the Great Tragedy, when ecological catastrophes, three pandemics, and civil wars decimated us, and only the tiniest group managed to survive. The survivors committed to preventing further social and ecological destruction and established a strong social contract. Among other concessions, they agreed to generation upon generation of voluntary, totally random breeding – meaning, all inhabitants of a generation across the entire globe were paired up in a fair lottery for reproductive purposes – in order to minimize future ethnic strife."

Call this *Planet Unity*.

Now, imagine the Galápagos Islands – the actual ones, on Earth. You may recall that Darwin examined Galápagos finches and tortoises while assembling his theory of evolution by natural selection. Here is the fantasy part: Drop identical small populations of early hominins onto different islands of this archipelago. Actually, add a few more dozen islands, make them impossible for the hominins to travel across, and give them starkly different environments. Maybe one of them is mostly sand and another is brimming with lush vegetation. Ask yourself: What would happen after untold millennia of evolution? Will the outcome be like Planet Unity, with the descendants of the early hominins remaining practically indistinguishable? Not at all. They will differ wildly in body, behavior, and probably also culture, eventually becoming different species.

This we shall call *Galápagos-Writ-Large*.

Now, let's take our thought experiment a bit further.[17] Within which of these two disparate scenarios should we place *Homo sapiens* today? Are

[17] I first presented this thought experiment in Winther (2018c), which was completed in 2015.

we more like the inhabitants of Planet Unity or more like those of Galápagos-Writ-Large? (Figure 2.1, chapter opener). Further, can the correct identification of the extreme scenario we more closely resemble tell us anything about our evolutionary origins, or about what we are like, not just in body but also in mind and society?

Given the manifest, observable differences in humans across the world, it may seem natural to identify with the scenario on Galápagos-Writ-Large. After all, we do not all have the purple skin of the inhabitants of Planet Unity – far from it – and we appear, at first glance, quite different from one another. Yet, appearances can be deceiving. The truth is, we are much more like the inhabitants of Planet Unity: We look and act much more alike than not. In fact, genomic models suggest that one of the most important answers to the question of *Who am I?* can only be found by asking *Who are we?* Genomic models do not let us rest in the superficial antagonism favored by us–versus–them politics. Instead, genomic models and methodologies show that we share common origins, common ancestry, and common migration patterns, and that we are all quite similar indeed. Our young species originated in Africa from highly related populations. Small groups left Africa – several times – expanded in size and range, and were modified somewhat by natural selection, random genetic drift, and mutation. (There was also some migration *back* to Africa.) In this way, we populated the world, occasionally interbreeding across significant distances throughout our history, more so in the relatively recent past.

In fact, the genomics of the past few decades have challenged and confirmed, often in surprising ways, earlier beliefs about the global history and distribution of human genetic diversity. We now know that *Homo sapiens* has relatively little genomic variation compared to many other species, that non-African genomic variation is something like a subset of African variation, and that most genomic variation is found within local human populations, with only about 7% found across continental groups.[18] In a crucial sense, *we are all Africans.*

Planet Unity and Galápagos-Writ-Large are, of course, extreme points at the ends of a spectrum of population structure. Yet, when we examine *Homo sapiens* in general and on average, as if surveying a map of the entire world on a single sheet of paper, our species fits the Planet Unity model surprisingly well. And, in at least some places on Earth – especially big

[18] This is *Lewontin's distribution*, as I call it, and is one of six empirical patterns fleshed out in Chapter 6.

multicultural cities, such as London and San Francisco – our children are becoming even more Unity-like. It is also true that if we zoom in to a finer grain, as if peering at a map of a neighborhood of Mexico City or New Delhi, to explore specific genes or compare just a few populations, our species exhibits some Galápagos-Writ-Large properties. For many genes, there are at least small allele frequency differences in populations from Africa, Central and South Asia, East Asia, Europe, the Middle East, Oceania, and the Americas. And for some genes, there are enormous allele frequency differences in different parts of the globe. This smidgen of Galápagos-Writ-Large is what makes the evolutionary genomics of our species so fraught, and so rich with investigative possibility: It makes us ask if, for example, alleles associated with cancer, sprinting speed, or intelligence might be unevenly distributed across the globe. If so, what consequences would knowledge of this have for tailored medical therapies or for social control and surveillance? Our Planet Unity characteristics mean we are all quite similar, but our Galápagos-Writ-Large characteristics indicate important heritable differences that could, potentially, be politically and socially controversial, particularly if social or medical interventions are discovered or identified that could influence or even change the effects of such differences.

Genomic Journeys

Although we can formulate answers to the question of *Who am I?* by looking to the developments of the genetic paradigm and the field of statistics, and while we gain further clarity by asking *Who are we?*, we learn even more by asking questions about who we *were*. In fact, we can find nuanced answers to questions about origins, identity, mutability, and futurity by reaching much further back in time and applying modern genomic models, informed by statistics, to our earliest human ancestors. The narrative for the remainder of this chapter moves deeper and deeper in time. In so doing, we find fundamentally related answers to our identity-based questions: Where and how did *Homo sapiens* emerge? Is there an original, earliest mother and father? Which other species and subspecies are on our same branch of the tree of life?

You may already have an inkling that genomic models are helpful in tracing lineages and ancestry. When a mutation, or a set of larger-scale genetic variants, occurred generations ago in a set of individuals that tended to reproduce together in a population, the mutation or genetic variant is traceable in the DNA within tissues and cells we might find in ancient human remains such as bones or teeth (or frozen bodies or body parts), or from their presence in humans of present-day populations. Genetic-level studies from developmental biology and medicine help us track and understand the mechanistic effects that given mutations or genetic variants have within an individual's lifetime. Consequently, identifying your or my population(s) of origin can provide information relevant not just for identity formation, but also for potential predictions about, and interventions in, our bodies and minds.

Although less reductionist approaches to the biological sciences exist, much if not most scientific practice today on the origin, population structure, and biomedicine of *Homo sapiens* is performed at the genomic level.[19] In part this is because genetic information is understood by many to be transmitted across generations and to be causally effective during the ontogeny of individuals. Given the current state of science, there can be little doubt that genes are important – but exactly how and to what extent remains open for debate.

Questioning European Origins

As illustrated above, although genomic evidence indicates we are very much like the inhabitants of Planet Unity, there is enough systematic difference in the human gene pool – that is, sufficient Galápagos-Writ-Large in our genes – to trace and tell robust origin stories. We know from lived experience that questions about origins are fraught in this hyperpolitical age. National, ethnic, and cultural origin narratives often play essential social and political functions by supplying some form of an answer to questions of origin and identity. Consider examples from across Europe: Romulus for the Roman Empire; or Arminius (or Hermann), a contemporary of Jesus, known as a unifier of the Germanic tribes in his victory over the Romans at the Battle of the Teutoburg Forest; or narratives of deep prehistoric uniqueness among the Basques, or Finns.[20] Such particularizing and ethnocentric origin

---

[19] Complementary perspectives to so-called *genetic reductionism* include the following: self-organization and autopoiesis (Maturana and Varela, 1980); gene–organism–environment co-interaction and niche construction (Levins and Lewontin, 1985); levels of selection (Maynard-Smith and Szathmáry, 1995); developmental systems theory (Oyama, 2000a, 2000b); general systems theory (Winther, 2008); and the extended synthesis (Pigliucci and Müller, 2010; Laland et al., 2015).

[20] For the cultural narrative of Finnish uniqueness, see the Finn national treasure epic poem *Kalevala*. On "the origin of the Jews" see Weitzman (2017).

myths remain strong in different European countries (and ethnicities). As is common with communal and socially normed historical mythologizing, these stories serve multiple purposes, including perhaps naïve identity formation or more pernicious political us-versus-them, anti-immigration rhetoric.

Despite their ongoing ubiquity, genetic studies of archaic human remains cast significant doubt on such stories and on the answers they provide to questions about origins.[21] Evidence from Neanderthals, or from Bronze Age skeletons in Europe, Central Asia, and the Middle East, indicates that humans cross-mated significantly, migrated dramatically, and consisted of many more groups than myths represent. Indeed, evidence of migrations consequent to the spread of farming technologies (Neolithic Age) or bronze metallurgy (Bronze Age) suggests that the Europe we know today was born during the Neolithic Age (roughly 6500–3000 B.C. for southeastern Europe) and, in particular, during the Bronze Age (roughly 3000–600 B.C. for southeastern Europe).[22]

Mythological origin stories strive for simplification, but the history, informed by evidence from the genetic paradigm, is more complex: To take just one example, consider the Yamnaya peoples who moved into Europe on horses from the Caspian steppes and mixed with Neolithic farmers. According to the Yamnaya hypothesis (Figure 2.2, top), in (roughly) northeastern Europe, this mingling gave rise to the Corded Ware culture, within which the Proto-Germanic and Proto-Slavic languages likely arose and spread. This basic pattern of migration and mixing in Bronze Age Europe has been corroborated using DNA evidence sampled from 101 ancient individuals from a variety of archaeological sites across Europe and Central Asia.[23] Danish geneticist Morten E. Allentoft and the coauthors who analyzed the evidence wrote: "Our results imply that much of the basis of the Eurasian genetic landscape of today was formed during the complex patterns of expansions, admixture

---

[21] For a genomic window on Basque roots, see Günther et al. (2015).

[22] Bellwood (2013, 2017) and Gamble (2013) discuss these and many other cases of early or ancient human migrations. The gradient between the Neolithic and Bronze ages is marked by introduction of bronze metallurgy and tool-making, from the Near East, diagonally up into Europe. It is perhaps hyperbolic to say this is the birth of Europe, but the cultural and biological foundations of Europe were indeed established with all these geographic expansions and migrations. This was the key "peopling" of Europe (or "Europe 2," according to Manco, 2016). Interested readers could also consider what role, for example, the Roman Empire, Slavic expansion in the first millennium, and nation-state building associated with post-1492 colonialism subsequently played in European identity – and culture and biology.

[23] Allentoft et al. (2015) and Haak et al. (2015).

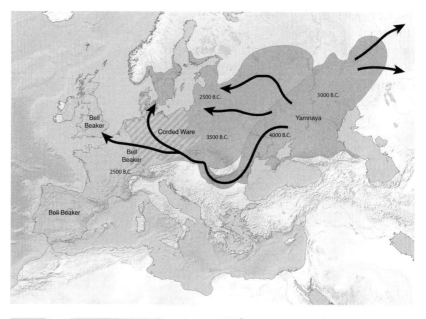

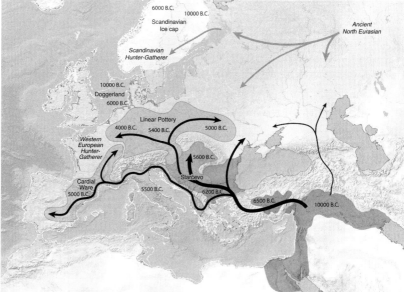

Figure 2.2. Making Europe

Maps showing two main hypotheses for the origin, migration, and expansion of European peoples and languages. (bottom) First, the Anatolian hypothesis postulates migration of farming peoples out of the Fertile Crescent, though Anatolia, and into Europe in the late Neolithic. This helps explain the origins of Linear Pottery and

and replacements during this [Bronze Age] period."[24] Indeed, the Yamnaya hypothesis for the making of modern Europe seems increasingly important compared to Renfrew's Anatolian hypothesis (Figure 2.2, bottom),[25] though the two need not contradict. After all, they cover different, overlapping periods, and each plausibly captures part of the migration patterns into Europe, even if the Yamnaya hypothesis is the likely picture for the origins of the Proto-Indo-European language.[26] This example illustrates the reality that simple historical myths about single "peoples" forming homogeneous and flat nations are rarely accurate or predictive.

Evolution within Africa; Migrations Out of Africa

The genomic evidence indicates that populations to which you or I belong are less particular and unique than cultural origin stories suggest, and that the *origins* of the populations to which you or I belong are less particular and unique than many of us would perhaps like to believe.

←

Caption for Figure 2.2. (cont.) Cardial Ware cultures found in the archaeological record, whose approximate extensions are shown here. The Anatolian hypothesis also embodies the spread of one of the three ancestral archaeogenetic components of European peoples, the Early European Farmer (EEF). Along with two other inferred ancestral European components, the Ancient North Eurasian (ANE) and Western European Hunter-Gatherer (WHG), EEF emerged around 22,000 B.C. (ANE remains are found near Lake Baikal in Siberia; Scandinavian Hunter-Gatherer emerged around 6000 B.C., with WHG partial ancestry; see also "Ancestral European" inset in Figure 2.10.) Glaciation extent and changing coastlines at 10,000 B.C. and 6000 B.C. are indicated, especially for Northern Europe. (top) Second, the Yamnaya hypothesis postulates the origin of Proto-Indo-European people between the Black and Caspian seas, and their migration across the Eurasian steppes, during the Bronze Age. Genomic evidence suggests that Yamnaya provided at least partial ancestry to Corded Ware culture. Bell Beaker culture arose through significant migration of Corded Ware into the British Isles ("peoples, not pots"), although particularly in the Iberian Peninsula, the Bell Beaker culture likely originated through cultural diffusion rather than migration (cf. Kadrow, 2018, and Olalde et al., 2018). (Sources include Allentoft et al., 2015, Haak et al., 2015, and Reich, 2018.) Illustrated by Mats Wedin and Rasmus Grønfeldt Winther. © 2021 Rasmus Grønfeldt Winther. (A black and white version of this figure will appear in some formats. For the color version, please refer to the plate section.)

---

[24] Allentoft et al. (2015, p. 170). See also Haak et al. (2015, p. 210).   [25] Renfrew (1990).
[26] For a review from a linguistic point of view, see Anthony and Ringe (2015).

It can be hard to imagine that most of Earth's landmass was not host to humans for almost the entirety of our planet's existence. But it is true. It was from Africa, and Africa only, that modern *Homo sapiens* spread across the world, and we did so quite recently indeed (Figure 2.3).[27] There is some agreement that the last significant *Out of Africa* dispersal event occurred between approximately 55,000 and 65,000 years ago, and Figure 2.3 represents this temporal range. Even so, there is evidence for previous dispersals, including into Asia as early as 120,000 or even 130,000 years ago and to at least Greece as early as 210,000 years ago.[28] Migration routes out of Africa, and then across Asia and/or into Europe, are not straightforward. For instance, two migration paths into Eurasia have been substantiated with archaeological and genomic findings: either from Africa's horn to the Arabian Peninsula (the "southern route" or "southern dispersal hypothesis"), or up through Egypt and into the Levant (the "northern route" or "northern dispersal hypothesis").[29] While non-exlusive, the latter seems the more likely candidate route, simply because it does not involve a water crossing.[30] Much remains to be learned about the timing and the spatial organization of Out of Africa migrations across the globe. Regardless of routes taken and destinations reached, the earliest populations originated in Africa.

The origin of human populations within Africa is much less clear. The debate about the precise site(s) of *Homo sapiens* evolution is one of continued controversy and, as such, also points to the current potential and limitations of genomic models. Let's turn first to the Great Rift Valley. According to genomic models, this area, more precisely known as the East African Rift System, is a key contender for the geographic region in which early humans and related primates evolved (Figure 2.4). A strong fossil record of early humans back to at least 160,000 years ago or so in this system supports this narrative.[31] Two key (and likely familiar) sets of players did fieldwork that led to relevant discoveries in this part of the world. First, the Leakey family – Louis and Mary Leakey and their son Richard – undertook fieldwork in the 1930s.[32] Later, in the late 1950s, Jane Goodall joined an expedition to

---

[27] See Maslin (2017) for a readable account. See empirical pattern #6 in Chapter 6.

[28] 120,000 or even 130,000 years ago: Bae et al. (2017); 210,000 years ago: Harvati et al. (2019).

[29] See, e.g., Pagani et al. (2012, 2015) and Bae et al. (2017). For further details, including relevant archaeological, paleontological, and linguistic evidence, consult Bellwood (2013, 2017).

[30] Bae et al. (2017, p. 2).

[31] See, e.g., Stringer (2012, 2016), Hublin (2013), and Hammond et al. (2017).

[32] Morell (1995).

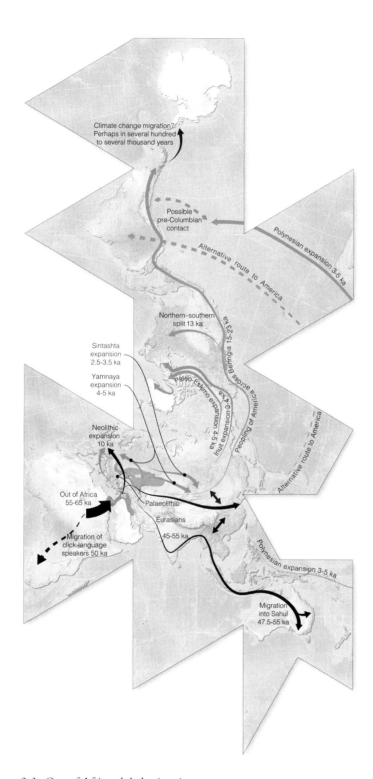

Figure 2.3. Out of Africa global migrations

Olduvai Gorge with Louis and Mary Leakey, after Louis Leakey asked Goodall if she was "willing to tackle [a] job."

According to Goodall, Louis was "particularly interested in the behavior of a group of chimpanzees living on the shores of [Lake Tanganyika] – for the remains of prehistoric man were often found on a lakeshore and it was possible that an understanding of chimpanzee behavior today might shed light on the behavior of our stone age ancestors."[33] At the time, the Leakeys basically discovered one hominin fossil after another. Meanwhile, Goodall did ground-breaking primatology research on chimpanzees in their natural environments, discovering phenomena such as tool-making and tribal warfare. Although these early studies were based on anatomy and archaeology (Leakeys) and comparative behavior (Goodall) rather than genomics, which emerged as a tool for exploring human evolution only decades later, the studies helped place the story of early human evolution on the appropriate corner of the map – East Africa.

The East African Rift System is thus an important and historically influential candidate for a single, coarse location for our shared African origin and subsequent evolution. However, plausible reasons also exist for seriously considering a southern Africa contender, given the distinct possibility that contemporary Khoikhoi and San are fairly direct descendants of ancient peoples.[34] Pagani et al. frame the issues at play:

Caption for Figure 2.3. (*cont.*) Key averaged routes and best temporal ranges of significant human global migrations, on an unnamable yet inspirational map projection. Both return movement and stable settlement occurred. Alternative routes across the Pacific to the Americas remain interesting possibilities. Changes in coastal landscapes due to changing sea levels (from +8 to −140 meters compared to present day), and changing ice cover especially in the Northern Hemisphere, including of Beringia, are visually suppressed. Will humans one day migrate to Antarctica? FC = Fertile Crescent; CA = Central Anatolia (per the Anatolian hypothesis); YR = Yamnaya Region (per the Yamnaya hypothesis); ka = thousand years ago. Illustrated by Mats Wedin and Rasmus Grønfeldt Winther. (Sources include Cavalli-Sforza and Feldman, 2003 and Nielsen et al., 2017.) © 2021 Rasmus Grønfeldt Winther. (A black and white version of this figure will appear in some formats. For the color version, please refer to the plate section.)

---

[33] Goodall (2010 [1971], p. 6).

[34] See, e.g., Schlebusch et al. (2017), who also use contemporary genomic data to place the origin of *Homo sapiens* between 260,000 and 350,000 years ago; see also Henn et al. (2011).

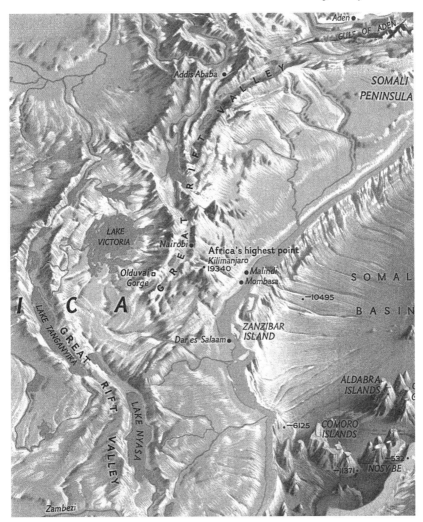

Figure 2.4. The East African Rift System
The strong fossil record in this area indicates that this was a crucial region for some hominin and some early human evolution, one species of which was to significantly impact the future of Earth as an evolving system (as hypothesized by James Lovelock in the Gaia paradigm of Lovelock, 1979), some few million years hence. As this is a cut-out of the 1967 *Indian Ocean Floor* panorama map, note cartographer and oceanographer Marie Tharp's masterful touch (in collaboration with artist Heinrich C. Berann and fellow oceanographer Bruce Heezen) in connecting marine and terrestrial geological features (see Winther, 2019). The eastern and western branches of the East African Rift can be seen clearly in the map on either side of Lake Victoria, and the triangular Ethiopian Rift can be seen toward the top of the map and to the

[Genomic] analyses pointed to click [Khoe and Sandawe] speakers, Pygmies, and a Nigerian-Congolese group as all having a deeper population history than both the whole genome and the African component of the East Africans sampled. Although this result might seem inconsistent with the outstanding fossil record available from Ethiopia, it may illustrate that genetic diversity assessed from modern populations does not necessarily represent their long-term demographic histories at the site. Alternatively, the rich record of human fossil ancestors in Ethiopia, and indeed along the Rift Valley, may reflect biases of preservation and discovery, with more fossils being exposed in regions of geological activity. Fluctuations in effective population size in the past and dispersals within Africa may have further confounded our analyses and their correlation with the fossil record.[35]

The truth is, despite the informed interpretations offered above, there may be no unique, geographically isolated, proverbial Garden of Eden in which modern humans evolved from our direct ancestors.[36] Also possible is an African multiregionalism hypothesis, which asserts that modern *Homo sapiens* evolved within Africa, but not in a single place. Africa, the second-largest continent, is immense, and no "linear progression towards later *sapiens* morphology" can be found; rather, various "'archaic' and 'modern' morphs" overlap temporally, producing "many potentially interfertile subdivisions of the evolving *sapiens* species across Africa."[37] The multiregionalism hypothesis is taken further by Scerri et al., who suggest a "Pan-African Cultural Patchwork."[38] Here, some Galápagos-Writ-Large structure gets built into a broader and more general Planet Unity model. Although the researchers respect and use genomic data, they also appreciate the limitations of genomic inferences, correctly warning that "all such models are necessarily abstractions and simplifications of the true population histories."[39] They therefore complement

---

Caption for Figure 2.4. (cont.) east of Addis Ababa. These rifts are found at the triple point (junction) where the Nubian African, Somalian African, and Arabian plates diverge from one another. This East African Rift System will eventually be inundated and become an ocean basin (Chorowicz, 2005). *Indian Ocean Floor* panorama map by Heinrich Berann/National Geographic Creative. Reprinted with permission. (A black and white version of this figure will appear in some formats. For the color version, please refer to the plate section.)

---

[35] Pagani et al. (2012, p. 93). [36] Hublin (2013, p. 525). [37] Stringer (2016, pp. 1, 7).
[38] Scerri et al. (2018, pp. 585–587). [39] Scerri et al. (2018, p. 588).

genomic information with data from archaeology and paleoecology, including material culture evidence. This involves tools and cultural artifacts such as heavy-duty axes, engraved ostrich eggshells, and bone points.[40] In the researchers' words, "the African archaeological record demonstrates the polycentric origin and persistence of regionally distinct Pleistocene material culture in a variety of paleoecological settings."[41] Paleontologist Louise Leakey, daughter of Richard Leakey, agrees.[42] What we do know is that the modern *Homo sapiens* globular brain shape, which "was not established at the origin of our species," appeared between approximately 35,000 and 100,000 years ago, across Africa.[43] We also know that environmental and ecological complexity, such as the intermittent appearance of freshwater lakes and rivers up and down the East African Rift System and "periodic extensions of arid and desertic zones," drove early human population structure and evolution toward biological and cultural complexity.[44] Thus, underdetermination of the plurality of our best geographic models of human evolution by limited genomic data is very much alive and well.

Our Ancestral Mother and Father

As we journey to our ancestral homes, we can trace our genealogy back to our shared and unique ancestral parentage. Doing so offers another means for building a genetic case for our similarities and, thus, for Planet Unity. This is because inside the bodies of every single one of us – right now – there is DNA we have inherited, with subsequent changes, directly from our common mother: the mitochondrial DNA from *Mitochondrial Eve*, the "lucky mother" of everyone on Earth, estimated to have lived 99,000–148,000 years ago.[45] In the bodies of roughly half of us, there is DNA we inherited, with subsequent changes, directly from our common father: the Y chromosome from *Y-Chromosomal Adam*, the father of all men, estimated to have lived 139,000, 254,000, 275,000, or

---

[40] Scerri et al. (2018, pp. 584–585).
[41] Scerri et al. (2018, p. 582). See also Skoglund et al. (2017) and Mounier and Lahr (2019).
[42] Yong (2018).   [43] Neubauer et al. (2018, pp. 1, 3).
[44] Hublin (2013, p. 525); see also Maslin (2017) and Scerri et al. (2018).
[45] See Poznick et al. (2013). Earlier work, such as Cann et al. (1987), had the range a bit further back in time. Per Wilkins (2012), one of the pioneers of these studies, Allan Wilson, preferred the term "lucky mother" since it emphasized the role of chance and the random pruning of all maternal lineages except one.

even 338,000 years ago.[46] Both Mitochondrial Eve and Y-Chromosomal Adam lived in Africa.

To be clear, Mitochondrial Eve and Y-Chromosomal Adam were not the first human woman and man. They were only the most recent matrilineal and patrilineal ancestors common to all modern humans. And even then, they are what evolutionary geneticists call the "most recent common ancestors" of only a small part of our genomes – respectively, mitochondrial DNA and Y chromosomes. To call them "Adam" and "Eve" should neither imply that they reproduced together nor even met – they almost certainly didn't live during the same millennium. That Y-Chromosomal Adam likely lived out his days long before Mitochondrial Eve simply means that during his lifetime there were several mitochondrial lineages (meaning, female genealogies), most of which were eventually pruned out of existence. Only one mitochondrial lineage survived: Mitochondrial Eve's, millennia downstream from Y-Chromosomal Adam.[47] And, because males also have daughters, some male descendants of Y-Chromosomal Adam had daughters *not* in the matriline of Mitochondrial Eve.

Coalescent theory stands as the basic mathematical theory and method used to make these kinds of inferences. It involves the premise that "for any pair of individuals in [a] population, their genealogy can always be traced back to a common ancestor."[48] If we generalize this to many individuals, we can infer a most recent common ancestor (MRCA) to all individuals (Figures 2.5–2.7).[49] If the theory sounds too simple (and it will indeed require some explication and qualification), consider the marvel of mitochondrial DNA and the DNA of the Y chromosome: They represent genealogies of, respectively, mother–daughter *matrilines* and father–son *patrilines*. Mitochondrial DNA and Y chromosomes are remarkably useful for tracing ancestor–descendant relations because they

---

[46] 139,000 years: Table 1 in Poznick et al. (2013, p. 564); 254,000 years: Karmin et al. (2015, p. 460); 275,000 years: Mendez et al. (2016, p. 731); and 338,000 years: Mendez et al. (2013, p. 454). See also Jobling and Tyler-Smith (2003).

[47] Are we to conclude that females or males are "more evolved," whatever exactly that means? In case it is not obvious, women and men have always evolved together from the very beginning of modern *Homo sapiens* approximately a quarter of a million years ago (Table 2.1), with almost all genetic mutations, selective pressures, and physiological and morphological adaptations affecting both sexes. See also Richardson (2013) and Fausto-Sterling (2020).

[48] Hartl and Clark (1989, p. 134); cf. chapter 3 in Nielsen and Slatkin (2013).

[49] The evolutionary genetics of coalescent theory was first worked out by Sir John Kingman (1982a, 1982b). For a historical perspective, and the importance of the interaction between Kingman and Warren Ewens, see Kingman (2000). See also Wakeley (2008).

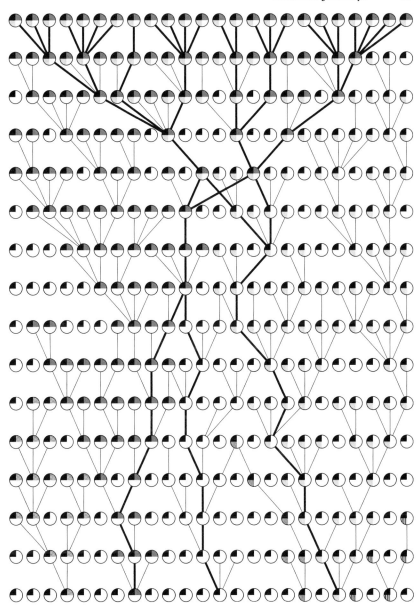

Figure 2.5. Four Gene Coalescent (Incomplete and Idealized)
An incomplete, idealized, and distorted projection of an actual full phylogeny
embedding complex pedigrees. A variety of evolutionary processes are represented
here simultaneously. Each of four autosomal genes is represented by a distinct
quarter of a circle (upper-left quadrant: blue; upper-right: reds; lower-right: yellow;

are both inherited as "blocks" (technically, *haplotypes* or, more generally, and when a set of similar haplotypes have a common ancestor, *haplogroups*). Unlike all non-sex nuclear chromosomes (known as *autosomes*) and unlike X chromosomes, mitochondrial DNA and Y chromosomes do not have a pair with which to exchange pieces of DNA during the making of eggs or sperm cells. They thereby do not mix and match across lineages. In technical language, they do not recombine. They act like a single unit – the alleles of the different genes on the haplotype ride together, if you will.[50] We can thus coalesce all

---

Caption for Figure 2.5. (cont.) lower-left: greens). Alternative gene allelic lineages are shown with different textures (black and white image) or color shades (reds and greens, color plate image). Time moves up in idealized generations. For a given gene, there are always alternative lineages, where all but one must go extinct (e.g., only the solid lower-left quadrant or the mid-tone green of the three lower-left lineages does not go extinct; Kingman, 1982a, 1982b; Wakeley, 2008; Felsenstein, 2004). That is, for each gene, all contemporary individuals, at both loci on the relevant chromosomal pair, must share a single common gene ancestor ("identity by descent" of all current alleles of a gene, e.g., Thompson, 2013; see also Malécot, 1969 [1948]). Furthermore, while tricky to show, sexual reproduction is idealized as haploid-like, analogously to an ancestral recombination graph (see Figure 2.7). While each gene allelic identity by descent lineage moves in a "pinball" fashion through the full sexual reproduction *inclusive* demography (with potential mutation), in effect the lineages per locus must be *exclusive* (cf. Felsenstein, 2004, Figures 26.1–26.3, pp. 451–453). (Mutation is not shown here.) Only a few lineage mergers in acts of sexual union and subsequent recombination are depicted here, although this must occur frequently in the full phylogenetic pedigree. That is, the figure highlights the moments of union and recombination when two lineages that ultimately prevail merge. Not all demographic relations are portrayed (nor is horizontal gene transfer). Indeed, while circles can be interpreted as representing individuals, it is advantageous to imagine them as idealized, multi-individual pedigree segments of the gene allelic lineages. An individual is diploid, and thus it and its gametes can be part of two different circles in the diagram (for an alternative visual strategy for diploids, see Felsenstein, 2004, Figures 26.1 and 26.2). The upper-left quarter coalesced a long time ago and is a lineage present in every individual proxied in this incomplete representation. Blank quarters imply either alternative lineages not represented or incomplete information. Concept by Rasmus Grønfeldt Winther, illustrated by Mats Wedin and Rasmus Grønfeldt Winther. © 2021 Rasmus Grønfeldt Winther. (A black and white version of this figure will appear in some formats. For the color version, please refer to the plate section.)

---

[50] Returning to the three laws of the early genetic paradigm, there is no segregation, no independent assortment, and pure linkage of genes in haplotypes of mitochondrial DNA and Y chromosomes.

mitochondrial DNA or Y chromosomes back to two, respective ancestors: Mitochondrial Eve and Y-Chromosomal Adam.[51]

More precisely, if we know how much genomic diversity or difference there is today among mitochondrial DNA or Y chromosomes, we can then estimate age backward by approximating how many average mutations per generation per nucleotide occur. The basic empirical method for telling time according to this "molecular clock" requires gathering mitochondrial DNA and Y chromosomes from individuals native to many different areas of the globe, and then using statistical phylogenetic methods to build phylogenetic trees showing which mitochondrial DNA (or Y chromosomes) from which individuals are most alike, and how and at which nucleotide sites they differ, estimating the timing both of branching events and of the most recent common ancestor, also calibrated against generation time of the species, or independent paleontological data.[52] To tell time in this way, a broad variety of mitochondrial DNA (or Y-chromosomal) haplotypes and haplogroups are compared, and, based on their relative similarity and difference, they are placed on different branches of an overarching tree.

To help us better grasp the method, we can turn to a useful analogy gifted to us by geneticist Spencer Wells. Visualize a southern French village in the eighteenth century, Wells advises, where recipes for a signature soup are passed down from mother to daughter. Even if the recipe is *only* shared orally, and even if it is *only* shared between mother and daughter, diversity in the soup will increase with each generation. This is because, as every home cook knows, each daughter will make

---

[51] As we shall see, coalescent theory also involves the premise that "for any population sampled at a particular time, the genealogy of an allele in the sample can be traced back to a single common ancestor" (Hartl and Clark, 1989, p. 134). Interestingly, since humans have approximately anywhere from 20,000 to 40,000 genes, each of these genes could give us different coalescents, although some of the coalescing lineages for different genes almost certainly go back to the same individual. The set of these original individuals for the sum total of all genes could perhaps be thought of as an *ideal original coalescent population*, though this is a theoretical, rather than a natural, field, population, in senses to be discussed in Chapter 3, and the hypothetical individuals of it would almost never coexist, whether in space or in time. Moreover, the species status of at least some of these hypothetical individuals would be nebulous, as they would predate modern *Homo sapiens*. Related to my concept of an ideal original coalescent population for the total set of all genes, Rosenberg and Nordborg (2002) write: "unlinked or loosely linked loci can often be viewed as independent replicates of the evolutionary process. In the absence of recombination, the entire genome would correspond to a single genealogical tree, and we would never have more than a single independent replicate" (p. 381).

[52] On phylogenetic trees, see van Oven and Kayser (2009) and Mendez et al. (2013); on molecular clocks, see Moorjani et al. (2016) and Bridget and Moorjani (2017).

small modifications to her soup, and then communicate those changes to her own daughter: "the longer the village has been accumulating these changes, the more diverse [the soup] is. It is like a clock, ticking away in units of rosemary and thyme – the longer it has been ticking, the more differences we see."[53] If we could taste the diverse soups and identify the number of additions and modifications, and if we knew roughly the average number of changes an individual made to their soup, then we could count backwards, and specify the approximate generation in which the woman who first made it belonged. With this picture in mind, we see that as estimation methods of the molecular clock continue to improve, we will one day be able to refine estimation and "approximation" (and perhaps even our analogies).[54]

A few further biological details can shed light on our common mother and father by helping explain why lineages are patrilineal or matrilineal (Figure 2.6). The first is easy: Fathers pass on their Y chromosome only to sons, but not to daughters. (They do not pass on mitochondria to either daughters or sons.) The second is a bit more challenging to grasp. DNA is housed in chromosomes located in the cell's nucleus. That is where the X and Y sex chromosomes that operate during development reside (females with an XX combination and males an XY).[55] But a *mitochondrion* – an organelle in the cell (analogous to an organ in the body) – is far away from the nucleus, yet, interestingly, it has its own circular DNA strand. This is a relic from 2 billion or so years ago, when independently living bacteria started living inside larger bacteria, perhaps first as parasites. Then, after many generations, the interactive system became stable. That is, the system eventually evolved a better harmony, by which the previously weakening parasite became mitochondria that *had* to live inside the larger, now eukaryotic, cell. In so doing, mitochondria also came to contribute to the welfare of host cells by producing ATP (adenosine triphosphate, a trinucleotide), a ubiquitous, biological energy-carrying molecule. All proper multicellular life, including humans, are descendants of such ancestral eukaryotic cells.[56]

[53] Wells (2003, pp. 32, 39).   [54] See Poznick et al. (2013).

[55] But see Fausto-Sterling (2020).

[56] This is *endosymbiosis* or *symbiogenesis*. The theory that some evolution occurs through this sort of combination of previously independent cells was influentially developed by biologist Lynn Margulis. Such evolution is especially central to the origin of eukaryotes. Dismissed at first, the theory is now widely accepted (e.g., Sagan, 1967; Margulis, 1975; Sapp, 1994; Maynard Smith and Szathmáry, 1995; Winther, 2009b; Lynn Margulis had been married to Carl Sagan, hence her earlier surname.)

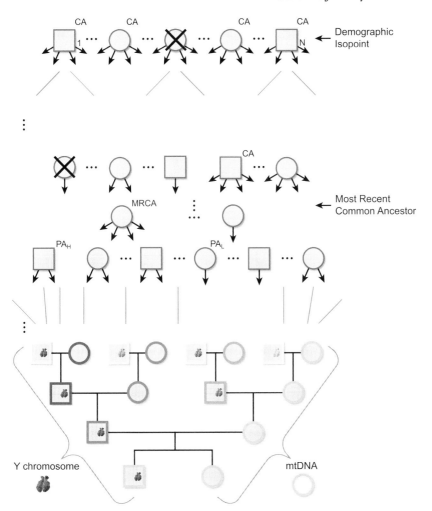

Figure 2.6. Deep Demography
Mitochondrial DNA ("mtDNA") matrilineality and Y chromosome patrilineality
represented at the bottom. Ignoring extreme inbreeding and other complicating
scenarios, every human has two parents, four grandparents, eight great-grandparents,
etc. But as we go back in time, ancestors are "duplicated" since they start being
shared by contemporary individuals, either partially (PA; unlabelled) – which can
interweave many (H, for high) or few (L, for low) demographic lineages – or with
everybody (CA), both within and across populations. There is a single, demographic
most recent common ancestor (MRCA). Further back, at the "demographic
isopoint," every individual is either the common ancestor of all contemporary
individuals (CA) or the ancestor of no one (X). Concept by Rasmus Grønfeldt
Winther. Partly illustrated by Bryant F. McAllister, University of Iowa (bottom
portion, with mitochondrial DNA and Y chromosome). Reprinted with permission.
Adapted by Rasmus Grønfeldt Winther.

Even in their intracellular location, mitochondria still divide autono-mously, employing not only their own DNA but also much of their own gene expression – replication, transcription, and translation – machinery to make their own proteins. Because all we get from our father's sperm are nuclear chromosomes, and because each of us started, as a cell, from our mother's egg, which was filled with her organelles, every mitochon-drion in every individual comes from the mother's egg.[57] Despite this, and perhaps counterintuitively, the mitochondrial genome is much smaller than the Y chromosome: (approximately) 16,000 nucleotide base pairs compared to (approximately) 59 million. Consequently, and given the current limits of comparative sequencing data, our estimates of how long ago Y-Chromosomal Adam lived are broad.[58]

For any one gene, the alternative descent lineages, only one of which coalesces, constitute an *exclusive* network. (The same is true for one larger nonrecombining unit such as mitochondrial DNA or the Y chromosome.) Since for a given locus there are only two spaces, one on each chromo-some, the genic genealogy over downstream generations is haploid-like, even if sexual reproduction can mix and match two lineages within single, diploid individuals (Figure 2.7). In essence, for any gene, the different gene allelic lineages of "identity by descent" neither mix nor intertwine.[59] Instead, they must compete in a zero-sum game, always eventually leaving just one lineage to be represented in all descendants (Figure 2.5).

Chang offered another, statistically informed, sense of shared human ancestry – an *inclusive* network based on demography (Figures 2.6 and 2.7).[60] Every human on the globe can trace their demographic ancestry back to a most recent single individual who lived some 3400 years ago.[61] She or he was your and my "great-great-great . . . great$_n$ grandparent."

---

[57] The egg (ovum) was itself made by an oogonium that grew (and exactly duplicated its nuclear DNA) into a primary oocyte, which divided twice within the ovary follicles, during meiosis, alluded to above. The oogonium and primary oocyte have the diploid number of 46 (nuclear) chromosomes (the primary oocyte's chromosomes have the "proper" X-shape, each with two identical sister chromatids); the egg has only the haploid number of 23 – one from each chromosomal pair, with recombined chromosomal segments. In fertilizing the egg, the sperm contributes 23 chromosomes, so that the offspring proceeds with the full complement of 46.

[58] Poznick et al. (2013) note how the mitochondrial genome "has been resequenced thousands of times" but full sequences of Y chromosomes "have only recently become available" (p. 562). For comparative perspectives on the global diversity of mitochondrial DNA and Y chromosome genomes, see, e.g., Jorde et al. (2000) and Underhill and Kivisild (2007).

[59] See Thompson (2013, pp. 305–308, 320–321); cf. "A caveat to forestall potential misunderstand-ing: this paper is not about genetics" paragraph in Chang (1999, p. 1006).

[60] Chang (1999).   [61] Rohde et al. (2004, p. 565).

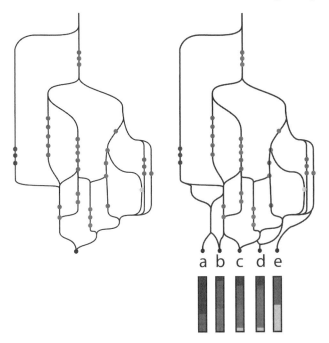

Figure 2.7. Ancestral recombination graph
(Left) Genetic ancestry in the form of the ancestral recombination graph (ARG) for a single individual. This representation shows the combination of different gene ancestries in graph form, idealizing away from all possible genetic ancestry paths. The ARG is embedded in the larger, actual pedigree. Only the nodes of ancestors where there was a coalescence or recombination event are portrayed; lines or edges depict genetic paths of descent, and linkage can facilitate the heredity of more than one gene together. (Right) The composite ARG of multiple, related individuals, also representing the ancestral sources of each individual's genome. From Mathieson and Scally, 2020. Published by *PLoS* under CC BY. (A black and white version of this figure will appear in some formats. For the color version, please refer to the plate section.)

We are united by this most recent common ancestor, even if we differ with respect to our *other* such great–great–great . . . great$_n$ grandparents in that generation. Then, before this point in history (at least n + 1 generations back), your and my inclusive demographic networks start overlapping to even *more* common ancestors. If we go even further back, we eventually reach a point where every single one of us on the globe will have "exactly the same set of genealogical ancestors."[62] For Rohde et al.'s conservative model simulation, this was some 7400 years ago.

[62] Rohde et al. (2004, p. 562).

Thus, in our globally shared, inclusive demography, there is a most recent common ancestor point (e.g., 1400 B.C.), as well as a *genetic isopoint* (e.g., 5400 B.C.), to use coinage by Rutherford,[63] though I would prefer to call it a *demographic isopoint* (Figure 2.6).[64] You and any full siblings you might have have a demographic isopoint only one generation back, with your parents. Double cousins – where a full sibling pair from one family couples with the full sibling pair from another family and both couples have children (e.g., two full sisters and two full brothers form two distinct parental couples) – have a demographic isopoint two generations back since all the children (cousins) have the same grandparents. More generally, and with any large family tree, including the one for all humans alive today, as we move back in time (generations), more partial ancestors accumulate, until we reach the most recent common ancestor, and then these ancestors start accumulating too until, eventually, we arrive at the demographic isopoint (Figure 2.6). Rohde et al. offer "a remarkable proposition: no matter the languages we speak or the colour of our skin, we share ancestors who planted rice on the banks of the Yangtze, who first domesticated horses on the steppes of the Ukraine, who hunted giant sloths in the forests of North and South America, and who laboured to build the Great Pyramid of Khufu."[65] These inclusive topographies, while informative, should not be confused with the exclusive topographies of the coalescing lineages of nonrecombining genetic units, such as Mitochondrial Eve and Y-Chromosomal Adam. Inclusive topographies inform genetic models with demographic evidence, helping to refine but also complicate our pictures of our pasts.

The preceding discussion should make it clear that, despite their significant limitations, genomic models have helped point out possible answers to questions about the origins of human populations.

Our Family Tree

All of the preceding also makes clear why it no longer makes sense to ask *Who am I?* Here, on this pale simulacrum of Planet Unity, the evidence suggests that we learn more about ourselves as individuals when we ask about the collective – not just *Who are you?* but *Who are **we**? Who were **we**? Where did **we** come from? From whom did **we** descend?*

Our answers depend, in some ways, on whom we are *not*. *Homo sapiens* today is not, for example, *Homo erectus* or *Homo heidelbergensis*, let alone *Australopithecus (Praeanthropus) afarensis* (Figure 2.8). But perhaps

[63] Rutherford (2020).   [64] Cf. Chang (1999, p. 1006).   [65] Rohde et al. (2004, p. 565).

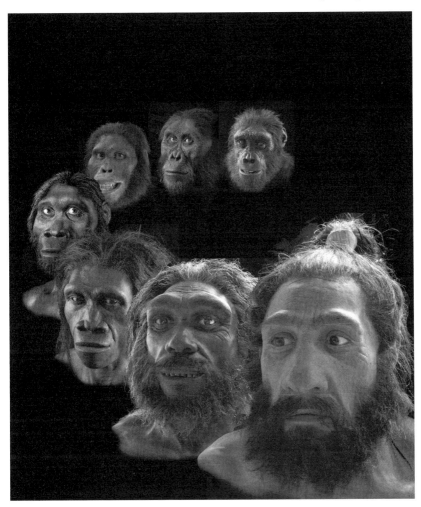

Figure 2.8. Hominin time spiral
Reconstructed models of a wide variety of hominins arranged chronologically from top to bottom along the spiral, without presuming to capture the branching phylogenetic relations (see Figure 2.11). From oldest to youngest: *Sahelanthropus tchadensis, Australopithecus (Praeanthropus) afarensis* (Lucy), *Australopithecus africanus, Homo habilis, Homo erectus, Homo heidelbergensis,* and *Homo neanderthalensis.* © 2021 John Gurche. Reprinted with permission. (A black and white version of this figure will appear in some formats. For the color version, please refer to the plate section.)

we are exceedingly similar to Neanderthals or Denisovans? There are consistent morphological and postulated behavioral differences between all these hominin species and subspecies, including differences in average brain size, presence or absence of air sacs in the vocal tract, the production of particular forms of tools and weapons, and foot morphology and function (Figure 2.9). Anthropologist Richard Klein usefully summarizes the prevailing view on the evolutionary history of hominins: *Homo erectus* evolved around 2 million years ago, within Africa, and spread throughout Eurasia.[66] It, or a subspecies closely related to it, is probably the ancestor of *Homo heidelbergensis*, which is likely the ancestor to *Homo sapiens* and our sister species (if not subspecies) Neanderthals and Denisovans (Figures 2.10 and 2.11).[67]

Importantly, and as carefully documented by evolutionary anthropologist Jean-Jacques Hublin, there are reasonably good African *Homo* fossils after 260,000 years ago and before 400,000 years, but a gap in between.[68] Recent discoveries of early *Homo sapiens* from some 315,000 years ago, found at the Jebel Irhoud archaeological site in Morocco, ameliorate this gap somewhat.[69] Europe, meanwhile, has a solid Neanderthal fossil record, even during this gap. While Neanderthals and Denisovans both likely emerged within, and migrated out of, Africa, Neanderthals also continued evolving in Europe and western Asia, and Denisovans called eastern and southern Asia and Oceania their later home.[70] Indeed, "African populations [of *Homo sapiens*] followed an evolutionary pathway clearly distinct from those found in Europe [Neanderthal], which lead to the emergence of a modern morphology."[71] These two subspecies or

---

[66] Klein (2009).

[67] I leave it as an exercise for the reader keen in taking this journey further to research *Homo floresiensis*, *Homo antecessor*, *Homo habilis*, *Australopithecus*, and so forth. Again, Klein (2009) is a good starting place, as is Kimbel and Villmoare (2016).

[68] Hublin (2013, p. 531).   [69] See, e.g., Hublin et al. (2017).

[70] Incidentally, by almost all estimates, as we saw above, Y-Chromosomal Adam, and certainly Mitochondrial Eve, lived many dozens of millennia after the evolution of modern *Homo sapiens*. This is as it should be, according to coalescent theory. Moreover, we still have no evidence of Neanderthal (or Denisovan) mitochondrial DNA or Y-chromosomal haplotypes in any modern human population. Perhaps male fetuses with Neanderthal fathers and human mothers were inviable (e.g., Mendez et al., 2016, p. 732). Or maybe any such haplotype lineages in *Homo sapiens* died out due to selection or even chance (see Prüfer et al., 2014; Nielsen et al., 2017; and https:// humanorigins.si.edu/evidence/genetics/ancient-dna-and-neanderthals). Even so, portions of the rest of the Neanderthal autosomal genome remain in humans. I leave it to the interested reader to draw out various mating combinations, in order to see that both Neanderthal (or Denisovan) mitochondrial DNA and Y chromosomes can be lost in two generations, while autosomal DNA, which recombines every generation, easily introgresses into the human genealogy. Of course, absence of evidence is not evidence of absence. It is possible that we have yet to discover a new Neanderthal Y-chromosomal or mitochondrial DNA haplotype from archaic remains, perhaps present somewhere in a human population, either already sampled or not. (On the converse point of human introgression into the Neanderthal lineage, see, e.g., Raff, 2017.)

[71] Hublin (2013, p. 525).

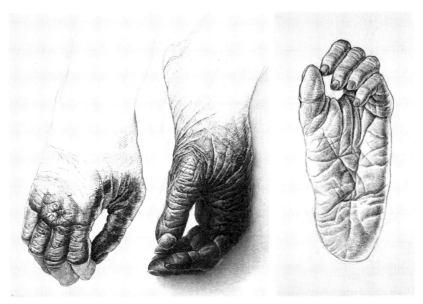

P. paniscus

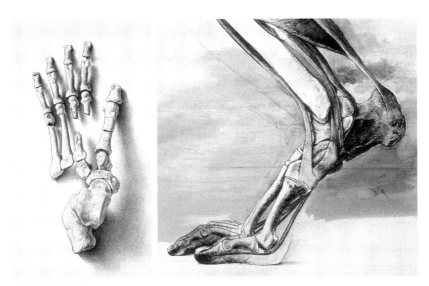

A. (P.) afarensis

Figure 2.9. Foot evolution
A topic of perennial fascination in human biology is the evolution of locomotion
and bipedalism. Foot morphology differs remarkably between *Homo sapiens* and our
nearest extant relatives. Yet Holowka and Lieberman (2018) argue that the foot

sister species of *Homo sapiens* elicit much interest (Figure 2.10). After all, most humans outside of Africa today typically have at least a few percent of Neanderthal DNA in their genomes, some of which is adaptive to, for example, fighting viruses.[72] At the same time, Oceanian peoples such as Papuans and Australian Aborigines often have more than 5% Denisovan DNA.[73] Very recent research finds Neanderthal DNA even in contemporary African individuals, suggesting that there were migrations back to Africa, perhaps even by European *Homo sapiens*, themselves carrying Neanderthal DNA, over the past 20,000 years or so.[74] We do have some answers but only very few: The history of *Homo* cannot help but raise questions about how to individuate DNA, "whose" DNA it is, and why Neanderthals went extinct approximately 40,000 years ago, among others.[75]

Out of Europe?

Other pressing questions related to our extended kin include which species gave rise to you and me, and to our nearest contemporary relatives, bonobos, chimpanzees, and gorillas. And where and how

---

Caption for Figure 2.9. (cont.) morphologies and functions of humans and contemporary nonhuman great apes (e.g., *Pan paniscus*, top; bonobos are both arboreal and terrestrial) show some astonishing likenesses. These researchers present a model of human foot evolution in three stages: (1) An ape-like foot evolved for use in trees but also became somewhat adapted to bipedal walking; (2) the foot evolved more for bipedal walking while keeping some capacity for arboreal grasping [e.g., *Australopithecus (Praeanthropus) afarensis*, bottom]; and (3) the foot became more human-like by adapting to long-distance running and walking, and lost its prehensility. © 2021 John Gurche. Reprinted with permission.

---

[72] See, e.g., Enard and Petrov (2018).

[73] See, e.g., Wall et al. (2013), Prüfer et al. (2014), Sankararaman et al. (2014), Racimo et al. (2015), Nielsen et al. (2017), Prüfer et al. (2017), Enard and Petrov (2018), Jacobs et al. (2019), Bergström et al. (2020), and Skov et al. (2020). The divergence between Neanderthals and modern humans, *Homo sapiens*, can be estimated. Mendez et al. (2016) use DNA from the Y chromosome of a Neanderthal to estimate divergence time to be around 588,000 years ago; Langergraber et al. (2012) find the temporal range for the split to be 400,000–800,000 years ago; Gómez-Robles (2019) uses fossil teeth, not genomics, to place the split at no less than 800,000 years ago. Typically, because of chromosomal recombination over many generations, the regions of the Neanderthal genome interspersed in our genome are less than 100 kilobases long. Regarding Denisovans, it is apposite to point out that many Central, South, and East Asian populations, as well as Indigenous Americans, typically have significantly more Denisovan DNA than individuals from Europe or Africa.

[74] Chen et al. (2020).   [75] Vaesen et al. (2019) and Chen et al. (2020).

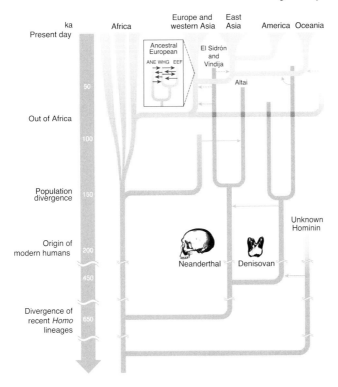

Figure 2.10. Evolution of *Homo sapiens*

Phylogenetic relationships among *Homo sapiens*. Corroborated admixture (solid lines) and hypothesized admixture (dashed lines), and direction of introgression, whether among modern humans or between modern and archaic humans, is shown. ANE = Ancient North Eurasian; EEF = Early European Farmer; WHG = Western European Hunter-Gatherer; ka = thousand years ago; see Figure 2.2 and text. The type, or representative, specimen of Neanderthals was a partial skull found in the mid-nineteenth century in the Neander Valley in Germany, from which archaic Neanderthal DNA was first extracted in the late 1990s. Initial Denisovan specimens that first yielded DNA were molars and a finger bone found in Denisova Cave in southern Siberia (Reich et al., 2010). At the top, place names indicate caves where Neanderthal fossils have been found and archaic DNA extracted: El Sidrón (Spain), Vindija (Croatia), and Altai (Denisova Cave, Russia) (e.g., Prüfer et al., 2014). From Nielsen et al. (2017), p. 304. © 2017 Nature Publishing Group, a division of Macmillan Publishers Limited, 2017. All Rights Reserved. Reprinted with permission from Springer Nature Customer Service Center GmbH. Adapted by Mats Wedin, Laura Laine, and Rasmus Grønfeldt Winther.

did *these* species originate? Such questions matter because they help expand our connections with our ancestors and contemporary relatives, inviting a sense of empathy and belonging across much larger branches of the tree of life. Additionally, for those of us interested in geography

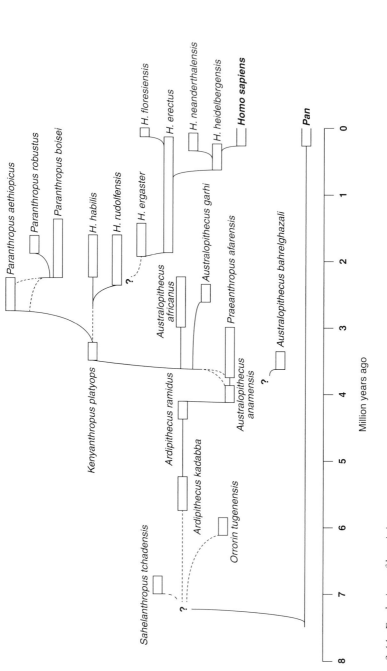

Million years ago

Figure 2.11. Evolution of hominins

The general consensus of the phylogenetic relationships among hominin species, as reconstructed from our best, contemporary fossil record. Many researchers prefer the name *Australopithecus afarensis* to *Praeanthropus afarensis*. From Strait et al., 2015, p. 2008. © 2015 Springer-Verlag Berlin Heidelberg. Reprinted with permission. Adapted by Mats Wedin, Laura Laine, and Rasmus Grønfeldt Winther.

and geology, these questions – because of the significant amounts of geological time involved – return us full circle to geographic Europe, with which we started this chapter's genomic journeys. Africa "lacks any reasonably complete hominoid crania between 17 and 7 million years ago (Myr), and no cranial specimens are known at all from between 14 and 10 Myr."[76] These facts might, on the surface, seem strange given that humans clearly evolved in Africa. Why shouldn't our distant ancestors also be found there? Perhaps this is due to sampling bias – after all, fossils of African great apes and hominins originate from less than 1% of the African continent area.[77] Yet, despite this small sample size, a few truly ancient, significant fossils have been discovered. In Kenya in 2017, for instance, a 13-million-year-old skull of the genus *Nyanzapithecus* was found – potentially, these primates "were stem hominoids [for these authors, the great apes (including humans), and their ancestors; gibbons are excluded] close to the origin of extant apes."[78]

While the ape fossil record during this period appears scant in Africa, it is correspondingly rich in Eurasia. Perhaps we should take the fossil record literally. To do so is to make another provocative and controversial set of theories possible. University of Toronto anthropologist David Begun suggests an *Out of Europe* paradigm, presenting ample evidence and reasonable arguments that "African apes evolved in Europe and moved to Africa, not the other way around."[79] Specifically, and in accord with Darwin's *The Descent of Man, and Selection in Relation to Sex*, Begun assumes that Dryopithecini, a tribe of Miocene apes widespread in Europe from something like 9 to 12 million years ago, is related to, and possibly an ancestor of, African apes.[80] Begun imagines dryopithecines, and in particular the genus *Dryopithecus*,

---

[76] Nengo et al. (2017, p. 169). On a right jaw mandibular fragment with three molar teeth of *Nakalipithecus nakayamai*, dated slightly less than 10 million years ago, see Kunimatsu et al. (2007).
[77] Cote (2004).   [78] Nengo (2017, p. 169).   [79] Begun (2016, p. 5).
[80] Darwin (1874 [1871]) declaimed: "In each great region of the world the living mammals are closely related to the extinct species of the same region. It is therefore probable that Africa was formerly inhabited by extinct apes closely allied to the gorilla and chimpanzee; and as these two species are now man's nearest allies, it is somewhat more probable that our early progenitors lived on the African continent than elsewhere. But it is useless to speculate on this subject; for two or three anthropomorphous apes, one the Dryopithecus of Lartet, nearly as large as a man, and closely allied to Hylobates, existed in Europe during the Miocene age; and since so remote a period the earth has certainly undergone many great revolutions, and there has been ample time for migration on the largest scale" (pp. 155–156, reference omitted). The dryopithecines consist of several extinct genera, including *Dryopithecus, Ouranopithecus,* and *Rudapithecus.*

to be a kind of combination of orangutan, in terms of its skill in the trees; chimpanzee, in terms of its diet; and gorilla, in terms of the anatomy of its face. Of course, this is exactly what we should expect of a fossil ape that lived just after the [Asiatic] orangutan and Africa ape lineages branched off from one another: a little bit of what would develop in each living species.[81]

Central to his narrative, Begun points out that "the ecological conditions in Eurasia selected for new adaptations in apes . . . related to diet and positional behavior."[82] According to his argument, geographic separation and adaptation to local conditions split the great ape phylogenetic tree into the orangutan-ancestor pongines in Asia, and the African ape hominines in Europe. Depending on the exact timing of the phylogenetic tree splitting, the last common ancestor of today's gorillas, on the one hand, and today's chimpanzees and humans, on the other hand, almost certainly lived in Eurasia, and the last common ancestor of *Pan* and *Homo* may have lived in Eurasia (alternatively, that split may have occurred in Africa). Many adaptations shared by all great apes, including orangutans (i.e., the hominids), likely evolved in Eurasia: longer lifespans and larger brains than monkeys, delayed onset of menarche, early bipedalism, and tool use, to name a few.[83] Using fossil morphology, Begun thus weaves a plausible narrative of *Dryopithecus* or its descendants accumulating adaptations and, likely, phylogenetic tree splitting in Europe, and then migrating to Africa some 7 to 9 million years ago.[84]

---

[81] Begun (2016, p. 154).

[82] Begun (2016, p. 5). On bipedalism locomotion adaptations evolving in Europe, see Böhme et al. (2019).

[83] Mithen (2015), Begun (2016), and Böhme et al. (2019).

[84] Regarding the *Dryopithecus* ancestry of African apes, and of hominins in particular, see chapters 7 through 9 of Begun (2016). Chen and Li (2001) use genomic data to date the Hominini split (humans and their closest ancestors, including, most likely, some *Australopithecus* species versus chimpanzees/bonobos and their closest ancestors) at 4.6 to 6.2 million years ago, and the Homininae split (Hominini, including chimpanzees/bonobos, and their closest ancestors versus Gorillini) at 6.2 to 8.4 million years ago. Langergraber et al. (2012) fine-tune generation times of chimpanzees and gorillas through field and genetic studies of parentage and kin structure in the wild, and use molecular clock-like mutation rates in humans to estimate lineage splitting, finding them to be older than Chen and Li (2001) do. The splitting times calculated by Moorjani et al. (2016) are even older (human and chimpanzee divergence at 12.1 million years ago, and human and gorilla divergence at 15.1 million years ago). To some interlocutors, the younger split times are inconsistent with Begun's Out of Europe hypothesis because at minimum the Hominini divergence would have taken place much later in time than dryopithecine evolution, thus suggesting that the European record is too old and irrelevant. The older dates are consistent with the presence of gorillas, chimps, and our ancestors in Europe between 7 and 15 million years ago, when few such fossils are known from Africa. Under either temporal range, ape adaptations accumulated in Eurasia, per the Out of Europe paradigm. David Begun provided constructive feedback here.

During their evolution, hominins left traces and tracks, some of which are found in the rocks and soils, and others in our contemporary genomes and languages. Moving backward in time – from *Out of Africa* to *Within Africa* to a potential, truly ancient, for primates, *Out of Europe* – illuminates the geographic and genealogical journey of *Homo sapiens*, that paragon of animals and quintessence of dust (see Table 2.1 for a summary). Out of Europe remains a fascinating paradigm for our deep geographic origin (Figure 2.12). Understanding the full story of our deepest primate ancestry helps refine geographic and temporal thinking. Moreover, analogous to the impressive progress made in isolating and sequencing DNA from Neanderthals and Denisovans (Figures 2.10, 2.11, and 2.13), especially as genomic material and computational technologies continue to improve, the potential for human evolutionary genomics-based findings on truly ancient apes could very much become real.[85]

Genomic data have strengthened – if not sealed – the case for a clear African origin of modern *Homo sapiens* (more on this in Chapter 6). Even so, it remains unclear whether the genomic signatures we continue discovering in contemporary humans as well as in DNA from ancient or archaic human remains will, ultimately or ever, tell us the *full* story about the migration and occasional interbreeding of very old human populations in Africa, and elsewhere.[86] After all, genomic data have real limits. As we have seen, other fields such as archaeology, paleontology, primatology, climate science, and linguistics contribute evidence, methods, and knowledge for our forays into the murky regions of human evolution. However, even with these other sources of evidence, many matters may never be completely and fully resolved.

---

[85] Every representation emerges according to the interests and abstraction practices of the representer. Consider the last three figures. Figures 2.10 renders recent evolution, informed by technologies available today for rescuing archaic human genomes from fossils, whereas Figure 2.11 zooms out temporally, thereby providing a much broader landscape of all hominin evolution. Figure 2.12 zooms out even more, with the two prior images as, respectively, only a small "dot" (under *Homo*) and a small end-branch (with *Ardipithecus*, *Australopithecus*, and *Sahelanthropus*). (Perhaps here there is some disagreement rather than only representational pluralism, with Figure 2.11 representing *Ardipithecus* as an ancestor of *Australopithecus*, whereas Figure 2.12 doesn't.) Together, the images tell of a branching history of some 35 million years of ape (and monkey) evolution.

[86] There is now an African Genome Variation Project; see Gurdasani et al. (2015).

Table 2.1. *Temporal and spatial information about key events in human evolution*

Human evolution event	Time (years ago)[a]	Place[b]
Out of Africa	55,000–65,000 (last significant dispersal) (Nielsen et al. 2017) 210,000 (at the earliest) (Harvati et al., 2019)	Northern Route (into the Levant) Southern Route (into what is Yemen today)
Modern *Homo sapiens*	260,000–350,000 (Schlebusch et al., 2017)	Africa (African multiregionalism; East Rift/South "Garden of Eden")
Mitochondrial Eve	99,000–148,000 (Poznick et al. 2013)	Africa (East Rift/South)
Y-Chromosomal Adam	139,000–388,000 (total range; see note 46)	Africa (East Rift/South)
Homo sapiens–Neanderthal split	400,000–800,000 (Langergraber et al., 2012) 588,000 (Mendez et al., 2016)	Africa
Hominini split (i.e., *Homo–Pan* split)	4.6–6.2 million (Chen and Li, 2001) 12.1 million (Moorjani et al., 2016)	Africa/Europe
Homininae split (i.e., Hominini–Gorillini split)	6.2–8.4 million (Chen and Li, 2001) 15.1 million (Moorjani et al., 2016)	Europe/Africa

[a] Discussions and sources for temporal ranges and points are given in the text. The events are presented in order of occurrence in the chapter and, except for the second row, in reality (assuming Y-Chromosomal Adam lived before Mitochondrial Eve).
[b] Discussions and sources for place are given in the text. Order matters: most likely region written first (e.g., Europe vs. Africa). "Africa" without specification in parenthesis means that not much information is known.

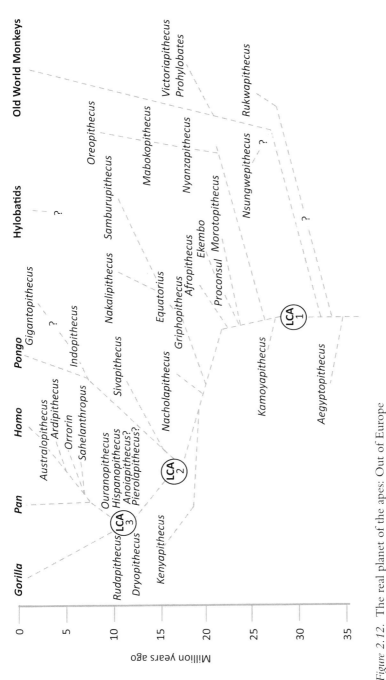

Figure 2.12. The real planet of the apes: Out of Europe
Phylogenetic relationships among hominoids (great apes and gibbons) and Old World monkeys – grouped as catarrhines (Catarrhini) – according to David Begun. LCA = last common ancestor. From Begun, 2016, p. 3, and personal communication. Reprinted with permission from Princeton University Press. Adapted by Mats Wedin and Rasmus Grønfeldt Winther.

Figure 2.13. Neanderthal Cave
Artistic rendition of fun times in a Neanderthal cave in southern France or
Spain – children chase each other while adults enjoy feast preparations. (top-right)
Cave art represents real, symbolic art in an Iberian cave, with a minimum age of
approximately 64,000 years ago. Concept by Rasmus Grønfeldt Winther, illustrated
by Larisa DePalma. © 2021 Rasmus Grønfeldt Winther.

The Path Ahead

Ultimately, the evidence and theory presented here shows that *Homo
sapiens* is a young species, with a cradle in Africa. Our young species
expanded quickly, and its genomic variation is fairly continuous and even
across its entire geography. We know we are more at home on Planet
Unity than we are on Galápagos-Writ-Large. Yet, we still do not know
who we are. Controversies proliferate. Exploring human genetic
variation and the existence (or not) of clearly distinct and individualized
human populations would be a wholly abstract and intellectual endeavor
except for the socially, politically, and morally relevant consequences of

Figure 2.14. Our cousins 1: Bonobos
Bonobos (*Pan paniscus*) are known for their empathy, matriarchal society, relative peacefulness, and rampant sexual activity across a range of relations (e.g., de Waal, 2013, 2019). Bonobo mothers frequently permit other group members to handle their infants; 96% of these interactions were positive in one study (Boose et al., 2018). Illustrated by Daphné Damoiseau-Malraux. © 2020 Rasmus Grønfeldt Winther.

such an exploration.[87] The questions asked in this chapter hint at the lessons, fears, and hopes – scientific and ideological – that abound for the future potential of precise, genomic biomedicine, and for social conceptualizations of ethnicity and race. Regardless of our political and moral stance on the origins and nature of human populations, clusters, or races, I suggest that we must gather and address all known evolutionary and genomic information about human origins, population structures, and biomedicine. Otherwise, we are arguing in the dark.

---

[87] See Lewontin (1970), Longino (2002, 2013), Hacking (2005), and Kitcher (2007).

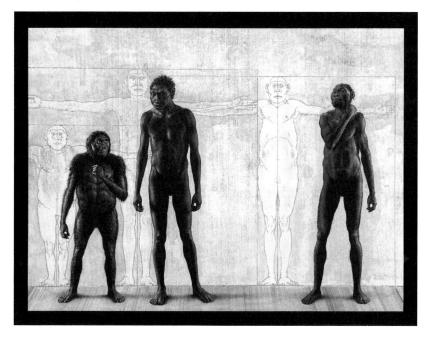

Figure 2.15. Vitruvian Hominins
Lucy [*Australopithecus (Praeanthropus) afarensis*] (left) juxtaposed with (center) *Homo erectus* (Turkana Boy) and *Homo naledi*. © 2021 John Gurche. Reprinted with permission. (A black and white version of this figure will appear in some formats. For the color version, please refer to the plate section.)

Truly interdisciplinary study helps us piece together various kinds of evidence of the human journey, including bones and fossils found across the world; whatever genomic information we can extract from these; the artifacts, technology, and even waste that humans have left behind; clues in spoken and written languages; and the presence and distribution of physiological and epidemiological adaptations in contemporary humans worldwide.

Wrangham discusses *the goodness paradox*, or the curious combinations – deeply rooted in our species – of selflessness and selfishness, virtue and violence, and peace and aggression.[88] By distinguishing two senses of aggression (reactive vs. proactive/planned), finding similarities and differences with (matriarchal, sexually prone) bonobos (Figure 2.14) and (patriarchal, violence-prone) chimpanzees, and postulating a hypothetical

---

[88] Wrangham (2019).

behavioral profile for Neanderthals (Figure 2.13), among other theoretical contributions, Wrangham provocatively tracks ethics and power in the evolution of *Homo sapiens*, and in comparative, contemporary societies. In earlier work, Frans de Waal had made analogous and resonant arguments.[89] Reconstructions of great ape evolution may provide insight into our moral and behavioral journey (Figure 2.15).

Given our increasingly globalized world, with a strong and accelerating flow of information, trade, and (voluntary and uprooted) migrants, we find ourselves exposed to an ever-broader variety of languages, behaviors, sights, and smells. While prejudice, misunderstanding, and violence grow, our best genomic science and findings in evolutionary genetics today suggest – in contrast to the essentialism of most eighteenth- and nineteenth-century anthropology – a deep biological connection among all peoples. We may exhibit some Galápagos-Writ-Large properties, but basic human physical and cognitive properties are universal. Indeed, it is one of my hopes that if we could only go deeper, in imagination and genealogy, we would realize how connected all life is, and how responsible humans are for our fellow creatures.

Genomics Glossary

Allele: one of various versions of a gene (or of any relevant segment of DNA) in a population, differing from alternative alleles at one nucleotide or more.

Chromosome: a structure of DNA double helix wrapped around histone proteins, and folded up and supercoiled; the human genome is composed of 23 pairs of homologous chromosomes, 22 pairs of standard autosomes, and one pair of sex chromosomes.

Codon: a sequence of three nucleotides that corresponds to an amino acid during gene expression.

Crossing over: when parts of two paired homologous chromosomes break apart and recombine during gamete formation (meiosis).

DNA (deoxyribonucleic acid): a molecular structure with a double helix ladder-and-rung structure, where the backbone of the ladder consists of alternating sugar (deoxyribose) and phosphate groups covalently linked, and the ladder rungs are pairs of four different kinds of nucleobases: adenine (A), cytosine (C), guanine (G), and thymine (T).

---

[89] See, e.g., de Waal (2013, 2019).

The rung pairs are linked to each other, inside the ladder, by hydrogen bonds, according to base-pairing rules: A pairs with T with two hydrogen bonds, and C pairs with G with three hydrogen bonds. The stability of the double helical structure is maintained by stacking forces attracting adjacent bases along the DNA strand. A nucleobase (e.g., G) covalently linked to a deoxyribose is known as a nucleoside, and, when the sugar of the nucleoside is also covalently linked to the phosphate group, it is known as a nucleotide. A nucleic acid (DNA, RNA) is made up of many nucleotides.

Evolutionary genetics (or population genetics): the mathematical study of the genetic composition and evolutionary changes of populations first developed by R.A. Fisher, S. Wright, and J.B.S. Haldane.

Gamete: an egg or a sperm (in humans and other animals).

Gene: a unit or part of DNA that codes for a particular protein serving a role in the biochemistry or micromorphology (and emergent physiology, morphology, and behavior) of an organism; can be defined structurally (in terms of consecutive or collated DNA segments) or functionally (in terms of how the gene affects the phenotype).

Gene expression: the process of mapping a DNA strand onto an RNA strand (transcription) and then cutting, pasting, and transforming that RNA strand before it maps onto an amino acid strand (translation), which may itself be post-translationally modified before the final, functional protein is produced via, for instance, protein folding.

Genetic code: the representation capturing the specificity of codon-to-amino acid mapping during gene expression (see Figure 7.5).

Genetic paradigm: the theoretical representations, observational data, experimental practices, and ontological assumptions associated with the overarching view that the structure, physiology, and behavior of organisms can be at least partly – and usefully – explained by the transgenerational inheritance of genes, including attention to the interaction of genes and environment during organismic development.

Genetics: the investigation of the role of genes in transgenerational heredity; in variation across individuals; and in individual development, physiology, and morphology.

Genome: an individual's complete DNA information; in humans, it consists of just over 3 billion unique nucleotide sites or base pairs (e.g., in our haploid gametes).

Genomics: an interdisciplinary field involving geneticists, biologists, statisticians, computer scientists, and doctors, among others, studying the architecture, function, and evolution of entire genomes of species.

Also used for mapping and experimentally editing genomes; for storing nucleotide information in databases; and for statistical analyses of many facets of the genome, including identifying the signature of selection and extracting knowledge about historical migration patterns of the species under study.

Genotype: the genetic constitution of an individual organism, at one or more loci (genes).

Heritability: the amount of the phenotypic variance correlated with (or explained by) the additive genetic variance.

Human evolutionary genomics: the genomic and evolutionary genetic study of humans, which has demographic, genealogical, forensic, and medical applications.

Law of independent assortment: the principle that alleles at one locus sort independently of alleles at another locus when a parent creates a gamete. This occurs either because the loci (genes) are on different chromosomes or because they are so far apart on the same chromosome that recombination effectively scrambles, reshuffles, or randomizes the two alleles at one locus (which may or may not be actually two different allele types) with respect to the alleles present at the other locus.

Law of segregation: the principle that a sexually reproducing individual has two alleles (which may be the same or not) for a given trait, one on either of the two chromosomes of a given homologous pair, and that those alleles separate when that individual creates a gamete.

Linkage: the phenomenon of genes that are close together on a chromosome tending to be inherited together; that is, they recombine less than genes that follow the law of independent assortment.

Locus: a chunk of the nucleotide sequence in the genome (though it can be composed of noncontinuous nucleotide sequences), sometimes used coextensively with "gene"; often but not necessarily functional in the sense of being causally or mechanistically associated with some part of the phenotype.

Microsatellite (aka short tandem repeats, STR; simple sequence tandem repeats, SSTR; or simple sequence repeats, SSR): a segment of tandemly repeated short DNA motifs (typically two to six base pairs repeated as many as 50 times) with a relatively high mutation rate and with many possible alleles (typically up to 20 or 30), which can fall within or outside of genes; microsatellites are generally not under direct selection (i.e., they tend to be neutral) and, hence, can be particularly informative of the history and migrations of populations of *Homo sapiens*.

Mutation: a single nucleotide change in the genome at one site; or a small insertion or deletion at (effectively) one site (see Figure 7.6); or a large (>50 base pairs) insertion, deletion, inversion, duplication, or translocation (e.g., a structural variant) that can, ideally, be anchored at one site; a mutation may or may not make a difference to the protein produced during gene expression, which in turn may or may not make a difference to the phenotype.

Natural selection: the phenomenon of individuals of a certain genotype leaving, on average, relatively more offspring in subsequent generations than individuals of another genotype, and this occurring because of systematic differences in the phenotypes associated with the genotypes such that the environment favors the former genotype. (The genotype that systematically leaves more offspring has a higher relative fitness than the other genotype.) Natural selection occurs when a trait or character is heritable, there is variation in the trait at the population level, and there are reliable differences in reproductive success associated with differences in the trait. Natural selection can lead to adaptation.

Nucleotide: a unit of DNA (or RNA) combining a sugar molecule; a phosphate molecule; and one of four molecules containing nitrogen – adenine, thymine, cytosine, or guanine (A, T, C, and G), or, in the case of RNA, thymine is replaced by uracil (U).

Phenotype: an organism's observable traits, which include physiological, morphological, cognitive, and behavioral features.

Population: an assemblage, an aggregate, or a collection of individuals, potentially with emergent properties and processes. Various kinds of populations can be identified and investigated, including theoretical, laboratory, and natural populations.

Phylogeny: the evolutionary history of an organism or species in relation to others.

Recombination: in eukaryotic organisms, during meiosis, chromosomes pair up appropriately before each pair is separated; when the 23 homologous chromosomes of humans are paired, different sections of the chromosomes can be broken, swapped, and reattached correctly in a process known as "crossing over." In so doing, alleles are reshuffled or recombined, and the produced gamete contains new recombinant chromosomes that have stitched together segments from the mother and the father of the individual producing the gamete (equivalently: the grandmother and grandfather of the individual to be produced by that gamete).

RNA (ribonucleic acid): a molecular structure with one-half the DNA double helix ladder-and-rung structure, where uracil (U) substitutes for thymine (T).

SNP (single-nucleotide polymorphism): a substitution or difference of a single nucleotide at a particular site in the genome, present in sufficiently many individuals of the population (e.g., >1%), which can exist within or outside of genes; most SNPs have only two alleles, where the least frequent one has a minor allele frequency.

Variant: an allele present in a certain percentage of individuals of a population (typically at least 1% or 5% of the population; alternatively, a "genetic polymorphism"); if an allele is only present in extremely few individuals, population geneticists prefer to call it a mutation.

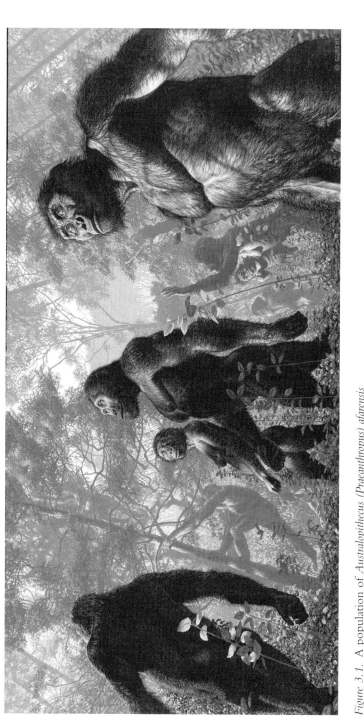

Figure 3.1. A population of *Australopithecus (Praeanthropus) afarensis*

In 1994, *National Geographic* asked Gurche to create a painting closer to our own ancestry, and closer to his training in anthropology. The Lucy fossil had made her debut without her head 20 years earlier, and the new *National Geographic* story was about the discovery of the first reasonably complete skull (a large male) of Lucy's species, *Australopithecus (Praeanthropus) afarensis*. Using the new skull, as well as a composite adult female skull and a child's skull known for the species, Gurche made three-dimensional anatomical reconstructions, subsequently wielding these as references for this painting showing a group or population of Lucy's species traversing a forest. (Co-written with Gurche.) © 2021 John Gurche. Reprinted with permission. (A black and white version of this figure will appear in some formats. For the color version, please refer to the plate section.)

3 · *The Mind, the Lab, and the Field: Three Kinds of Populations*

Clearly, when we stand at the intersection of human evolutionary genomics and philosophy and ask *Who are we?* we implicitly acknowledge that in body, mind, and history, we are fundamentally relational. Perhaps paradoxically and certainly provocatively, we are not really individuals at all: Connections, relations, and larger holistic and pluralistic systems are as important to your and my reality as the internal subjectivity many of us feel. In terms of the genetic paradigm, a single individual out of relationship with others is of little evolutionary or ecological significance. A single individual needs other individuals for behavioral and psychological interactions, competition and cooperation, and reproduction (Figure 3.1, chapter opener). Populations are a unique organizational level because most, if not all, local interaction happens at this level.

Even so, populations are also abstractions. Why? Because the constitution of the population into individuals, as well as the composition of its gene pool, are dynamic. Populations change via natural selection, mutation, and other evolutionary forces; populations have unstable spatial and temporal extents and exist at varying scales; and what constitutes a population depends on multiple definitions and interpretations. Differently put, scientists use the fundamental concept of *population* to select specific attributes of genetically, evolutionary, and ecologically relevant communities in which they are interested. Features are chosen based on the particular goals, assumptions, and practices scientists adopt in three research contexts: the theorist's mindscape, the experimenter's laboratory, and the fieldworker's landscape.[1] Paraphrasing the biologist Jean Rostand's quip, "populations pass; the frogs remain."

Context shapes the populations postulated in idealized theoretical models, set up in experimental laboratories, and surveyed and sampled in the wild. This chapter investigates such contextual influence,

---

[1] On the latter two, see Kohler (2002).

distinguishing *theoretical*, *laboratory*, and *natural* populations as three qualitatively and conceptually different kinds of populations at the heart of distinct styles of scientific practice in evolutionary genetics and ecology. Indeed, researchers frequently cycle through investigations into theoretical, laboratory, and natural populations, expressing genuine interest in each population type, while respecting the differences among them.

Because scientists can study all kinds of populations, a bird's eye view of each population type is useful, particularly in a field as potentially open to interpretation as evolutionary genetics. This chapter begins our charge to analytically understand population types, as living and embedded in the vibrant and ever-growing theories and models, experimental practices, and field data protocols of human evolutionary genomics. I focus on exemplary students of each population type: R.A. Fisher in the theoretical domain, Thomas Park (1908–1992) in the laboratory, and David Lack (1910–1973) in the natural domain. Each of these biologists influentially articulated and promoted the respective population type and thus serve as particularly useful levers or mini universes for comprehension. They were trailblazers, and while the populations they investigated may initially appear unrelated to our questions about identity, ancestry, and connection, their work informs our efforts to understand to which populations we belong, from genomic, evolutionary, and ecological points of view. Furthermore, two case studies in this chapter illustrate the value of distinguishing theoretical from natural populations, the two kinds of populations particularly central to *Our Genes*: the concept of *effective population size* as deployed in theoretical evolutionary genomics, and the use of model-based genetic clustering algorithms such as *Structure*.

The trichotomy of theoretical versus laboratory versus natural population is not intended as a rubric for determining how much a model does or does not correspond to reality. While an overarching aim of evolutionary biology and ecology is to understand the complex structure and dynamics of populations "in the wild," the multiple ontologies of scientific practice are complex: There is perhaps a world in a theoretical model (e.g., Morgan, 2012) or in an experimental system (e.g., Leonelli, 2007) or in a map of the natural field (e.g., Winther, 2020a). Consequently, this chapter does not intend to provide a singular, complete, and strict delimitation of the *population* concept. Other classifications and analyses of the concept of population, so central to the biological sciences and beyond, are compatible with the views expressed here. Further, my familiarity with both the limitations and the flexibility of measures of genetic variation – explored further in Chapter 4 – makes me a pluralist about population definitions and concepts, about the kinds of complex objects and processes one could delimit as populations, and even about distinct classifications of populations.[2]

What ultimately follows is an attempt at "taking a look" at differing styles of practice of working biologists.[3] Although they bring a variety of metrics and models to the study of populations, I ask which kinds of populations biologists believe they study. Which figures in the history of biology might shine through as exemplars of distinct styles of practice regarding populations?[4] Which critical tools allow biologists to avoid conflating different kinds of populations and to perform important work internally, within each style of practice? We may not be able to arrive at a single concept of population, but by exploring the study of distinct kinds of populations, we gain insight into the abstractions of which we are a part.

Three Kinds of Populations

As I started describing above, human evolutionary genomics bases its inquiry on the three kinds of populations used across many biological and biomedical sciences:

- *Theoretical populations* are groups of abstracted individuals (or genes) whose properties and behaviors are studied in formal mathematical and statistical models constructed with idealized assumptions.
- *Laboratory populations* are collections of actual organisms – or parts of organisms, such as stem cells and other forms of cell lines – assembled in an experimental setting.
- *Natural populations* are collections of actual organisms living in the wild – settings that are not constructed expressly for studying the organisms (although researchers might modify the habitat).

Because of higher-level processes and relations, populations are not identical to the concrete set of individual organisms composing them,

---

[2] For some of the philosophy of science literature on this topic, consult Matthen and Ariew (2002), Godfrey-Smith (2009), Millstein (2009, 2010), Stegenga (2010), and Earnshaw-Whyte (2012, 2018).

[3] On "taking a look," see Hacking (2007c, pp. 36–38). In a qualified manner, I interpret theoretical populations as belonging within the "axiomatic style" of Crombie (1994), while experimental populations are subsumed under the "experimental style," and natural populations live within the "taxonomic style."

[4] On exemplars, see Kuhn (1970); on styles of practice in biology, see Winther (2012a).

whether in the mind or theory, the lab, or the field. Although I describe the three kinds of populations as distinct types of collections of – and relations among – objects, the three kinds of populations entail distinct population concepts. That is, laboratory and natural populations are also, in some sense, theoretical. In practice, researchers may modify their use of the concept of population to suit the questions they pursue.

Given their simultaneous fluidity and fixedness, all three types of populations have received philosophical attention. For example, Margaret Morrison shows which assumptions and idealizations were necessary to overcome conflicting notions of theoretical populations in the biometrician–Mendelian debate in the early twentieth century.[5] As we saw in Chapter 1, the Mendelians argued that most phenotypic characters were associated with a few genes with large and robust effects. By contrast, the biometricians focused on continuous phenotypes such as height and weight, and believed there was not an easy story to tell – certainly not one told in terms of discontinuous, Mendelian factors – for the heredity of such characters. Rachel Ankeny and Sabina Leonelli address laboratory-based work on model organisms, and the role of such work in experimental and biomedical contexts.[6] Roberta Millstein focuses on delineating natural populations in terms of individuals who causally interact.[7] James Griesemer simultaneously speaks to the three kinds of populations in drawing out a "data journey" of abstracted gene sequence data moving from global natural human populations to immortalized HGDP-CEPH cell line laboratory populations in Paris (explored further in Chapter 5),[8] to the theoretical populations (clusters) of Rosenberg et al.[9] Ultimately, each type of population has a rich history of use in biology and developed in its own way, as shown by each of the historians and philosophers of biology cited in this paragraph.[10]

Fisher on Theoretical Populations

According to one (if not *the*) founding father of modern statistics, R.A. Fisher, the investigation of theoretical populations provides important insights for investigating evolution. In the preface to the first edition of

---

[5] Morrison (2000, 2002).

[6] Ankeny and Leonelli (2011). On animal models in medicine, see Swindle et al. (2012), Denayer et al. (2014), and Pedersen and Mikkelsen (2019).

[7] Millstein (2009, 2010, 2014, 2015).   [8] Griesemer (2020).   [9] Rosenberg et al. (2002).

[10] See also, respectively, e.g., Kingsland (1995), Kohler (2002), and Mitman (1992).

The Genetical Theory of Natural Selection, Fisher observed that "practical biologists" may deem it ludicrous to "work out the detailed consequences experienced by organisms having three or more sexes," but this is precisely what biologists should do if we "wis[h] to understand why the sexes are, in fact, always two."[11]

As Fisher understood, the creative power of mathematics lies partially in its capacity for generalization, abstraction, and idealization. Very roughly, *generalization* concerns the increasing breadth and scope of situations to which a mathematical structure applies, while *abstraction* relates to the paucity of assumptions and axioms of the structure. The sparser the set of assumptions and axioms under which a theorem is derived, the more abstract it is, and the more concrete cases it can subsume, perhaps incompletely.[12] *Idealization* is reasoning about representations that may not be physically realized, such as infinitely long lines in geometry.[13] Mathematical activity involves *proofs* and *applications* of general, abstract, and idealized mathematical structures, deductively hitched.[14]

Fisher argued that certain properties of groups of organisms could be understood without detailed knowledge about individual organisms.[15] Specifically, Fisher considered the effects of selection in the aggregate, "borrow[ing] an illustration from the kinetic theory of gases." Just as the physicist specializing in statistical thermodynamics investigated the behavior of idealized gas particles in a theoretical aggregate, the population geneticist studied the behavior of abstracted and idealized organisms in a theoretical population, a theoretical aggregate that was "independent of particular knowledge about individuals."[16]

Indeed, Fisher constructed a novel, *theoretical* notion of population in part by likening selection laws to gas laws. By 1918, Fisher assumed that a population consisted of many "random[ly] mating" individuals, later clarifying that "the [fundamental] theorem is exact only for idealized

---

[11] Fisher (1958 [1930], p. ix).

[12] See, e.g., Cartwright (1983), Levins (2006), and Winther (2020a).

[13] See, e.g., Cartwright (1989), Ohlsson and Lehtinen (1997), Weisberg (2013), and Winther (2021a).

[14] See, e.g., Hacking (2014).   [15] Fisher and Stock (1915).

[16] Fisher and Stock (1915, pp. 60–61). Indicative of the pervasive scientific context of the times, the article speaks of "eugenists" rather than "geneticists," and of a "general theory of eugenics." It appeared in *The Eugenics Review*, a journal that changed its name to *Journal of Biosocial Science* only in 1969. Other influential contemporary scientists published in that journal, especially in its early days, including Raymond Pearl, R.C. Punnett, and Charles Spearman.

populations."[17] Just as "laws of gases" ensure averaged behavior across individual particles, so the fundamental theorem of natural selection ensures averaged behavior across individual organisms.[18] Indeed, according to Fisher in *The Genetical Theory of Natural Selection*, despite certain differences – for instance, "entropy changes lead to a progressive disorganization ... while evolutionary changes ... produc[e] progressively higher organization in the organic world" – both the second law of thermodynamics and the fundamental theorem of natural selection, are, in Fisher's view, "properties of populations, or aggregates, true irrespective of the nature of the units which compose them."[19]

Nevertheless, the physics–biology analogy pertained less to content than to method, especially that of averaging across the properties and processes of individuals to identify statistical, central tendencies of the population.[20] The fundamental theorem of natural selection and the analogy behind it show that general, abstracted, and idealized theoretical populations were Fisher's object of study.

In developing ground-breaking statistical machinery, Fisher followed the mathematical method of generalization, abstraction, and idealization. For instance, the eventually ubiquitous analysis of variance (ANOVA) method was first developed by Fisher in the hearth of evolutionary genetics: Fisher's poster-boy ANOVA model was a phenotypic decomposition model attributing (variation in) the phenotype to a combination of (variations in) multiple factors – genetic, environmental, and other components, including genotype-by-environment interaction (i.e., a given genotype's causal effect depends on the environmental background), and genotype–environment covariance (i.e., certain genotypes tend to be associated with certain environments).[21] More generally and foundationally to statistical language and conceptualization, Fisher

---

[17] Respectively, Fisher (1918, p. 401) and Fisher (1958 [1930], p. 38). To be clear, the individuals in Fisher's populations are abstractions and do not literally mate (randomly or otherwise), although they do join their genetic factors randomly to give rise to the next, idealized, generation.

[18] Fisher (1958 [1930], pp. 39–40).

[19] Fisher (1958 [1930], pp. 39–40); cf. Edwards (1994, 2014), as reprinted in Winther (2018a), and Morrison (2000, 2002).

[20] On what has been called the "averaging strategy," see Welsh et al. (1988), Sober and Lewontin (1982), Wade (1992), Sterelny and Kitcher (1988), Okasha (2004, 2006) and Winther et al. (2013). There are various ways to understand the fundamental theorem of natural selection. See, e.g., Frank and Slatkin (1992), Edwards (1994, 2014), as reprinted in Winther (2018a), Plutynski (2006), Okasha (2008), and Ewens (2011, 2018). However, the power of Fisher's overarching mathematical, theoretical style is widely accepted.

[21] Fisher (1918).

baptized the statistical distinction between *population* and *sample*, faulting previous researchers for:

apply[ing] the same name, *mean, standard deviation, correlation coefficient*, etc., both to the true value which we should like to know, but can only estimate [i.e., population parameters], and to the particular value at which we happen to arrive by our methods of estimation [i.e., estimates or sample statistics].[22]

The distinction between population and sample became foundational to statistics. Fisher also encouraged statisticians to proceed "by construct-ing a hypothetical infinite population, of which the actual data are regarded as constituting a random sample."[23] Indeed, in addressing the basic problems of statistics (per Fisher, these were *specification, estimation,* and *distribution*),[24] Fisher repeatedly appealed to infinite and hypothetical (theoretical) populations.

In the context of evolutionary genetics, Fisher used the formal, math-ematical concept of theoretical population in providing analytical argu-ments for, among other things, the evolution of sex and why the stable sex ratio for many mating systems was 1:1. Simultaneously with Sewall Wright in the late 1920s and early 1930s, Fisher modeled the evolution-ary consequences of repeated binomial sampling at many single loci across generations – meaning both of them cared about random genetic drift. Today, this model is referred to as the Wright–Fisher or Fisher–Wright model. Fisher's formal methods provided ways of simultaneously building analytical models of evolutionary processes and testing these statistically against real data.

Park on Laboratory Populations

The scientific research of ecologist, evolutionary biologist, and ento-mologist Thomas Park illustrates the use of laboratory populations. Park spent much of his career at the University of Chicago developing, modifying, and observing the *Tribolium* flour beetle laboratory system. Together with statisticians Jerzy Neyman and Elizabeth Scott, Park wrote about two kinds of models that can be employed when

---

[22] Fisher (1922, p. 311).

[23] Fisher (1922, p. 311). On the history of "representative sampling," see Kruskal and Mosteller (1980), replete with instructive diagrams for nine different meanings of the term. Interestingly, I have thus far been unable to find a history of the *population* concept in statistics.

[24] Fisher (1922).

populations are too challenging to study in the field: "*mathematical* or *laboratory-experimental*."[25] This is consonant with my three-way differentiation of populations. Both mathematical and laboratory-experimental models "depict the workings of at least a part of nature" and, moreover, "enhance the interaction of certain factors," while "diminish[ing]" or "eliminat[ing]" others.[26] Most generally, both kinds of models "are abstractions of nature designed to illumine natural phenomena."[27] Since mathematical theoretical populations were investigated above in relationship to Fisher, I now set aside this type of model, focusing instead on laboratory populations.

The first of two parts of the 1956 paper "Biological Aspects," written primarily by Park, presented "a laboratory-experimental model":

A population exhibiting a relatively rapid life cycle in a not too artificial laboratory habitat; cultured under easily controlled, yet manipulatory, environmental and trophic conditions; for which repeated censuses of all stages can be taken with negligible disturbance, and for which adequate replication is feasible.[28]

This definition compresses a list of 10 characteristics that make a laboratory system optimal "for study of population phenomena": census, generation time, age and sex distributions, replication, physical environment, trophic environment, stock cultures, observation of behavior, genetic situation, and combination of models.[29] *Tribolium* satisfies many of the listed characteristics. As applied to flour beetles, the laboratory system is ideal. Flour can easily be sieved to retrieve eggs, larvae, pupae, and adults, facilitating censuses. Moreover, flour is simultaneously the "climactic," trophic, and spatial habitat, simplifying a potentially complex environment. Finally, different *Tribolium* species can be mixed in the same flour, enabling studies of interspecific behavior (Figure 3.2). Park used *Tribolium* to study processes such as cannibalism, interference among individuals of the same and of different species, oviposition, and rate of food-depletion.[30] The *Tribolium* model was influential because it was an outstanding laboratory system for addressing ecological and evolutionary questions of, for instance, how populations of different species competitively interact, and how populations grow and crash under conditions of

---

[25] Neyman et al. (1956, p. 42).   [26] Neyman et al. (1956, p. 42).
[27] Neyman et al. (1956, p. 43).   [28] Neyman et al. (1956, p. 45).
[29] Neyman et al. (1956, pp. 44–45).
[30] Neyman et al. (1956, pp. 43, 48); cf. Mitman (1992) and Winther (2005).

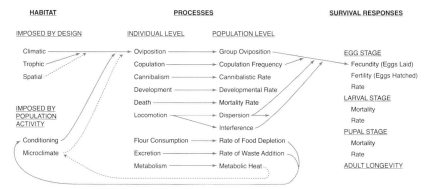

Figure 3.2. Major components of the *Tribolium* model
Habitat features, individual- and population-level processes, and survival responses
for *Tribolium* flour beetles kept under experimental conditions. Lower-level
individual phenomena such as copulation and locomotion interact to create higher-
level population phenomena. These in turn affect processes such as egg fecundity
that feed back to individual-level processes, including future amount of flour
available during development. From Neyman et al., 1956, p. 48. Reprinted with
permission from University of Chicago Press – Journals. Redrawn by Mats Wedin.

limited resources. Consequently, this model system continued through
the work of Park's students (and their students), including Monty
B. Lloyd, David B. Mertz, Michael R. Nathanson, and Michael
J. Wade.[31]

In their paper, Neyman, Park, and Scott expressed worry that their
laboratory model would be "criticized as being 'artificial.'" They
accepted that laboratory models "though not simple, are simplified,"
but they rejected the implication that artificial and simplified models
are trivial.[32] Rather, laboratory populations are *abstract* compared with
natural populations. For example, many features of natural populations
are eliminated in laboratory populations (e.g., rain, presence of predator
species). Constructing a laboratory population is also an *idealization*:
A previously nonexistent entity, the laboratory population, is granted

---

[31] I am indebted to Michael J. Wade, my mentor for my master's degree in evolutionary biology,
with whom I worked on *Tribolium* from ecological, evolutionary, and genetic angles (e.g.,
Winther 2004, 2006a). Wade was also a co-advisor for my doctorate in history and philosophy
of science, for which Elisabeth Lloyd was my main advisor. I started learning population genetics
and ecology in my undergraduate days from Marcus Feldman, Joan Roughgarden, and Deborah
Gordon. Around the same time, I first learned philosophy of biology and history of science from
Peter Galison, Peter Godfrey-Smith, and John Dupré.

[32] Neyman et al. (1956, pp. 45–46).

reality in the counterfactual – or better yet, "counternatural" – experimental setting. In addition, the model is *general*, and the authors cite Park's 1955 claim that the "unrealistic aspects" of laboratory models "may be a virtue instead of a vice" and "can contribute to the maturation of ecology, at least until . . . they are no longer needed."[33] In truth, they are often needed, or at least they are useful. For instance, interspecies competition experiments or studies of sexual selection in the laboratory can shed light on more complex and interactive ecological dynamics in nature. Laboratory and natural populations are not identical, and while the former represent the latter, they do so imperfectly and can, moreover, also only serve as limited instantiations of theoretical populations. However, the simplified, abstract, ideal, and general results from a two-species laboratory population competition system can be usefully compared and generalized to studies of natural populations.

Indeed, Neyman, Park, and Scott's work shows how assumptions holding neither for idealized laboratory populations nor for natural populations can lead to broad conclusions that may still be correct. It also shows how false models can point to further experimental work.[34] It is powerful, desirable, and crucial that Neyman, Park, and Scott's article makes assumptions clear in part by permitting and reminding the reader to consciously move between levels of abstraction – natural population to laboratory, and laboratory population to theoretical – always aware of which assumptions have been made and which information has been lost. In this way, their work employs *assumption archaeology* and attests to the reality that even false models can be a means to truer theories.[35]

Lack on Natural Populations

The ornithologist David Lack, known for studying Darwin's finches in detail, was an evolutionary ecologist and a student of natural populations. In the preface to the first edition of his influential *Darwin's Finches*, he sketched the nature of the book:

[33] Neyman et al. (1956, p. 46); cf. Park (1955, p. 69).   [34] See, e.g., Mertz et al. (1976).
[35] On assumption archaeology, see Winther (2020a) and the end of this chapter; on the uses of false models, see Levins (1966, 1968) and Winther (2006c). For studies of laboratory populations of human cell lines pertinent to human evolutionary genomics, such as the HGDP laboratory populations in Paris, see Cavalli-Sforza (2005), M'charek (2005), Reardon (2005), Griesemer (2020), and Chapter 5.

This is a work of natural history, based on a study of living birds in the Galapagos and of dead specimens in museums. The evidence is circumstantial, not experimental, so that theories must be presented cautiously. They should not, however, be excluded.[36]

Natural populations are the basic unit of Lack's investigations. Indeed, by studying field populations, including remnant models stored in the museum, Lack was able to investigate plumage in the context of sexual selection, beak size differences among finches on different islands, and hybridization.[37] Lack circumscribed his populations using features of the Galápagos finches that interested him: few competitors for food; few predators; and, crucially, "owing to geographical conditions," division "into a number of partly, but not completely, isolated populations, some of which are of very small size."[38] Accordingly, Lack identified natural populations of a species using geographic isolation.

Although Lack's units of study were natural populations, complementing our two other exemplars, Fisher and Park, his investigations drew on insights from other types of populations, including theoretical populations. Indeed, he changed his theoretical interpretation but repeatedly referred to the same data from natural populations found in the field or stored in the museum. For example, in his earlier work, Lack hypothesized that most variation across populations was nonadaptive, attributable to the "Sewall Wright effect," and emphasized random genetic drift as a key evolutionary force, at least on a par with natural selection.[39] By 1947, however, and in part due to the influence of Julian Huxley (1887–1975) – a proponent of the ubiquity and power of natural selection, a eugenicist, brother of Aldous Huxley, and author of the book that baptized "the modern synthesis" – Lack's views changed significantly. In *Darwin's Finches*, he postulated that interspecific competition fine-tuned the heritable variation of each isolated population of Darwin's finches. Lack now understood variation among populations as the result of natural selection and adaptation, with each population (and, eventually, species) adapted to its local environmental conditions, to its niche. Both earlier and later, Lack used data models, including data tables,

---

[36] Lack (1983 [1947], p. xiii).
[37] On remnant models, see Griesemer (1990, 1991). The suggestion here is that specimens in a museum are better thought of as samples, potentially unrepresentative, of natural populations rather than as constructed laboratory populations.
[38] Lack (1945, p. 116).   [39] Lack (1945, pp. 119, 135).

histograms, and maps, to abstract the properties of individuals from natural populations.

Peter and Rosemary Grant have taken Lack's research program further in their study of repeated bouts of selection in natural populations of Galápagos finches. More specifically, they studied the evolution of finch morphology and beak size as a consequence of food availability and environmental conditions (such as drought) over many decades on the Galápagos Islands. Discussing their precursor, Peter Grant observed in 1977 that "Lack himself was not a tester of ideas so much as an explainer of observations and hence a generator of ideas" and "Lack's field work strategy was to be a generalist, sacrificing some depth for breadth."[40] The Grants maintained Lack's focus on natural populations, adding new insights about evolutionary, ecological, ethological, genetic, and physiological processes while remaining close to populations in the wild.[41]

Of course, many others have also investigated natural populations from an evolutionary point of view. Among the more famous in the twentieth century was the Ukrainian American Theodosius Dobzhansky (1900–1975), whose work "took him into the field and caused him to abandon his beloved *Drosophila melanogaster*, the standard fly, for a wild cousin, *D. pseudoobscura*"; this inspired historian of science Robert Kohler to later write an entire book on the cultural border joining and separating the lab and field in biology.[42] Other mid-century evolutionary work examined genetic and phenotypic variation in natural populations of the snail *Cepaea nemoralis* in England (Cain and Sheppard, 1950) and France (Lamotte, 1959). Similar examples could be multiplied almost without end, as Endler does in a table presenting more than 100 "direct demonstrations of natural selection."[43] In all of this work, natural populations are simultaneously assumed, abstracted, constructed, and investigated by those doing fieldwork and interested in natural processes.

Fisher, Park, and Lack, and the respective work that they each inspired, demonstrates that theoretical, laboratory, and natural populations can be distinguished from one another and are each important to various fields of study (see Table 3.1). To say that these types of population can be distinguished is not to say that they cannot also be integrated. In fact, it is often necessary to invoke multiple types of population in the course of a single inquiry. For example, when we ask

---

[40] Grant (1977, p. 299).
[41] Grant and Grant (1989), Weiner (2014 [1994]), and Winther (2022b).
[42] Kohler (2002, pp. xiii–xiv).   [43] Table 5.1 in Endler (1986, pp. 129–153).

Table 3.1. *Key contrasts among theoretical, laboratory, and natural populations*

	Theoretical population	Laboratory population	Natural population
Core definition	Groups of abstracted individuals (or genes) studied in formal models	Collections of actual organisms or organism parts assembled in an experimental setting	Collections of actual organisms living in the wild
Worlds studied	All conceivable worlds	All materially possible worlds	Actual world
Chapter examples	Wright–Fisher model populations	Park's *Tribolium*	Darwin's finches

questions about ourselves and each other, about our origins and our identity, we must examine the population types most suited to our questions. In some cases, this means taking genetic samples from blood or saliva from human "field" populations across the globe and feeding the gene sequence data into theoretical models, such as the clustering algorithm and computer program *Structure*. In other cases, such as in biomedicine, it means ethical and careful experimental set up of human stem cells in laboratory populations, which may then be used as therapies for cancer treatments in individuals. The boundaries between population types can also be fluid. This can be beneficial for seeking similarities between population types and making inferences from one kind of population to another. But we must be clear in our assumptions and definitions, and must not automatically transfer and impose contextual and hidden presumptions from one population type to another.

Distinguishing Population Types

Although the boundaries of these population types may be fluid, the fluidity must be managed. When unmanaged, it is possible to conflate and confuse the three types, which can lead to errors of various sorts. For instance, theoretical assumptions appropriate for natural populations of one taxon of organisms may bias away from appropriate measurements or delineation of the natural populations of another taxon of organism. Or, as we shall see below, theoretical assumptions about population sizes may lead to inappropriate phylogenetic reconstructions. That is, theoretical population properties must be specified, especially when we move from one theoretical context to another (e.g., population genetics to phylogenetics).

More generally, abstract and necessarily simplified theoretical models should not be conflated and confused with a more complex and multifarious reality – a kind of "diseased ontologizing" or pernicious reification – lest seemingly accurate theoretical predictions be inappropriately attributed to rich nature by scientists or the public at large. In the two examples that follow, I describe some of the trouble resulting from conflating or confusing population types. I first show what population geneticists since Sewall Wright have recognized: the mistake of conflating the census size, N_c, of a natural population with its *effective*, or theoretical, population size, N_e. Second, I turn to *Structure* analyses to show how this clustering analysis tool, which works perfectly well for identifying certain kinds of theoretical populations, can fail to ground claims about natural populations.[44]

"Effective Population Size" as a Barrier: Distinguishing Theoretical from Natural Populations

Effective population size, a concept clarified by distinguishing theoretical, laboratory, and natural populations, highlights the translations that researchers must make between statements about natural and theoretical populations. As evidenced by the example of Fisher above, evolutionary genetics was primarily a theoretical discipline a century ago. With relatively little genetic data, evolutionary geneticists studied the ways in which evolution unfolded in theoretical populations.[45] As more genetic, and subsequently genomic, data became available, evolutionary genetics became more empirical. Evolutionary geneticists thus needed methods

---

[44] Winther (2014, 2020a, 2020b). Distinguished biologists Marlene Zuk and Mike Travisano critiqued an earlier version of this chapter that appeared as a journal article (Zuk and Travisano, 2018; see Preface). Although Zuk and Travisano take issue to some extent with the practice of bounded population types, I argue that much of our work in the natural sciences depends on this kind of – often implicit – separation. If we fail to acknowledge the boundaries we draw, we lose an essential means of interpretation and analysis. This is not to say boundaries should not be interrogated rigorously: Rather, it is to say, to repeat the purported words of my fellow Dane Niels Bohr to physicist Edward Teller, "If I cannot exaggerate, I cannot talk" (Teller, 1985, p. 182).

[45] See Haldane (1964) and Lewontin (1974) for commentary on this situation.

for translating insights between their rich theoretical heritage and their empirical pursuits. Effective population size is one such bridging method.

Today, contemporary evolutionary genomicists carry on the work of early evolutionary geneticists in studying theoretical populations in which the following properties, by and large, are assumed true for each population:[46]

1. random mating among sexual, diploid individuals;
2. constant number of breeding individuals across generations;
3. equal numbers of females and males, all of whom can reproduce;
4. no migration among ideal, theoretical populations;
5. no selection; and
6. no mutation.

These assumed features describe and capture a simplified, abstract theoretical population of sexually reproducing individuals. As such, these assumptions enabled geneticists to gain insights into idealized theoretical populations. On the basis of their interests, researchers might relax some of these assumptions or add other assumptions.[47]

For example, the Wright–Fisher model starts from assumptions 1 through 6. It ignores mutation, selection, unequal numbers of breeding individuals across generations, and other nonidealized properties.[48] With the resultant formal model, we can determine the rate at which random genetic drift occurs. If we add assumptions about the existence and degree of mutation and selection, we can go even further, determining, for example, the expected heterozygosity of the population or the approximate probability that all individuals in the population will

---

[46] See Hartl and Clark (1989), Hedrick (2005), Winther (2006a), Kliman et al. (2008), Ewens (2009), Nielsen and Slatkin (2013), and Hofrichter et al. (2017). For didactic ease, a sexual species is here considered; but even more simply, we could assume a hermaphroditic species. Some modelers do this: Ewens (2009) "consider[s], as the simplest possible case … [that] there is no concept of two separate sexes" (p. 15, cf. pp. 23–24). For Hofrichter et al. (2017), "the distinction between female and male individuals is irrelevant for the basic model" (p. 17). Indeed, once we assume random mating and equal numbers of females and males, and also try to assume that the two sexes in the theoretical population have the same allele frequencies, intergenerational allele sampling from a sexual or a hermaphroditic species are formally equivalent for the purposes of the Wright–Fisher model. For various formal reasons, it is best, easiest, and simplest to assume a hermaphroditic species, however.

[47] When considering data, the researcher might be forced to add or relax assumptions because of the features of her data. For example, she might have data that rule out the possibility that mating is random with respect to traits she studies. The focus here is on theory rather than empirical work.

[48] See, e.g., Hartl and Clark (1989, pp. 66ff., Table 1, p. 64), Gillespie (2004, pp. 47ff.), and Equation 35 in Ewens (2009).

eventually carry a naturally selected allele. Additionally, in the Wright–Fisher model, the size of a theoretical population affects its evolution. For example, other things equal, the larger the Wright–Fisher population, the smaller the influence of drift, the greater the influence of selection, and the greater the expected heterozygosity. Such results can shed light on patterns of evolution observed in species in the field, such as the historical populational bottlenecks that cheetahs have gone through in the wild, and in laboratory populations of, for instance, the genetic workhorse *Drosophila*.

Outside the Wright–Fisher model, early researchers also noticed that deviations from assumptions 1 through 6 affected the evolution of a theoretical population in many of the same ways as changing the size of the population. For example, modifying assumption 3 so that the population consists of different numbers of breeding females and males decreases the heterozygosity just like decreasing the population size does. Although the populations fleshed out in this case are theoretical and idealized, they can lead to discoveries that are more practically applicable. One such example is the population size needed for a healthy amount of genetic diversity to protect multiple populations of particular endangered species in conservation biology.

One way we relate models of theoretical populations (where each model adopts a distinct set of assumptions) is through Wright's concept of *effective population size*, denoted as the parameter N_e.[49] Whereas the census size, N_c, denotes the number of organisms in the population, whether in nature, in the laboratory, or in theory, an effective size N_e represents the number of organisms with genetic characteristics that match a Wright–Fisher model theoretical population satisfying assumptions 1 through 6 with a population size equal to N_e. In other words, the effective population size is the size of a Wright–Fisher idealized population that would be expected to have a value of a statistic, or a theoretical property, identical to the one calculated or observed for the natural, laboratory, or theoretical population of interest.[50]

This is important, because these numbers frequently differ (the effective population size is almost always smaller than the census size), and scientists are often tasked with translating the theoretical effective population size into field census sizes to achieve the results calculated by

---

[49] Wright (1931a, 1938).

[50] See, e.g., Li (1955, pp. 320–321), Crow and Kimura (1970, pp. 109–111), Hartl and Clark (1989, p. 82), and Hedrick (2005, pp. 318–319).

theory. Their work is sometimes aided by choosing different properties on which to base the correspondence of the two populations, leading to different effective population sizes. Population geneticists often distinguish among *variance*, *eigenvalue*, and *inbreeding* effective population sizes, based on these three properties of theoretical population models.[51] For instance, the inbreeding effective population size is the number of idealized individuals that would generate the same level of inbreeding as measured in the natural population of interest.

Effective population size illustrates the positive benefits of distinguishing theoretical, natural, and laboratory populations, as well as the risks of not doing so. When evolutionary geneticists study laboratory and natural populations empirically, they use the effective population size to relate natural and laboratory populations to theoretical populations. However, and as mentioned above, because many natural populations do not meet the assumptions of theoretical models, the effective population size is sometimes strikingly different from the census population size. For example, though the census population size of humans is today close to 8 billion, under many measures, gene types, and models, the effective population size of humans has been estimated to be on average somewhere between roughly a few thousand and somewhat close to 30,000 over the past million years or so and has likely not been less, during this period, than approximately 1000.[52] However, the picture is more complicated during the past 10,000 years, and especially during the past few thousand years. Depending on models and assumptions used, an effective population size of as much as 1.1 million has even been found for contemporary Europe.[53] After all, super-exponential growth has occurred in recent human history, which "skews patterns of genetic variation and distorts basic principles of population genetics."[54] Discrepancies between effective population sizes and census sizes throughout our history and today may look enormous, but they are potentially less meaningful to findings in evolutionary genetics than may

---

[51] Crow and Kimura (1970, pp. 345–365), Hartl and Clark (1989, p. 82), Hedrick (2005, p. 319), and Ewens (2009).

[52] See Takahata (1993) and Table 1 in Hawks et al. (2000, pp. 9, 17). Again, this is not to suggest that the human population meets all the assumptions of a Wright–Fisher model with a theoretical population size of around 10,000, or, more precisely, ranging between approximately 2000 and 28,000. Rather, the empirically observed heterozygosity of humans is approximately the heterozygosity expected in a Wright–Fisher population with such effective population sizes.

[53] Keinan and Clark (2012, p. 741). Note that this is still much less than the census size.

[54] Keinan and Clark (2012, p. 740).

initially appear to be the case. After all, clear methods exist for mathematically translating parameters from theoretical to natural populations.

One necessary consequence of the use of the effective population size concept is that population geneticists are reminded that the estimates of population sizes obtained should not be interpreted as the actual number of individuals in the natural population.[55] The concept acts as a barrier to conflating theoretical and natural populations.[56] Evolutionary geneticists are keenly aware that simple theoretical models, such as the Wright–Fisher model, may be poor descriptions of natural populations. Nonetheless, the focus of much population-genetics research is to relate predictions from theoretical populations to natural populations. By fitting theoretical population models to data from natural populations, geneticists obtain estimates of parameters such as migration rates, divergence times, and population sizes.[57] This, in turn, provides insight into the dynamics of evolution (e.g., speed of evolution, possibilities of adaptative changes, and potential extinction rates) in natural populations.

The Danger of Conflation: Theoretical and Natural Populations in *Structure* Analyses

In the past two decades, population geneticists have used model-based clustering methods to assign individual organisms to distinct statistical clusters using genetic data.[58] *Structure*, discussed in more detail in Chapters 5 and 6, is an influential algorithm and computational program for genetic clustering.[59] It produces analyses that have sparked insight into human genetic variation, especially as an exploratory tool for describing population-level patterns of genetic variation.[60] *Structure*-based analyses have been controversial, however. As we shall explore further in Chapter 9, proponents of biological and biogenomic race

---

[55] Slatkin (1991) proposed an analogous usage of the concept of *effective migration rate* for estimates of migration rates between populations.

[56] Interestingly, the effective population size concept itself can also be perniciously reified. In fact, Messer and Petrov (2013b) argues that coalescent effective population size, which is sensitive to neutral evolution over long time scales, has been incorrectly used by some evolutionary geneticists as a single description of population size, leading some researchers to ignore the dynamics of natural selection in populations whose sizes fluctuate rapidly.

[57] See Slatkin (1985), Neigel (1997), and Beerli and Felsenstein (1999).

[58] See Pritchard et al. (2000), Falush et al. (2003, 2007), Tang et al. (2006), Hubisz et al. (2009), Brisbin (2010), Maples et al. (2013), and Stift et al. (2019).

[59] See Pritchard et al. (2000), Falush et al. (2003, 2007), and empirical pattern #5 in Chapter 6.

[60] See Rosenberg et al. (2002), Friedlaender et al. (2008), Tishkoff et al. (2009), and Chapter 5.

concepts have argued that the classifications resulting from running *Structure* analyses on human data mimic race classifications traditionally used in the United States and thereby validate the reality of race.[61]

How do *Structure* analyses fit my population trichotomy? In Pritchard, Stephens, and Donnelly's original 2000 paper describing *Structure*, the clusters produced by the algorithm are repeatedly and consistently called "populations." But it was unclear to what kind of populations the word referred. To answer this question, we must consider *Structure*'s input, model, and output. The input to *Structure* is genetic data. Pritchard et al. used three examples: simulated data, data sampled from three geographically distinct groups of Taita thrush birds, and data from African and European humans. Thus, the genetic data used by *Structure* has come from theoretical populations (e.g., simulations), laboratory populations, or natural populations.[62]

Structure uses genetic data to *estimate* populations.[63] That is, *Structure* estimates the population membership of each organism in the sample (organisms might be assigned to more than one population) and estimates the allele frequencies in each population at each genetic locus in the data set. To make these estimates, *Structure* compares the provided data to a model. In the model, individual organisms have only two properties: population memberships, which may be fractional; and multilocus genotypes, or the individual's collection of alleles. Populations, in turn, have only two properties: allele frequencies and Hardy–Weinberg equilibrium.[64] In sum, *Structure* finds population memberships and allele frequencies that lead to the closest fit between the provided data and the model.

Using the trichotomy of population concepts, *Structure* outputs clusters that are estimates of theoretical populations. These clusters are the groupings optimizing the fit between, on the one hand, a model of a theoretical population and, on the other hand, data that may come from theoretical, laboratory, or natural populations.

Treating clusters from a *Structure* analysis as populations with properties that are not part of *Structure*'s model can lead to inferential errors.[65]

[61] See Sesardić (2010, 2013), Spencer (2013, 2014, 2015, 2019), and Wade (2014).

[62] On laboratory populations in this context, see, e.g., Whiteley et al. (2011); Griesemer (2020).

[63] The concept of population can apply to groups at different levels of a nested structure. Thus, the targets of estimation in a *Structure* analysis are called populations rather than subpopulations, even though they may be subsets of a larger population.

[64] The assumption of Hardy–Weinberg equilibrium can be relaxed; see Gao et al. (2007).

[65] See Anderson and Dunham (2008), Weiss and Long (2009), Gilbert et al. (2012), Putnam and Carbone (2014), and Chapter 6.

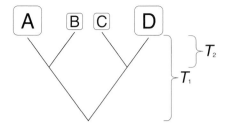

Figure 3.3. The true history
Four populations, A, B, C, and D, with their true historical, phylogenetic
relationships. T_1 and T_2 are divergence times, as described in the main text. From
Winther et al., 2015, p. 18. (See Preface). © 2015 Elsevier Ltd. All rights reserved.
Reprinted with permission. Redrawn by Mats Wedin.

Consider four (theoretical) populations with a true history, as depicted in
Figure 3.3, making the following two assumptions:

- The populations undergo random genetic drift but are not subject to
  natural selection.[66]
- Populations A and D and their ancestral populations are very large
  compared with their divergence time in generations (denoted T_1),
  meaning that the allele frequencies in the two populations will be very
  similar to the original ancestral population from which they diverged.
  This is because allele frequencies in large populations such as A and
  D drift very slowly, that is, their allele frequencies do not change much
  by chance, per generation. In contrast, if populations B and C are
  smaller, with sizes similar to their divergence times in generations
  (denoted T_2) from, respectively, populations A and D, then the allele
  frequencies in populations B and C will change appreciably due to
  random genetic drift.[67]

In this example, populations A and D will have similar allele frequencies, and will each maintain much of the same genetic variation.
However, populations B and C will have distinct and unique allele
frequencies across the genome, in part because they will have different
loci fixed for different alleles. A *Structure* analysis assuming three clusters
($K = 3$) would likely infer and ontologize the three clusters as {A + D},
{B}, {C} (Figure 3.4). No inferential errors have been made yet – this is

---

[66] See chapters 7 and 8 in Crow and Kimura (1970), chapter 2 in Hartl and Clark (1989), chapter
6 in Hedrick (2005), and chapter 2 in Nielsen and Slatkin (2013).
[67] See Figure 4 in Hartl and Clark (1989, p. 68), which doubles as the book cover.

Figure 3.4 Structure ontologizes
Hypothetical results of *Structure* clustering applied to the populations in Figure 3.3.
From Winther et al., 2015, p. 18 (see Preface). © 2015 Elsevier Ltd. All rights
reserved. Reprinted with permission. Redrawn by Mats Wedin.

the algorithmically and formally appropriate clustering if we are considering allele frequencies alone. However, depending on our interests, it
will make little sense to interpret cluster {A + D} as a population that
does not include populations B and C. For example, if a geneticist is
interested in questions about which natural populations of organisms
freely interbreed today, then the *Structure* clusters do not correspond
with the researcher's purposes – populations A and D do not freely
interbreed and have not done so for a long time.

If a geneticist is interested in phylogenetic studies and in exactly the
type of groupings found by *Structure* in the data, conflating and confusing
theoretical populations derived in the context of clustering analysis with
the theoretical, or even natural, populations pertinent to phylogenetic
studies can create problems. Shared ancestry is a fundamental component
of research into phylogenies, through which insights are gained into
genealogy, the divergence of populations, and the origin of subspecies
and (eventually) species. But *Structure* does not directly inform us about
shared ancestry. After all, {A + D}, {B}, {C} is not a phylogenetically
accurate grouping (Figure 3.4); {A + B}, {C + D} is.[68]

Structure can also err systematically. Suppose we were to perform a
Structure analysis on data from populations B and C from the example
above, excluding populations A and D. *Structure* analyses find a clustering
scheme in which individuals in the same cluster have maximally similar
genotypes and individuals in different clusters have maximally different
genotypes, subject to certain formal constraints on all clusters, especially
that they satisfy Hardy–Weinberg assumptions. With enough genetic
data, *Structure* would likely be able to distinguish between populations
B and C. But *Structure*'s assignments will be imperfect, and to the extent
that *Structure* errs, it tends to err systematically, assigning "B-like" individuals from population C partial membership in population B, and
assigning "C-like" individuals from population B partial membership in

[68] Kalinowski (2011) reaches a similar conclusion.

population C. Thus, the *Structure* clusters corresponding to populations B and C will likely be *more* genetically differentiated than the theoretical populations used or predicted by phylogenetic studies, or than the natural populations themselves. *Structure*-inferred clusters might not even approximately or remotely represent ancestral groups. Estimates of population parameters such as divergence times or migration rates might thereby be inaccurate for studies in which *Structure*-inferred clusters are treated as natural or theoretical populations for phylogenetic studies. If properties of *Structure* clusters are analyzed as if they were properties of natural populations, as is sometimes done, then this potential bias must be kept in mind.[69]

In all of this, the statistical methodology cannot be faulted. It has done just what we asked: produced groupings maximizing the fit between data and a model of populations as groups or clusters meeting Hardy–Weinberg equilibrium that also differ in allele frequencies. If our purposes extend beyond the minimal and highly simplified theoretical population concept expressed by the model underlying *Structure*, then we have more work to do after running the analysis. For one, we need supporting information to make a case that a *Structure* cluster corresponds to the type of entity in which we are interested. Users who conflate *Structure* clusters – which are perfectly reasonable theoretical populations in one context – with ecological or phylogenetic, natural or theoretical, populations will make confused interpretations.[70]

Such conflations and confusions are, alas, pervasive. Weiss and Long argued in 2009 that "architects of *Structure* . . . are well aware of the limitations of the method and state them clearly in their papers [references omitted]. However, applications of such programs are often made without heeding caveats or recognizing the limitations of the underlying models with respect to the questions and data at hand."[71] Weiss and Long were concerned with scientists reifying the output of *Structure* and similar programs, and they gave examples of such conflations – or pernicious reifications – of different population types. Lawson and colleagues voiced analogous concerns in 2018: "most researchers are cautious but literal" in how they interpret the results of *Structure*. In particular, Lawson and colleagues argued that it can all too easily be assumed that "the chosen value of K [cluster number] is the true value of

---

[69] For one such analysis, see Jeong et al. (2014).
[70] Putnam and Carbone (2014) consider similar issues from a more technical point of view.
[71] Weiss and Long (2009, p. 704).

K," that "each of the *K* ancestral population[s] existed at some point in the past," or that "modern individuals were produced by recent mixing of these ancestral populations."[72] Of course, these assumptions do not always hold – and Lawson et al.'s program *badMIXTURE* helps avoid some of these reifications.[73] But similar errors can occur when the media reports on *Structure* analyses, such as when science journalist Nicholas Wade described a *Structure* analysis by Tishkoff et al. (2009) as revealing "14 different ancestral groups."[74] The mistake here in a popular science context resulted in the lay public being told that *Structure* automatically and transparently identifies ancestral, phylogenetic, natural populations – populations with which we identify.

In these cases, mindful attention to the population trichotomy presented in this chapter, as well as simple philosophical reflection on ambiguities of meanings, and on the importance of considering research context, might benefit scientific practice. By demanding clarity about the aims and nature of studies, the kinds of populations used in studies, and the inferences legitimately supported by those populations, we can discourage the use of methods for questions they cannot legitimately answer.

Assumption Archaeology and Population Pluralism

Distinguishing three kinds of populations assumed and used in scientific practice – in theoretical, laboratory, and natural contexts – sheds light on the exact nature and content of the kinds of populations we wish to use, and of which we are a part. In this way, the trichotomy can be considered a tool of and for *assumption archaeology*, or the attempt to study the system of assumptions underlying a family of scientific models and theories.[75] One way of finding our way in a system of representations – a set of models or a scientific theory – is by excavating and locating impactful, even if hidden, assumptions. In Newtonian physics, for instance, space and time are God given, linear, and absolute; gravity is a universal force; and energy and mass are distinct and nonconvertible. By diagnosing such presumptions, we are less likely to miss when we make an inferential

---

[72] Lawson et al. (2018, p. 2).

[73] Novembre and Peter (2016) also review statistics and modeling strategies for how to correctly identify genetic differentiation among populations, admixture, and the local, spatial distribution of genomic variation, including with the program *teraSTRUCTURE*.

[74] Wade (2014, p. 100).

[75] See Winther (2006a, 2006c, 2020a), Godfrey-Smith (2009), and Servedio et al. (2014).

mistake (e.g., conflating one population with another), and we remain more flexible and receptive to critique from others. We can block model–world conflation and de-reify.

Scientists engaged in assumption archaeology ask questions such as these: Which kinds of assumptions are at play in modeling and theorizing, including methodological and ontological assumptions? What are the functions of each assumption, and what happens when we replace a given assumption, or add new ones? Which mistaken inferences could be drawn when we forget that certain assumptions have been made? Many working scientists consider such questions carefully.[76]

The *population* concept itself is used in human evolutionary genomics in addition to countless other fields in the natural and social sciences, as well as statistics. Close attention to this central concept of science shows the breadth, depth, and richness – not to say useful ambiguity – a scientific concept can possess. But what we learn about one type of population does not always apply to another. Theoretical populations need not describe all aspects of natural populations in order to provide insight. Indeed, careful users of *Structure* compare *Structure* results with a plurality of other kinds of evidence related to their interests and concerns. Finally, the analysis of populations found in this chapter may also be applicable to other biological fields. We shall see this in more detail in Chapters 4 and 5, where we turn to a variety of metrics and models for analyzing population-level genetic phenomena.

---

[76] Philosophical sources for framing and addressing these questions include Kuhn (1970), van Fraassen (1980), Levins and Lewontin (1985), Hacking (2002), Longino (2002, 2013), Wimsatt (2007), Elwick (2012), and Winther (2012b, 2021a).

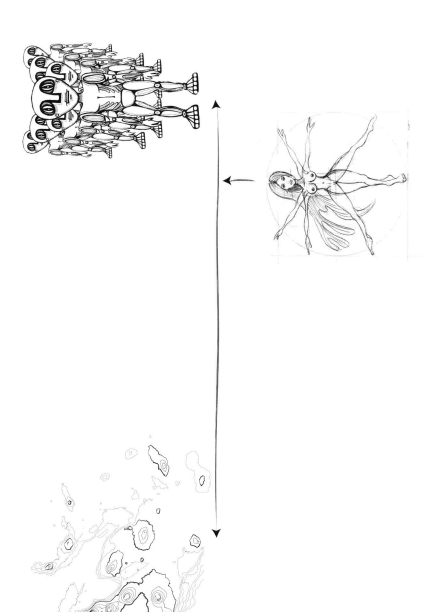

Figure 4.1. From Galápagos-Writ-Large to Planet Unity 2: Vitruvian Woman

Multiple lines of argument in *Our Genes* suggest that humans, here represented by Vitruvian Woman, are closer to Planet Unity than to Galápagos-Writ-Large. Concept by Rasmus Grønfeldt Winther, illustrated by Larisa DePalma (aliens, Vitruvian Woman) and Mats Wedin

4 · *Metrics and Measures*

At their most basic, the answers provided by modern genomic science to questions about populations depend on data. Data offer such a crucial lens through which to observe and begin to understand our similarities to, and differences from, one another that one of my major arguments in *Our Genes* is simply this: *Data matter.* Field data, yielded by measuring features of natural populations, help make evolutionary theory mathematically and statistically explicit by viewing evolution as the change of allele frequencies across generations. Evolutionary geneticists use many kinds of data to develop and select among mathematical models, and to subject explanatory and predictive evolutionary genetic theory to empirical tests.[1] To test and improve such models, evolutionary geneticists depend on measurements and data provided by field samples and laboratory experiments, fed through statistical machinery.

This chapter provides background information necessary to understand exactly what evolutionary geneticists measure and why they measure it. However, and as in my analysis of the trichotomy from Chapter 3, I also show that this information is not as straightforward as it seems. A plurality of measures and metrics of genetic variation exist, including diversity measures such as Jost's D, and heterozygosity-based measures such as Wright's F_{ST} and, to an extent, the Shannon information-theoretic measure. Each measure was designed for certain purposes, holds under certain assumptions, and provides certain kinds of knowledge. While these measures help point to the differences among theoretical, natural, and laboratory populations, which then inform various scientific inferences and conclusions, the methodological and theoretical assumptions behind such genetic variation measures can be unclear and hidden. That is why I also argue that data, even when arrived at through measurement and "naked mathematics," are not only the objective result

---

[1] See Lewontin (1974), Hartl and Clark (1989), and Nielsen and Slatkin (2013).

of transparent and precise measurements. They are also, as I claimed in Chapter 3, the result of contextual and local abstraction and idealization. In other words, data emerge when some measures or metrics are selected over others. Choices (and subsequent losses) are always involved.

A quick primer on statistical nomenclature is in order. While the two concepts overlap, and the same equation for genetic variation can be used as a measure or a metric, the former – *measure* – refers to data defined and shaped via well-motivated mathematical equations and, moreover, made meaningful by their role in informing and confirming some models or interpretations over others. The latter – *metric* – refers to an equation or concept used primarily in theoretical contexts, such as in modeling. The former is closely related to what statisticians refer to as a *sample statistic* or *summary statistic* or *estimator*, the latter to what they would call a *population parameter*. To be specific to our purposes, we can even say that a *measure* is the instantiation of the relevant genetic variation equation (and sampling protocol associated with it) in natural – and, to a lesser extent, laboratory – populations, while a *metric* is the instantiation of the equation as it is used in theoretical modeling efforts on theoretical populations. In this chapter, I tend to refer to theoretical metrics, although the corresponding empirical measures are closely associated, even if they are typically more complex.

Complex data are open to interpretation. The data also find themselves within the working context of particular empirical and theoretical research programs. In the process of deploying measures and metrics, we must keep well in mind that data rarely provide an unequivocal answer to the question of which model or theory is correct. Models and theories are underdetermined by the data; assumptions and epistemological values are also crucial to model selection or theory choice. To grasp the broader personal and philosophical interpretations – about populations and individuals, origins and identity, similarities and differences – we must understand not only what is measured by evolutionary geneticists but also the centrality of the theoretical, disciplinary, and work contexts in which measurements are made.

Useful background material in this empirical key is David Houle and colleagues' 2011 article "Measurement and Meaning in Biology." Whereas these researchers appreciate close attention to data, they also argue that "the key principle of measurement theory is that theoretical context, the rationale for collecting measurements, is essential to defining appropriate measurements and interpreting their values," where the cardinal sins of measurement include "not paying attention to theoretical context, inappropriate transformations of data, and inadequate reporting of units, effect sizes, or estimation error."[2] This chapter builds on their effort. I define and characterize measures and metrics pertinent to genetic variation, in order to make sense of their respective formal properties, implicit assumptions, and theoretical contexts, and thereby help to explain findings about *Homo sapiens*. By deploying assumption archaeology introduced in Chapter 3 – that is, by unearthing and evaluating the assumptions that evolutionary geneticists make – we get closer to understanding the multifarious meanings of genetic variation within and among populations (even as populations are also defined variously). Careful philosophical and conceptual analysis of what *genetic variation* and *population* are illuminates the instability, interpretative openness, and context dependence of data and models in human evolutionary genomics and, hence, the care we must take in the narratives and interpretations about our minds and bodies drawn from such data and models. In this, I follow the injunction of Houle and colleagues to remember theoretical context; interpret the numbers; respect scale type; know what the parameters mean; and make meaningful measures vis-à-vis a plurality of metrics, and a variety of theoretical – or modeling – contexts, of genetic variation.[3]

The Meanings of Metrics and Measures: Realism or Constructivism?

To conceptually analyze and provide an assumption archaeology of genetic variation that can shed light on, for example, Galápagos-Writ-Large and Planet Unity patterns, I propose a classification of its metrics and measures, distinguishing *genetic diversity*, *genetic heterozygosity*, and *genetic differentiation*. When basic definitions from genetics and genomics are required, the reader is invited to refer to the glossary at the end of Chapter 2.

Because the meaning of measurements in evolutionary genetics is determined by the theoretical contexts within which metrics and measures are embedded, we must study measurements through the disciplines making use of them. This means we investigate not just the metrics and measures themselves but also the relevant disciplines in the biological

---

[2] Houle et al. (2011, p. 4). Krantz et al. (1971) reviews measurement theory in general, and across the sciences.
[3] Houle et al. (2011, pp. 19–28).

sciences – including *conservation biology*, *taxonomy*, and *phylogenetics* – in which population structure and genetic variation are objects of study. Combined with population genetics, these fields flow together with – and are part of – human evolutionary genomics. Within each disciplinary and theoretical context, the relevant metrics and measures are a step toward telling us whether the species we look at satisfies a Galápagos-Writ-Large or a Planet Unity population structure pattern (Figure 4.1, chapter opener). This, of course, helps us get at whether there are real subdivisions within the species studied or not, and – if addressing humans – what populations you and I might be a part of, and what our genes might say about our future potential and perils.

Presaging ontological themes addressed in Chapter 9, there are both realist and conventionalist (alternatively, constructivist) ontological interpretations of the value and significance of metrics and measures, in their living theoretical and disciplinary contexts. The realist believes that given the appropriate measures and discipline, there *must be* a fact of the matter about whether the given species or set of populations under study best fits a pattern such as Galápagos-Writ-Large or Planet Unity (or a pattern, specifically, in between). In this, the realist fits populations according to a *subdivision* determination (Galápagos-Writ-Large) or a *continuum* determination (Planet Unity). According to the realist, one of these, and only one of these, must be true for a particular species at a particular time.

Of course, given the complexity of the theoretical machinery and a plurality of modeling assumptions that exist even within the same disciplinary context, a more conventionalist or constructivist interpretation is also possible. According to this philosophical position, there is rarely a single fact of the matter about whether we can absolutely and objectively judge a given species (or set of populations) as subdivided *or* continuous. Thus, argues the constructivist, evolutionary geneticists meet the data having already adopted one of two perspectives: *Splitters* typically subdivide and *lumpers* typically find continuity, and the preexisting commitment to – or the judicious and case-based choice of – either perspective determines different valuations of complex data, measurements, and models.[4] But, unlike the realist, the conventionalist or constructivist does

---

[4] Zerubavel (1996) and Bowker and Star (1999) provide useful articulations of splitting and lumping perspectives. Sociologist Eviatar Zerubavel (1996) explores the "mental process" that allows "grouping 'similar' things together in a single mental cluster" (i.e., lumping) and, conversely, "perceiving 'different' clusters as separate from one another" (i.e., splitting) (p. 421). A similar

not believe that there is a single, objective fact of the matter about whether the splitter or lumper is correct. In general, the constructivist argues that taxonomists reasonably disagree about whether identifying, naming, and publishing articles about human populations below the species level can tell us anything meaningful about robust and predictive genetic differences among groups in terms of, for example, cognitive capacities or disease susceptibility. According to the constructivist, split-ting and lumping perspectives are both reasonable, if somewhat narrow, interpretations informing the inferential consequences of each perspec-tive's use of genomics and conceptualizations of genetic variation. There can thus be genuine, reasonable, and irreducible disagreement between the two perspectives. This is the general ontological interpretation that I somewhat cautiously defend, at least with respect to human evolution-ary genomics. Let's now turn to the metrics and measures of genetic variation, and those theoretical, disciplinary contexts within which they are embedded.

Three Metrics and Measures of Genetic Variation

As three kinds of metrics and measures of genetic variation, genetic diversity, genetic heterozygosity, and genetic differentiation encompass much of our current knowledge of data-driven genomic analyses of the similarities and differences between individuals and populations. They are used, primarily and respectively, in conservation biology, taxonomy, and phylogenetics.

Genetic Diversity

Diversity is most naturally thought of as a measure of a system's hetero-geneity. Genetic diversity is assessed according to the general lack of *identity* of elements within and across populations. Genetic diversity might therefore be measured by how many different alleles there are in

characterization is found in Zerubavel (1991, p. 21). In developing the metaphor of the "mental distance" between objects classified within a category or classification, and between categories of a classification, Zerubavel (1991) argues that "whereas lumping involves playing down mental distances within entities, splitting entails widening the perceived gaps between entities so as to reinforce their mental separateness" (p. 27). In their investigation of the politics and ethics of "classification and its consequences," sociologists of science Bowker and Star (1999) note that lumpers are "those who see fewer categories and more commonalities," while splitters are "those who would name a new species with fewer kinds of difference cited" (p. 45).

a population, either at some particular locus or on average across all (or just some set of) loci.

Metrics of genetic diversity are broadly useful. In conservation biology, for instance, genetic diversity helps diagnose the genetically most diverse species (or populations) on which limited conservation effort resources should be focused. All other things being equal, a population is more (rather than less) diverse if it has more alleles per locus, and the frequency of alternative alleles are more rather than less equal.

A number of particular metrics illustrate this general characterization of genetic diversity. As an example, consider the formal metric known as *Jost's D* (D_J)[5]:

$$D_J = \left(\sum_{i=1}^{u} p_i^q \right)^{\frac{1}{1-q}},$$
[Eq. 4.1]

where there are u alleles at a locus, p_i is the frequency of the ith allele, and q controls the sensitivity of the metric to the allele frequencies.[6] Equation 4.1 yields important information. According to Jost, "true diversity $[D_J]$ has an easily interpretable metric, unlike heterozygosity. A drop of, say, 20% in diversity is equivalent to a loss of 20% of the alleles of a perfectly even population (a population whose alleles are all equally common)."[7] But there are also unclear use cases, such as when two populations have an unequal number of alleles at a given locus (alternatively, the comparison could be between two different loci in the same population). There is no straightforward general answer to a question like "does a locus with three alleles of equal frequency have more or less diversity than a locus with one very common allele and 20 very low-frequency alleles?" Indeed, for cases like this, the diversity measure can *reverse*, depending on allele frequencies and on how much one discounts rare alleles as compared to common alleles.[8] In other words, this use case also clarifies

---

[5] See Jost (2006, 2008). While Jost's D is a convenient example of a diversity metric, my use of it should not be taken as an endorsement of it as *the only* or *the best* metric. For discussion, see, e.g., Hoffmann and Hoffmann (2008), Gerlach et al. (2010), and Whitlock (2011). For earlier formulations of this metric, see Hill (1973), Gregorius (1978, 1987), and Routledge (1979).

[6] See Jost (2006, p. 364) and Jost (2008, p. 4020).   [7] Jost (2008, p. 4020).

[8] When $q = 0$, Jost's D is equivalent to the allele number, and then the locus in the second population (alternatively, the second locus in the same population) clearly has the higher diversity, with 21 alleles. In contrast, when $q = 2$, Jost's D becomes the reciprocal of Nei's gene identity (which is $1 -$ Weir's D, which we shall meet again in the next section), and as long as the frequency of the very common allele at the locus in the second population (alternatively, the second locus in the same population) is sufficiently high, this locus's diversity will be *lower* than that of the first population. Jost (2006) and Jost (2008) explain Equation 4.1 in detail.

how data interpretation and statistical analysis depend critically on the metric and measure chosen, and the parameter values (e.g., q) one assigns to a metric.[9]

Genetic Heterozygosity

Heterozygosity measures and metrics are formulated explicitly in an evolutionary genetic theoretical framework, and are used in Wright's F_{ST} and Nei's closely related G_{ST}. Heterozygosity can be thought of as a special kind of diversity measure, but it has an important history and use in population genetics. Sewall Wright used it as a cornerstone for his F-statistics, as we shall see. Theodosius Dobzhansky appealed to heterozygosity in his influential argument that genetic variation was more common in nature than previously thought; his doctoral student, Richard Lewontin (1929–2021), showed this to be indeed the case in *Drosophila* and, with some qualifications, in humans.[10]

Measures of heterozygosity use allele frequencies to assess the fraction of heterozygotes, as opposed to homozygotes, in a population. For instance, the basic Weir's D (D_W) metric of heterozygosity is:

$$D_W = 1 - \sum_{i=1}^{u} p_i^2, \qquad \text{[Eq. 4.2]}$$

where p_i^2 is the homozygotic genotype frequency of the ith allele and u is the total number of alleles at the particular locus.[11] D_W is recalculated to give heterozygosity metrics depending on the level of analysis – whether we are assessing the heterozygosity of a subpopulation (H_S) or the heterozygosity of the total population (H_T). H_S uses Equation 4.2 based on allele frequencies for the respective subpopulation, whereas H_T uses Equation 4.2 based on grand average allele frequencies for the total population, across all subpopulations. Finally, H_I is the *observed* heterozygosity of individuals within respective subpopulations.[12]

---

[9] See Houle et al. (2011).

[10] See, e.g., Dobzhansky (1955), Hubby and Lewontin (1966), Lewontin and Hubby (1966), and Lewontin (1972); for further context, see Beatty (1987) and Winther (2022a).

[11] Alternatively, $D_W = \sum_{i,j=1}^{u} p_i p_j$, where p_i and p_j are the frequencies of, respectively, the ith and jth alleles of a total of u alleles, and $i \neq j$ [Eq. 4.1n]. Because heterozygote and homozygote frequencies must add up to 1, this latter equation is equivalent to Equation 4.2, although population geneticists prefer Equation 4.2 because it is easier for calculations and tracking genotypes.

[12] From the observed heterozygosities for each population, allele frequencies can be extracted for calculating the two higher-level genotypic, heterozygosity measures, assuming random mating

In nature, in the laboratory, and in theoretical models, there tends to be a general heterozygosity reduction of, first, H_I with respect to H_S, in a subpopulation (or of the average H_I with respect to the average H_S, both averaged across subpopulations) and, second, of the average H_S with respect to H_T. With respect to the first general heterozygosity reduction, the more *inbreeding* there is in a subpopulation, the more the same alleles tend to be recycled, as it were, within broad family lineages across generations. That is, inbreeding increases the probability that at a given locus, two downstream individuals will have the same allele due to shared ancestry (this same allele across the two, related, individuals is said to exhibit *identity by descent*). Inbreeding will cause reduced heterozygosity, within any given population, compared to the Hardy–Weinberg heterozygosity expected under random mating. Conversely, populations can outbreed and migrate with respect to one another, and can thereby have increased heterozygosity, compared to Hardy–Weinberg expectations. Inbreeding can occur to various degrees but, except for cases of self-fertilization in some plants as well as other unique and interesting mating systems, is relatively rare in nature and in humans.[13] Setting aside other evolutionary forces – which can themselves cause deviations from Hardy–Weinberg heterozygosity proportions within subpopulations (e.g., selection) – inbreeding effectively results in a lower H_I than the H_S of a subpopulation (or, when there is more than one sub-population, in a lower average H_I than the average H_S); conversely, outbreeding can result in a H_I higher than H_S (or in an average H_I higher than the average H_S). Wright's F-statistic of F_{IS} captures inbreeding in the sense thus far discussed, as deviations from Hardy–Weinberg expectations within subpopulations, and can in principle take on values between -1 and 1 (positive values indicate inbreeding).[14]

---

and Hardy–Weinberg equilibrium at each of the two levels. Recall the distinction between sample statistics and population parameters above. Strictly speaking, H_I, H_S, and H_T are still sample statistics unless the sample size is sufficiently large (typically greater than 100 individuals, with their respective loci), in which case we are also finding – that is, reliably estimating – parameters, even for H_I. In other words, with sufficiently large sample sizes, sample statistics accurately give parameters.

[13] On the relative rarity of inbreeding with the meaning discussed here (i.e., often $F_{IS} \approx 0$; mating within subpopulations is effectively random), see, e.g., Hartl and Clark (1989, p. 298) and Holsinger and Weir (2009, p. 645).

[14] I cannot in this chapter do full justice to F-statistics, including a deeper analysis of F_{IS} (or of inbreeding, even within just one large population), or of the overall F-statistic, F_{IT}, the three of which are related by this equation: $(1 - F_{IT}) = (1 - F_{IS})(1 - F_{ST})$ [Eq. 4.2n]. Hartl and Clark (1989) provide a pithy, powerful, and precise explication of F-statistics (pp. 293–296). See also Wright (1965, 1969), Weir (1996), and Nielsen and Slatkin (2013).

However, in order to explore population structure in humans and other organisms, I set aside F_{IS} and instead focus on Wright's F_{ST}. F_{ST} is a classic measure and metric of the sheer and almost ineluctable reduction in heterozygosity in subpopulations relative to the total population, due to evolutionary forces such as random genetic drift and even selection:[15]

$$F_{ST} = \frac{H_T - \overline{H}_S}{H_T}.$$ [Eq. 4.3]

This is the second general heterozygosity reduction – of average H_S (i.e., $\overline{H}_S$) with respect to H_T. In principle, F_{ST} takes on values only between 0 and 1. Notice that, according to Equation 4.3, if there is no population structure (if the subpopulations are identical in allele frequency composition), then $F_{ST} = 0$ (i.e., no excess of homozygotes across subpopulations). If, by contrast, the two populations are fixed for two different alleles ($p_1 = 1$ and $p_2 = 0$; $q_1 = 0$ and $q_2 = 1$, for alleles p and q at a biallelic locus in two populations, 1 and 2), then $F_{ST} = 1$ (i.e., $\overline{H}_S = 0$ and $H_T = 0.5$; excess of homozygotes is at a maximum), and the populations are maximally diverse genetically.

The reduction of $\overline{H}_S$ with respect to H_T, per locus, is the *Wahlund effect*, discovered by the Swedish geneticist Sten Wahlund early in the twentieth century, which describes the "excess homozygosity" of subpopulations relative to the total population.[16] In particular, the averaged heterozygosity of two (or more) populations (i.e., $\overline{H}_S$) is almost always *lower* than the total heterozygosity (i.e., H_T) calculated by *pooling* those populations into a total population and, for a given locus, calculating Equation 4.2 using the grand average allele frequency across every

---

[15] See Wright (1922 1931a 1943 1949 1965 1969). A remark by Hartl and Clark (1989) is worth bearing in mind: "Wright developed his *F* statistics in terms of one locus with two alleles, and treated multiallelic data by keeping the most common allele and pooling the other alleles. An explicit multiallelic formulation [i.e., G_{ST}] was presented by Nei (1973)" (p. 293). Moreover, for clarity about the distinction between population parameter (metric) and sample statistic estimator (measure) in the case of F_{ST}, see, e.g., Weir (1996), Weir and Hill (2002), Holsinger and Weir (2009), and Willing et al. (2012). Provine (1986) provides an intellectual history of Wright's development of his *F*-statistics, a formal, population genetic framework that reached maturity in the 1940s.

[16] Wahlund was concerned with questions about eugenics and the evolution of different human groups. Säll and Bengtsson (2017) narrate: "[Wahlund's] interest in the gradual decline in genetic differentiation over time as a result of migration and intermarriages can be seen as a way to predict how long it will take before the Sami people [Indigenous peoples of Northern Scandinavia] disappear" (p. 147). These are clearly controversial – indeed, distasteful as well as tragic – matters. We must be wary of the dangers and problematic history of human evolutionary genomics, which *Our Genes* does its best to face head-on.

subpopulation. Conversely and equivalently put, the averaged homozygosity of many subdivided populations is almost always *higher* than the average homozygosity of the total population. The one exception (and the reason for the "almost always" in the previous two sentences), occurs when the *same* allele frequencies for a given locus are found across all populations – in that case, the averaged heterozygosity of the subdivided populations and the heterozygosity of the pooled, total population will be equal. But the averaged heterozygosity of the subdivided populations will never be higher than that of the total population.[17]

Wright developed these *F*-statistics in the service of his interest in population subdivision, migration, selection, and evolution in hierarchical populations.[18] More specifically, Wright studied guinea pigs and livestock. Among the major questions he sought to answer was whether, at the level of groups (or demes or populations), there would be sufficient intergroup variation on which group or interdemic (interpopulation) selection, a phase of his shifting balance process, could act.[19] He created *F*-statistics in conjunction with his *shifting balance theory*.[20] This theory postulated that evolution occurs both within small populations, subject to a combination of random genetic drift and mass selection, and *across* populations, with populations that send out more migrants and boast higher fitness against changing and distinct local environments having a relatively larger influence in terms of their alleles and allele frequencies on the overall, total population. In other words, according to Wright's work, populations with higher mean fitnesses exerted a stronger influence on the evolution of the overarching species.

While Lewontin was more influenced by Dobzhansky's thinking about heterozygosity than by Wright's models and theories, he was clearly knowledgeable about Wright's work.[21] And although he used a somewhat different metric (and partitioning method) in his landmark 1972 study, as we shall see below (and in Chapter 5), Lewontin cemented the subsequent interpretation of heterozygosity-like metrics (and the hierarchical partitioning of heterozygosity) for genetic

---

[17] See Winther (2022a) for a fuller picture of the Wahlund effect, and the role of measure convexity and Jensen's inequality. The Wahlund effect also holds for more than two alleles at a locus, although with some complexities (e.g., Hartl and Clark, 1989, pp. 288–291; Nielsen and Slatkin, 2013, chapter 4).

[18] See Provine (1986).   [19] See Provine (1986) and Winther et al. (2013).

[20] See, e.g., Provine (1986), Wade (1992, 2002, 2016), Winther (2006a), and Winther et al. (2013).

[21] See, e.g., Lewontin (1967).

taxonomic purposes of identifying the reality and robustness, or not, of human populations as racial or ethnic groups.[22]

Genetic Differentiation

Differentiation should be thought of as how different two things are. For genetic differentiation, two populations that share most of their common alleles are relatively undifferentiated, whereas two that do not share many alleles will be highly differentiated. Thus, genetic differentiation is a metric of the lack of *similarity* between populations.

Differentiation, when used with assumptions permitting it to be interpreted as a measure of genetic distance, encourages the recursive nesting of populations. This is what makes it so useful for producing tree structures and phylogenies. All other things being equal, two populations are relatively more (rather than less) differentiated if they share relatively few alleles (rather than many) or if the *frequencies* of the alleles shared between the two populations are relatively more different (rather than relatively more similar), or both. As an example, given certain assumptions, the following equation is a valid measure and metric of genetic distance:[23]

$$T_{EU} = \sqrt{\left(\sum_{i=1}^{u} \left(p_{1,i} - p_{2,i} \right)^2 \right)}, \qquad \text{[Eq. 4.4]}$$

where two populations have u total alleles (the number of distinct alleles at that locus in both populations), $p_{1,i}$ is the frequency of the ith allele in population 1, and $p_{2,i}$ is the frequency of the ith allele in population 2, and

---

[22] That F_{ST} holds only under certain assumptions has been recognized by, for instance, Long and Kittles (2003), Long (2009), Long et al. (2009), Jakobsson et al. (2013), and Edge and Rosenberg (2014). Two such assumptions that Long and Kittles (2003) diagnose are that effective population sizes (a concept we met in Chapter 3) must be equal across all subpopulations and that there must be no extra or hidden hierarchical population structure, lest subpopulations do not evolve independently (Long and Kittles, 2003, pp. 455–457). They then calculate F_{ST} values and goodness-of-fit statistics for four models of human evolution and population structure, where each model successively "relaxes an important constraint [assumption] from the previous model" (p. 460), and the third model is the first of the four that is "consistent with a large body of molecular data" (p. 463). F_{ST} values and goodness-of-fit statistics differ between the four models, and Long and Kittles believe they have "demonstrat[ed] the shortcomings of F_{ST}" (p. 464). Perhaps. For three of the four models, F_{ST} values hover between 0.119 and 0.131, utterly consistent with standard findings. In Chapters 5 and 6, I further investigate assumptions of, and mathematical constraints on, F_{ST} and consider the range of F_{ST} estimates in *Homo sapiens*.

[23] See Nei (1987, pp. 208ff.), Excoffier et al. (1992, pp. 480–481), Weir (1996, pp. 190–194), and McDonald (2008).

we are measuring allele frequencies at one locus. Equation 4.4 is the simplest and most general form of a Euclidean genetic distance metric, which satisfies three crucial axioms in a standard Euclidean space: Distances are greater than or equal to 0 (i.e., not negative), distances are symmetric, and distances satisfy the triangle inequality (i.e., the length of the longest edge of a triangle is less than or equal to – for cases of triangles with area 0 – the sum of the lengths of the other two edges). There is no single one-size-fits-all genetic distance measure for assessing genetic differentiation between pairs of populations. Weir proffers: "Distances may be regarded simply as data reduction devices, or as a means of comparing pairs of extant populations, or as the basis for constructing evolutionary histories for the populations. These different goals require different distances and, especially for inferring phylogenies, require genetic models."[24] Indeed, the degree of differentiation of a single, given pair of populations will differ based on the measure used, of which T_{EU} is only one of many possible measures. Furthermore, the rank order of a total set of population pairs from most to least differentiated can change based on the genetic distance metric used, a fact that has deep implications for statistical phylogenetic inference. For instance, the same data with different measures can produce trees with different branching patterns and branching topologies.[25] Semantic and formal choices need to be made in selecting and using formal metrics and measures.

Contrasting (Genetic) Diversity, Heterozygosity, and Differentiation

To place these three conceptualizations of genetic variation and their respective uses in the context of the populations of our thought experiment from Chapter 2, we could assume that Galápagos-Writ-Large has high levels of diversity, heterozygosity, and differentiation, both within and across populations, while Planet Unity has very low levels. However, the measurements are frequently more complicated. In fact, if we compare these three metrics and measures of genetic variation, we begin to see just how multifarious and complex the seemingly simple central concept of *genetic variation*, indeed, *genomic variation*, is. We also come to understand the wildly varying theoretical contexts and paradigms, and their associated purposes and research questions, in which the three measures live and are deployed.

---

[24] Weir (1996, pp. 190–191).
[25] Cf. chapter 9 in Nei (1987) and chapter 11 in Felsenstein (2004).

For example, let's consider the distinction between genetic diversity and genetic differentiation. These metrics are especially pertinent to conservation biology and phylogenetics, respectively, because they illustrate amounts of standing genetic variation for further evolutionary change and even for bare survival given changing local conditions (e.g., changes in parasitic load, rain patterns, or human-caused habitat destruction), or represent the relative genetic difference and distance between different populations (and species). With information about relative difference and distance, a phylogenetic tree can be constructed to capture relative groupings, in terms of the branching genealogical patterns of the origin of populations and origin of species. But diversity and differentiation are broadly independent of one another, and two populations quite different from each other, according to a differentiation metric such as Equation 4.4, can maintain that particular level of differentiation for a significant range and kinds of internal diversity, according to a diversity metric such as Equation 4.1 (and diversity values may themselves differ between the populations).[26]

Conversely, two populations that are very diverse at some locus (or loci), meaning each population has a large number of alleles at the locus (or loci) in question at sufficiently significant frequencies, can vary from entirely differentiated (i.e., they share no alleles) to undifferentiated (i.e., they have the same alleles at the same frequencies at each locus of interest). Further, while it may be obvious that two populations differing with respect to their diversity (for instance when one population has 2, 10, or 100 alleles at a locus and the other has just 1 allele) cannot be entirely undifferentiated, their degree of differentiation can still vary from relatively low to complete. Diversity and differentiation are thus distinct measures (and meanings) of genomic variation: Diversity metrics indicate how much nonidentity of elements populations have (both within and among populations), and differentiation metrics indicate how dissimilar populations are to each other.

Heterozygosity metrics, by contrast, reflect (at least partially) *both* population diversity *and* differentiation among populations and also

---

[26] Here is one way to think of this: consider that we can use Equation 4.4 iteratively and calculate Euclidean genetic distances between populations for each of many loci, counting each locus equally; by contrast, Equation 4.1 can be extremely sensitive to the differing magnitudes of allele frequencies and is not democratic, as it were, per locus. And although the square function in Equation 4.4 also somewhat inflates the relative importance of loci giving larger between-population differences, this measure, unlike Equation 4.1, fundamentally assesses genetic differentiation.

represent one way of conceptualizing genetic variation. As such, they often cannot distinguish differences that are the result of different degrees of diversity from differences that are the result of different degrees of differentiation.[27] That said, heterozygosity is not just a blunt tool. Whereas metrics of diversity and differentiation allow us to draw conclusions, respectively, about standing genetic variation (relevant to conservation biology) and differences among populations (relevant to phylogenetic inference), heterozygosity metrics can be used to draw taxonomic conclusions about the robustness and reliability of group – especially population – identification and uniqueness.

Admittedly, there is some overlap and interaction among the three conceptualizations, and among them and their respective theoretical contexts. For instance, Reich et al. first applied an interpretation of F-statistics, in particular F_{ST}, to infer the phylogenetic relationships among two (f_2), three (f_3), and four (f_4) populations.[28] However, while important, this is a somewhat nonstandard and still emergent use of F-statistics. Although projects dependent on measuring diversity, heterozygosity, and differentiation themselves exhibit theoretical variation, there are strong links between, respectively, assessing the identity and diversity of a particular population or subpopulation and conservation biology, exploring heterozygosity and taxonomy, and describing the similarity and differentiation of populations and phylogenetics. Each of the three meanings and families of metrics and measures of genetic variation – *diversity*, *heterozygosity*, and *differentiation* – needs to be relativized to its appropriate background paradigm and modeling practices. Consequently, each measure yields clearer, richer meanings when kept to its appropriate research aims and theoretical frameworks.

Theoretical and Disciplinary Homes

To exaggerate, each type of metric and measure has a particular theoretical and disciplinary home. This paradigmatic home influences the meaning of the data and how it is used to produce theoretical representations, such as taxonomic groupings and phylogenetic trees, and even policy recommendations. As I have written in another context, abstract maps make concrete worlds.[29]

---

[27] See Jost (2008, p. 4019).   [28] Reich et al. (2009); see Peter (2016).   [29] Winther (2020a).

Table 4.1. *Three-way comparison of the three theoretical and disciplinary contexts considered in this chapter*

Biological discipline	Concepts and aims	Metrics and measures
Conservation biology	Examines biodiversity to preserve species and ecosystems in the context of limited public and private resources and awareness	Diversity metrics (e.g., Jost's D)
Taxonomy	Defines classes and groups of individuals on the basis of molecular (e.g., genetic or protein) information or diagnostic, phenotypic features in order to discover the ranked "order of nature"	Heterozygosity metrics
Phylogenetics	Investigates evolutionary, genealogical relationships among classes or groups of individuals (e.g., subspecies, species, genera, clades) through characters assessed and collected via molecular (e.g., genetic or protein) or phenotypic studies in order to reconstruct the tree of life[30]	Differentiation metrics (e.g., Euclidean genetic distance)

In this section, I illustrate the ways three theoretical and disciplinary contexts tend to use genetic diversity, genetic heterozygosity, and genetic differentiation measures (see Table 4.1 for a summary of the three contexts). I examine the objective and subjective factors involved in deploying each kind of metric and measure of genetic variation. Knowledge in each of the three disciplines requires a finely tuned ability to join and separate groups of different sizes, and potentially at different levels, according to particular distinctions of difference, including about genomic variation. Therefore, each discipline is informed by prominent discourse on how distinctions are and should be defined and, thus, about how groups should be sorted and understood.

---

[30] While tree topologies continue to be the standard, orthodox representation of phylogenies, for both humans and the general tree of life, there are calls for alternative trellis (e.g., Templeton, 1999) and network (e.g., Doolittle, 1999; Huson et al., 2010) topologies.

Conservation Biology

Comparisons of population-level genetic variation, as measured by genetic diversity (e.g., Jost's *D*), are most at home in conservation biology. For instance, Lou Jost works and publishes on birds, plants (especially orchids), and conservation of Ecuador's forests, in addition to working on mathematical biology, as he explains on his website (www.loujost .com). It is in conservation biology that the appropriate meanings and measures of diversity, genetic and otherwise, for ecological theory and practice are most ardently discussed – and where much of the most interesting work on these questions has been done.[31]

When encountering a wide-ranging species facing the destruction of much of its habitat, the conservation biologist often asks: "What would happen if we could save only *this subpopulation*? Or, *this one*?" Otto Herzberg Frankel offers one way to understand this task. He decried the loss of plant diversity and based his conservation arguments and priorities on the loss of gene-pool diversity, believing that in the interest of both humans and other species it was necessary to conserve as much genetic diversity as possible, of both domesticated and wild gene pools, "to keep evolutionary options open" and to respect the "intrinsic value" of nonhuman species.[32] In his work, Frankel poignantly painted the frightening, accelerating rate of habitat loss.[33] He also insightfully portrayed some of the temporal paradoxes of conservation biology: "Perhaps our greatest difficulty stems from the contradictions of time scales: evolutionary time is compressed into [human] historical time and made subject to decision making on a [proximal, immediate] sociopolitical time scale."[34] His argument continues to resonate: Operating at time scales many orders of magnitude larger, evolution – and associated ecological and geological processes, which often operate at even significantly longer time scales – simply cannot adapt to the whims and short-sightedness of human actions and policy making. Consequently, maximizing genetic diversity to the extent possible can help ameliorate this "contradiction" somewhat and provide a potential, partial solution.

---

[31] On genetic diversity in conservation biology, see Frankel (1974), Jost (2008), Funk et al. (2012), Coates et al. (2018), and Drori (2018). Their findings are disparate and distinct, but all of these researchers use metrics, measures, and meanings of genetic diversity to inform conservation efforts. Environmental DNA (eDNA) is an interesting, recent development in this context.

[32] Frankel (1974, pp. 62, 63).   [33] See also Ward (2004) and Bradshaw et al. (2021).

[34] Frankel (1974, p. 63).

In conservation circles, there is active discussion about what role is played by *conservation units* below the species level. Two kinds of conservation units are often distinguished. An *evolutionary significant unit* is "a classification of populations that have substantial reproductive isolation, which has led to adaptive differences so that the population represents a significant evolutionary component of the species." A *management unit* is "a local population that is managed as a distinct unit because of its demographic independence."[35] There are other units, and they are based on a variety of criteria – adaptive, functional, demographic, and historical – but these two serve my purposes of showing the cross-cutting and distinct sets of criteria used to define and characterize kinds of subspecies populations pertinent to conservation biology.

In general, the discourse about conservation units has increasingly focused on ways to use the modern measurement tools of genomics, including species barcoding and genetic diversity measures of genetic variation, such as Jost's D, to delineate the populations and associated habitats worth conserving. In a useful article reviewing the state of the art and implications of genomics for conservation, Coates, Byrne, and Moritz urge the "increasing need to recognize ... genetic diversity below the species level. The challenge is to ensure that highly structured populations and deeply divergent lineages that are of conservation concern are protected in the face of competing demands for conservation attention."[36] Others from the heartland of conservation management have mounted similar arguments.[37] Cards on the table – I believe, when it comes to answering questions about conservation biology, that it is crucial to think in terms of populations. The cause of the current sixth mass extinction (or seventh)[38] is not just the extinction of species, but the *decline of populations in endless retreat.*[39]

The arguments in conservation biology about the value that statistical measurements can or should offer with respect to individual- and population-level variation – as well as with respect to larger questions about the literal and formal uniqueness of populations, and about

---

[35] Funk et al. (2012, p. 489); see also Coates et al. (2018).   [36] Coates et al. (2018, pp. 9–10).
[37] See Allendorf et al. (2013), Hoban et al. (2013), and Koskela et al. (2013).
[38] Cf. Rampino and Shen (2021).
[39] Cf. Gaston and Fuller (2008). Conservation biologist Mark Thompson and I are collaborating on a project showing how his field is ultimately concerned with population-level phenomena. Although conservation biology is about connectivity – networks of people and the biota as a whole – the legislative and financial infrastructure for conservation tends to emphasize single species, and in a variety of other ways frequently misses the problem of population-level declines.

metaphorical, phenomenological identity and embedding within populations – link to arguments in other discourses, such as taxonomy and
phylogenetics. More broadly speaking, they are also some of the same
arguments investigated in *Our Genes*: What can statistical measurements
tell us? About ourselves? About how we are connected to one another?
About our world? Increasingly, we see the answer is, "It depends." What
exactly do the statistics in conservation biology tell us? What amount of
diversity, local and global, is helpful or necessary for robust and resilient
ecosystems and climate? How do we prioritize unfortunately limited
conservation resources? Total species number is not everything. As
Houle et al. argue, we need to search for "meaningful statistics."[40] This
injunction requires us to understand the appropriate, central senses of
genetic diversity in conservation biology and to grasp the role of statistics
more generally in biology – and beyond.

Taxonomy

Taxonomy tells us about the ranked, hierarchical order of populations
within species, within genera, within families, and so forth. It begins to
get at questions about the exact structural relationships of similarity and
difference among individuals from distinct populations and species. In the
case of *Homo sapiens*, it thus helps us pursue answers to the ontological
status of the reality of biogenetic race. This may look like a straightforward application of taxonomic practice to which there is a clear answer,
but looks can be deceiving. For instance, are the populations we call
"races" in ordinary social discourse the sorts of contemporary subpopulations that a reasonable biologist would also recognize as existing and
worthy of attention, and perhaps even of naming?

Diametrically opposed answers to this question can be found in recent
philosophical and biological literature, much of it building on, or at least
alluding to, Lewontin's findings. The philosophical answers will have to
wait until Chapter 9, but Lewontin's 1972 article will be discussed in this
chapter, as well as in Chapters 5 and 6. For the moment, I simply trace
Alan Templeton's argument that the degree of differentiation among
human populations does not rise to the level required by biologists to
recognize subpopulations as legitimate populations of interest.[41] He

---

[40] Houle et al. (2011, pp. 17–19).   [41] Templeton (1999); see also Templeton (1997, 2002).

interprets the human species as (roughly) a continuous genomic field, and uses F_{ST} metrics and measures, the bread-and-butter metric for assessing genetic heterozygosity, to make his case. In Templeton's taxonomic analysis of human populations, he draws on Table 1 in Barbujani et al. (1997, p. 4518) to show that our global F_{ST} is fairly low (0.156; cf. empirical pattern #4, Chapter 6), somewhere between impalas and waterbucks, and nowhere close to the highly differentiated species of North American deer and gray wolf.[42] Moreover, he argues that "human 'races' do not satisfy the standard quantitative criterion for being trad-itional subspecies."[43] Indeed, "a standard criterion for a subspecies or race in the nonhuman literature under the traditional definition of a subspe-cies as a geographically circumscribed, sharply differentiated population is to have F_{ST} values of at least 0.25 to 0.30."[44] Long and Kittles are critical of F_{ST} measures interpreted as "simple statistical partitions" that are taken to be informative of complex, actual human population structure; even so, they somewhat ironically find human F_{ST} values congruent with those of Templeton.[45]

Returning to our overarching ontological positions, the realist holds that there must be a fact of the matter about whether populations exist or not in *Homo sapiens*. Based just on the gray-zone F_{ST} values just explored, however, a conventionalist ontological position seems much more appropriate for our species. While Templeton and Long and Kittles can reasonably argue that races do not exist, opposing philo-sophical arguments are possible (see Chapter 9), and we shall revisit some technical details and empirical estimates of F_{ST} in Chapters 5 and 6. Add to this the assumptions – and epistemological and political value-judgments – in the theoretical machinery and science communi-cation process, and the conventionalist gloss on the existence of popu-lations in our species explains and accounts for the way genomics is discussed and handled by scientists, and by the public and by politicians, both today and tomorrow, with multiple and valid interpretations about whether biological populations exist or not. I suggest that there is no absolute answer to questions such as, "Does our species clearly consist of biogenomic racial groups or populations?" In the taxonomy of *Homo sapiens*, the genomic data underdetermine judgment about the reality of population structure in our species.

[42] Templeton (1999, pp. 633–634).   [43] Templeton (1999, p. 635).
[44] Templeton (1999, p. 633).   [45] Long and Kittles (2003, p. 466). Cf. note 22.

Lewontin (1972): Taxonomy Meets Conservation Biology

Lewontin's paper "The Apportionment of Human Diversity" garnered significant attention when it appeared in 1972. In it, Lewontin argued that the vast majority of the genetic variation of *Homo sapiens* exists *within* groups rather than *between* groups. Given the impact of Lewontin's article in framing a whole set of issues and methodologies around human evolutionary genomics, it seems apposite to zoom in on the ways Lewontin's paper engaged both heterozygosity and diversity metrics and was, effectively, a piece of both genetic taxonomy and conservation biology.

For the young Lewontin, the central question of experimental population genetics was, "At what proportion of his loci will the average individual in a population be heterozygous?"[46] His mentor, Theodosius Dobzhansky, had inspired this kind of question. As we have already seen, population genetic theory uses the measure and metric of heterozygosity. Lewontin was clearly familiar with heterozygosity.[47] Statisticians and ecologists often refer to this measure as *Gini* or *Gini diversity*.[48] However, Lewontin preferred a distinct metric: the Shannon information measure, H, "which bears a strong resemblance numerically to h [heterozygosity or D_W of Equation 4.1n]."[49] Formally, Lewontin defined this measure as follows:

$$H = -\sum_{i=1}^{u} p_i \log_2 p_i, \qquad \text{[Eq. 4.5]}$$

where p_i represents the frequency of the ith allele, there is a total of u alleles at the given locus, and the base of the (binary) logarithm is 2. After all, as Lewontin wrote, Equation 4.5 "is widely used to characterize species diversity in community ecology, and since I am performing a kind of taxonomic analysis here, I will use H [Shannon information]."[50] Importantly, both heterozygosity (Gini) and the Shannon information measure (i.e., entropy) satisfy various criteria required of a genetic diversity measure, including convexity and maximum value when all allele frequencies at a locus are equal,[51] and the two metrics formally resemble each other (Figure 4.2). Moreover, in his book *The Genetic Basis of Evolutionary*

---

[46] Lewontin and Hubby (1966, p. 603); cf. Hubby and Lewontin (1966, p. 577).
[47] See Lewontin (1972, p. 388).
[48] Gini (1912); cf. Simpson (1949). Lewontin was likely familiar with this work.
[49] Lewontin (1972, p. 388).   [50] Lewontin (1972, p. 388).
[51] Lewontin (1972, p. 388); see also Winther (2022a).

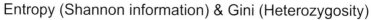

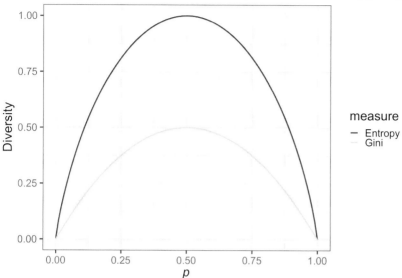

Figure 4.2. Entropy and Gini diversity
The actual diversity values of Shannon information, H, and heterozygosity, D_W, mapped against allele frequency, for a biallelic locus, per, respectively, Equations 4.5 and 4.2 (or 4.1n). Concept and draft by Rasmus Grønfeldt Winther, illustrated by Amir Najmi using ggplot2 in R. © 2021 Rasmus Grønfeldt Winther.

Change, Lewontin stated that Shannon information was "a nearly identical measure" to heterozygosity.[52] In this sense – as well as in his methodology of apportionment of genetic variation at three levels, which resembles F-statistics but which we shall not encounter until Chapter 5 – Lewontin's analysis is simultaneously diversity-like and heterozygosity-like.[53]

Lewontin found something astounding. Genetically speaking, you are almost as different from someone chosen at random from your population(s) of origin or ancestry as you are from a random person from anywhere on Earth. More precisely put, Lewontin's measure of

[52] Lewontin (1974, p. 155).
[53] Recall H_I, H_S, and H_T from the heterozygosity section above; H_O, H_{race}, and $H_{species}$ are the three basic Shannon information measures in Lewontin (1972); the other three measures that Lewontin uses are various averages of H_O and H_{race}. H_S is formally analogous to H_{race}, and H_T to $H_{species}$. H_I is the observed heterozygosity within single populations. Lewontin did not generally have access to observed heterozygosity data in humans, only to allele frequencies, which he thus had to use to calculate his within populations measure, H_O. Informally speaking, he could therefore assess population structure, but not inbreeding. See Winther (2022a).

genetic variation was, on average, 86% as large when calculated for a single population as when calculated for the world as a whole.[54] In fact, Lewontin's 1972 data showed that total global genetic variation of humans could, on average, be partitioned (or apportioned) thus, rounded to the nearest whole number: 7% (on average) among continental regions (Africans, Europeans, East Asians, South Asian Aborigines, Oceanians, Indigenous Americans, Australian Aborigines); 7% (on average) among populations, but within continental regions (for example, Kikuyu, Tutsi, Zulu); and 86% (on average) within populations (e.g., Icelanders, Yanomama, and Zulu).[55] This I shall call *Lewontin's distribution*, which corresponds roughly to saying that, for our species, $F_{ST} = 0.14$.[56] Chapter 5 explores the partitioning methodology underlying Lewontin's distribution and the relation of this methodology to F-statistics, and Chapter 6 unpacks it further as an empirical pattern.

The background assumptions and hidden theoretical contexts informing Lewontin's use and interpretation of the Shannon information measure he chose are most telling. On the one hand, he was clearly interested in taxonomy. The last paragraph of his 1972 paper states that he has shown "racial classification" to be "of virtually no genetic or taxonomic significance,"[57] and recall that he affirmed that he was doing "a kind of taxonomic analysis."

On the other hand, Lewontin's arguments carry more than a hint of the conservation biologist's focus. In his book *Human Diversity*, Lewontin wrote:

If, after a great cataclysm, only Africans were left alive, the human species would have retained 93% of its total genetic variation, although the species as a whole would be darker skinned. If the cataclysm were even more extreme and only the Xhosa people of the southern tip of Africa survived, the human species would

---

[54] The earliest published data explicitly giving just this statistical result can be found in Table 12, "World Variation of Gene Frequencies," in Cavalli-Sforza (1966, p. 367); cf. Bodmer (2018, pp. 322–323). However, it was Lewontin (1972) who unpacked the implications and meanings of this kind of data and mathematical calculations.

[55] The true apportionments of Lewontin's 1972 study are 7.2%, 7.1%, and 85.7% (Winther, 2022a). As we shall see further in Chapters 5 and 6, Lewontin (1972) had incorrectly written 6.3%, 8.3%, and 85.4% (p. 396).

[56] Think of this as adding the amount of genetic variation among populations, but within continents (7%) and among continents (7%); or, conversely and equivalently: 1 − within populations variation (86%).

[57] Lewontin (1972, p. 397).

still retain 80% [86%] of its genetic variation. Considered in the context of the evolution of our species, this would be a trivial reduction.[58]

This ecological and conservation biology focus resonates with the paper published a decade earlier. Indeed, let's take the apocalypse situation further. Imagine a situation akin to the history of Planet Unity, where the Great Tragedy forced harsh choices. In our modern context, we can imagine a cataclysm analogous to the current ecological crisis, but where humans bear the primary brunt – a pandemic such as one caused by a ramped-up coronavirus with disproportionate effects in different populations across the globe. According to Lewontin, if Earth faced an even more devastating apocalypse, and alien conservation biologists were forced – during such a Great Tragedy-like apocalypse – to save only a single, small subpopulation of *Homo sapiens*, the aliens could reasonably pick any small subpopulation, nearly at random, and be confident that they would not be losing much of the extant genetic diversity in doing so.

The alien conservation biologist thought experiment highlights the implications of the claim that practically any particular population could effectively stand in for the whole of humanity without great loss. The alien conservation biologist might well point toward the lack of genetic diversity in our species, and the immense fraction of within populations genetic variation of the total genetic variation, and lump all of humanity together as a single population. Effectively, we can be interpreted as a single species with no significant infraspecies genetic subdivisions. Call this alien a *lumper*.

According to my conventionalist or constructivist ontological interpretation, one could also easily imagine an alien more committed to saving particular aspects of genetic diversity, à la Frankel above. This alien might be less willing to sacrifice unique alleles – saving only 86% of the extant genetic variation would represent an unacceptable loss. Or consider another alien biologist intervening in our perhaps hypothetical Great Tragedy who is unwilling to sacrifice unique populations that, while genetically similar, each have particular sets of allele frequencies making them special; that is, a conservation biologist who is committed

---

[58] Lewontin (1995 [1982], p. 123). These percentages should be interpreted as averages. See also Lewontin et al. (1984, p. 126), where the 85% (i.e., 86%) quantity is stated correctly. Long and Kittles (2003) appropriately observe "that it would make a great difference which group is chosen" (p. 467).

to preserving genetic structure between populations as best she can. These latter two aliens adopt quintessentially *splitter* positions.

The lumper wanting to preserve more than the average 86% of the variation Lewontin calculated might note the following: First, since the majority of genetic diversity exists within Africa, any populations chosen should probably be African (see Chapter 6). Second, a collection of populations from Africa should be preferred to a collection of populations drawn from several continents. Certainly, if the goal were to reconstitute the species, this strategy has much to recommend it, as Lewontin, or Long and Kittles, imply. While we might lose many particular allele combinations, the vast majority of alleles themselves would be present, and combinations could of course be re-created (with enough alien time, interest, and funding).

Phylogenetics

Conservation biology tells us about the modern spatial structure of genomic variation, providing insight into which populations and species require protection. Taxonomy seeks to discover and name unique populations and species, and rank these in natural, classificatory groups within groups. In contrast, phylogenetics tells us about the origins, gross genealogy, and nested historical relationships of populations and species, including humans. The mission of phylogenetics is to reconstruct life's genealogy. To do so, a variety of kinds of data from the genomic and phenotypic levels are used, as are a plurality of computational algorithms and, frankly, many hidden and input assumptions about what constitutes the appropriate populations (e.g., human ethnicities) and species from which to sample and measure individuals. Perfectly valid reasons exist for a realist ontological interpretation of phylogenetic trees: Some seek to discover the "objective" and "true" tree of life.[59] However, as I argue here and elsewhere, we often get out of the models what we put in, perhaps more so in phylogenetics than in other biological subdisciplines.[60] That is, we can also adopt a constructivist or conventionalist ontological interpretation here, arguing that genomic data and models can be perfectly reasonably interpreted from either splitting or lumping perspectives. Our best data simply underdetermine which trees of life to adopt because so many different trees are consistent with the same set of data and metrics.

---

[59] But see Vergara-Silva (2009) and Morrison (2014). [60] See Winther (2009a, 2020a).

This is due in part to the complexity of the forms of genomic data useful for phylogenetics. When, in the 1960s, important work showed that molecular signals (genes or gene products – i.e., proteins) could be used to reconstruct phylogenetic history,[61] biologists were able to argue for a fairly early split of approximately 5 million years between humans and African great apes.[62] Since then, data have accumulated and been enriched, as nucleotide sequences have become data-analysis targets, including but not limited to copy number variants (CNVs), single-nucleotide polymorphisms (SNPs), and short tandem repeats (STRs) or microsatellites.[63] Some of these sequences refer to variation occurring at one nucleotide site (e.g., SNPs). Some are typically found in noncoding regions, meaning they usually have no functionality and are thus not subject to selection (e.g., STRs). Today, entire genomes are measured (e.g., genome-wide association studies, or GWAS), and mitochondrial (matrilineal) and Y chromosome (patrilineal) data have been added to the empirical mix, as we saw in Chapter 2.

These complex data serve as inputs to phylogenetic modeling. Over the course of the twentieth century, and in fulfillment of Darwin's argument that "all true classification is genealogical,"[64] the two most influential linages or families to emerge for modeling phylogenetic trees were cladism (developed by Willi Hennig) and probabilistic inference. (It is noteworthy that the only diagram or figure in *On the Origin of Species* was an abstract tree of life.[65]) According to *cladism*, parsimony or simplicity is the best method for inferring trees. A phylogenetic tree illustrates a nested branching order illuminating the relative genealogical relationships among different (theoretical or natural) populations, or species, which are placed at the top of the tree – the tip leaves. Furthermore, in order for taxonomic classifications to be natural and objective, they must make reference to systematizations of the order of nature captured

---

[61] See Zuckerkandl and Pauling (1965).

[62] Sarich and Wilson (1967) and Chapter 2. More generally, Hubby and Lewontin (1966) and Lewontin and Hubby (1966) presented an influential early use of molecular technologies (e.g., protein electrophoresis) to assess evolutionary genetic variation, in fruit flies. This is the research paradigm within which one should also place Lewontin (1972, 1974).

[63] For a fairly comprehensive classification of types of genomic variation or polymorphism, see Pollex and Hegele (2007). Recall also the glossary at the end of Chapter 2.

[64] Darwin (1964 [1859], p. 420).

[65] I improvise on this diagram with Figure 8.5, "Pondering Darwin's Forms of Life," in Winther (2020a, p. 236).

in our phylogenetic cladograms, which show a nested clade (genealogical unit) structure.[66]

In contrast, *probabilistic phylogenetic inference* employs, for instance, metrics of genetic distance (often derived from genetic differentiation metrics of genetic variation, such as Euclidean genetic distance, captured by Equation 4.4, among many others) to construct trees. Probabilistic inference was trailblazed by two students of R.A. Fisher: A.W.F. Edwards and Luigi Luca Cavalli-Sforza. To provide a flavor of their impact, Felsenstein noted: "Edwards and Cavalli-Sforza's paper of 1964 is remarkable in that it introduces the parsimony method, the [maximum] likelihood method, and the statistical inference approach to inferring phylogenies, all in one paper."[67] Whereas the parsimony method in the probabilistic context refers to using a "minimum evolution" algorithmic approach to building trees, the likelihood method refers to testing potential phylogenetic trees, inferred through particular statistical methods (including minimum evolution), with a maximum likelihood approach permitting selection of the tree(s) best matching the data.[68] In short, cladistic and probabilistic methods are today the standard modeling strategies in phylogenetics, though one might say that probabilistic phylogenetic inference won the day.[69]

Phylogenetics improves our understanding of the identification and hierarchical relationships of populations and groups, but even so, we cannot expect phylogenetics to provide unequivocal answers to human population structure identification and, in particular, for racial history and ontology. I focus here on data and topology, as widely used by phylogeneticists, and the ways choices about these impose problems and limits on identifying human population differentiation, and narrating the history of human populations. In general, data choice includes gathering from, and managing the complexity of, a multitude of gene sequence

---

[66] See also Sober (2008); cf. Winther (2012a, Section 6.2.2) and Nielsen (2018).

[67] Felsenstein (2004, p. 128); cf. Felsenstein (2018, pp. 330–331). See also Sober (2020).

[68] This is a slightly more modern way to make the contrasts. Interestingly, the synoptic paper Cavalli-Sforza and Edwards published, clearly laying out three models or methods of phylogenetic inference – minimum evolution, maximum likelihood, and additive tree – is their 1967 paper (Cavalli-Sforza and Edwards, 1967, which is paper 37 in Winther, 2018a), although the seeds were present in their 1964 paper (Edwards and Cavalli-Sforza, 1964, paper 27 in Winther, 2018a). Sommer (2015) analyzes trees and maps in the context of Cavalli-Sforza's work on "the great human diasporas"; cf. Winther (2012a, Section 6.2.3).

[69] See Nielsen (2018). Probabilistic phylogenetic methods that use genetic distance metrics, i.e., differentiation metrics, include UPGMA (unweighted pair group method with arithmetic mean), neighbor joining, and minimum evolution.

phenomena, as well as phenotypic characters, available to us for sampling and measurement. Meanwhile, topology choice includes the different genealogical topologies and whether a topology permits distinct genealogical units to hybridize or "reticulate": A tree does not, a trellis or network does.[70]

Unsurprisingly, data choice influences and informs phylogenetic work. For instance, different lineages will be traced – and distinct trees built – depending on, for example, whether data from mitochondria or Y chromosomes are collected. As ever, different inputs yield different outputs, and lineages of women and men indicate different population structures and migration paths, as we saw in Chapter 2.[71] Perhaps a consilience "total evidence" approach – according to which many strands of evidence, such as genetic, biochemical, morphological, and embryological (these latter two at the phenotypic level), are combined – could adjudicate a total pluralism, as well as data conflicts.[72] However, data variety at the genomic level often indicates a diversity of incongruous gene lineage signatures, with no single, evident neat phylogenetic tree.[73] Moreover, given war, invasions, and migrations, various methodological decisions have to be made, such as which kinds of populations should be sampled, including whether they should be considered "aboriginal" and "indigenous," and whether other kinds of data (e.g., archaeological, linguistic) need to be considered in identifying possible genetic mixing between populations and in establishing the potential autochthonous nature of the given population.

The decision to use a tree or trellis or broader, network topology to represent the evolution of *Homo sapiens* diversity is similarly fraught and can also limit population identification. While this is not the place to enter the territory of graph or network theory,[74] an important discussion remains about whether, given processes of admixture and migration, the evolution of *Homo sapiens* diversity can ever be represented by a well-behaved tree or whether we need a trellis or network model indicating repeated episodes of population isolation and population hybridization in *Homo sapiens* and closely related kin, such as Neanderthals or Denisovans (see Figure 2.10).[75] If migration and hybridization as evolutionary

---

[70] See Doolittle (1999) and Templeton (1999).   [71] Wells (2003).
[72] Kluge (1989) and Eernisse and Kluge (1993); but see Rieppel (2009).
[73] See Maddison (1997) and Nakhleh et al. (2009).
[74] See Huson et al. (2010) and Newman (2010).
[75] See Templeton (1997, 1999), Relethford (1998), Wells (2003), and Adams (2008).

mechanisms, and a branch reticulation topology, have been as common as Alan Templeton and others suggest, then indeed it becomes even less clear how to subdivide *Homo sapiens*, and how to tell a historical narrative of our species. Again, the antirealist argument need not carry the day. Legitimate modeling strategies using particular measures can still solidify a phylogenetically based racial ontology out of the indeterminate genomic field.[76] For instance, Andreasen and Brandon use the phylogenetic theory of cladism, in light of human evolutionary genomics, to argue that human races are biologically real entities.[77] For Andreasen, "cladistic races are ancestor–descendent sequences of breeding populations that share a common origin."[78] As we continue to witness, the particular research paradigms and purposes embraced are as important as the genomic facts in explaining topological model choice and the ontological conclusions reached.

That human phylogenetics has improved over time does not license all-out realism. There are good reasons to adopt a conventionalist perspective. A number of conventional choices need to be made in terms of data and measurements and of tree or trellis or network topologies. Our phylogenetic tree is underdetermined by the data. The appropriate topology of human evolution remains a bewildering question. In short, interpreting genomic data involves values and assumptions, and we need to select from a plurality of modeling methodologies, in order to gain clues about the pattern and dynamics of human evolution. Whatever the absolute fact of the matter about our evolution, it is almost certainly too complex to be fully represented. And as we shall see in Chapter 5, a certain circularity and conventionalism in phylogenetic reconstruction seems unavoidable.

Homo sapiens: A Special Case

By exploring three kinds of measures of genetic variation and placing them within their respective theoretical and disciplinary contexts, we begin to understand how to determine similarities and differences at the genetic level, and potentially why making such determinations is essential

---

[76] For instance, Figure 5 in Mountain and Cavalli-Sforza (1997, p. 712); Agrawal and Khan (2005).

[77] Andreasen (2000, 2004, 2007) and Brandon (2022).

[78] Andreasen (2004, p. 430). Andreasen also argues that races are probably on "their way out" due to increased migration and globalization in the modern world (p. 431).

for comprehending the existence or not of human populations, and for identifying, defining, and sampling them.

Conservation biology makes use of genetic diversity to determine which species, subspecies, and other groups require resources and protection. Taxonomy below the species level deploys heterozygosity metrics to identify robust populations and to assess population structure. Phylogenetic inference applies genetic differentiation to illustrate the historical relationships among species and groups. Further, each of these disciplines is informed by both splitting and lumping thinkers and perspectives, and, as we have seen through the conventionalist ontological interpretation that I advocate, such perspectives and paradigms greatly impact our concept and understanding of population structure.

The use of genetic variation and population structure to reconstruct the history of the human species, including our history of migrations and gene exchange within and between geographic areas, is of real intellectual and personal interest. To continue exploration, however, we need some clarity, following Houle and colleagues,[79] about what the statistics mean and what the appropriate genetic variation metric and measure are for the task at hand. A full analysis of the modeling strategies and theoretical context involved would require not only exploring the metrics above, but also tracking the role of ontological assumptions and formal methodology for a variety of cases (see Chapter 5). The *SMEO-P* model – that is, set up, mathematically manipulate, explain, objectify–pluralize – developed in Winther (2006a) might be useful here.

A qualification is worth making, by way of our ongoing thought experiment (Figures 2.1 and 4.1). If *Homo sapiens* were much less structured (e.g., characterized by a smaller range, even more admixture, or an even younger species), our best genomics would strongly argue for a continuum perspective, including knockdown arguments about the nonexistence of races.[80] However, if *Homo sapiens* were much more structured – perhaps in a world more like Galápagos-Writ-Large, with hundreds of islands separated by large distances, with hundreds of thousands of years of evolution – there would very likely be clear structure below the species level, and genomic data and models would converge on a subdivided, racial realist position. But *Homo sapiens* is neither kind of

---

[79] Houle et al. (2011).

[80] As long as there were some (even potentially very small) population structure, genomic methods would find differences quite effectively, and so splitters might still exist.

species: It is neither completely unstructured (i.e., thoroughly admixed and hybridized) nor totally substructured (i.e., living in extreme isolates).

Yet, as we have seen in the preceding pages, our species is *closer* to a continuum (Planet Unity) than to a subdivided species (Galápagos-Writ-Large). Ours is a fairly admixed and globalized species (Figure 4.1). It is also a relatively young one, evolving in a continuous and clinal world, genetically and ecologically. Although there are many other facets of conservation biology, taxonomy, and phylogenetics, genetic metrics and measures help inform the focus of *Our Genes*. That said, I should not wish the reader to misinterpret me as a "gene-bean counter." I fully realize the complexity of these disciplines. After all, that is why I have spent Chapter 3 and this chapter exploring some of this complexity and nuance by reflecting philosophically on the very concepts of *population* and *genetic variation*.

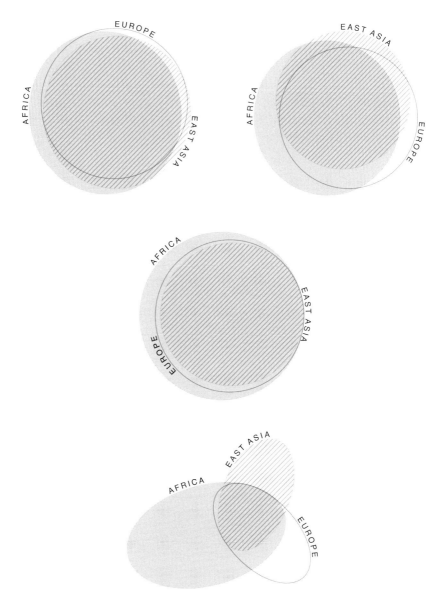

Figure 5.1. Africa–East Asia–Europe genomic overlaps
Overlap and relative uniqueness of genomic variation in Africa, East Asia, and Europe, represented in various ways and as drawn by the author using data from various sources: (top left) geographic distribution of 8290 microsatellite alleles at 783 loci, representing 87.96% of all the alleles of (the non-exhaustive) Table 2 in Rosenberg (2011, p 665); (top right) geographic distribution of 8516 microsatellite alleles at 783 loci at a sample size correction parameter $g = 40$ alleles per subsample,

5 · *Models and Methodologies*

Evolutionary genetics attempts to make evolutionary theory mathematically explicit through measuring change at the gene level, but once data have been gathered, modeling and interpretation – albeit contextual and thus limited – must be applied. In this chapter, I take a closer look at population modeling, moving from measurements and data to the investigation of two mathematical modeling strategies at the heart of genomic calibrations and analyses of human populations: variance partitioning and clustering analysis.

Variance partitioning assesses genetic differences among preestablished populations or groups using models and methodologies of partitioning genetic variation (itself often measured through heterozygosity), such as

---

Caption for Figure 5.1. (cont.) representing 95.13% of all variation per (the exhaustive) Table 1 in Szpiech et al. (2008 p. 2501);[1] (middle) geographic distribution of SNP variants that are common (>5%) to at least one continental region (including singletons) – these account only for approximately 11% of the total (singleton and nonsingleton) SNP variants (per Biddanda et al., 2020, Figure 3, figure supplement 2, A); (bottom) geographic distribution of SNP variants that are either common (>1%) or rare (but not undetected) in a continental region (*not* including singletons), accounting for just about 85% of the (nonsingleton) SNP variants that can be read from the source figure, which is only approximately 86% of the total (per Biddanda et al., 2020, Figure 3, figure supplement 1, B; all East Asian variation shared with either Africa or Europe is shared with both). (I used EulerAPE software [https://www.eulerdiagrams .org/eulerAPE/]. There was no "diagonal error" – i.e., area distortions – for the top three diagrams, but the bottom diagram has a small amount of error represented in the tiny East Asia–Africa and East Asia–Europe overlap regions; these two regions are actually nonexistent, i.e., 0.) © 2021 Rasmus Grønfeldt Winther.

---

[1] Table 2 of Rosenberg (2011) is still missing 10.12% of the alleles. (Only 1.89% of the alleles listed in Table 2 are not within Africa, East Asia, or Europe.) Since I was unable to secure the full table and hence do not know the geographic distribution of the remaining 10.12% of the alleles, the figure is somewhat incomplete. Table 1 of Szpiech et al. (2008) lists all geographic region combinations and hence is not missing any alleles.

analysis of variance (ANOVA) or other closely related methods. Among these methods is the one used by Lewontin (1972) and one suggested by Laurent Excoffier and colleagues in 1992. Figure 5.1 (the chapter opener) presents an informal visualization of different ways to apportion variation among populations of *Homo sapiens* aggregated at the continental level. *Clustering analysis* deploys clustering algorithms to form populations from the ground up. A Bayesian form of this modeling strategy was first presented by Jonathan Pritchard and colleagues in 2000 and influentially applied to human genomic data by Noah Rosenberg and colleagues in 2002. In a sense, variance partitioning and clustering analysis are two sides of the same mathematics coin. Both are legitimate and consistent probability theory and statistics modeling practices, but they answer different questions about population phenomena: On the one hand, of all the genetic variation in the species, how much is found within local populations, among populations but within continental aggregates (alternatively, "races"), and among continental aggregates or races? On the other hand, starting with multilocus data on individuals – that is, with knowledge of the alleles present at different loci in the same individual – how can we best pool together individuals into small numbers of clusters, such that individuals in the same cluster exhibit maximally similar multilocus genotypes while individuals in distinct clusters exhibit maximally different multilocus genotypes, subject to Hardy–Weinberg assumptions?

Despite their mutual consistency and their respective roles in answering different questions, the two modeling strategies often seem to pull in opposite directions. Results from variance partitioning are often interpreted as if human groups or races do not exist, while results from clustering analysis in *Homo sapiens* are typically interpreted as if human groups or races are real. However, both serve as important case studies for the potential pernicious reification – that is, conflating and confusing model and world – that can accompany, and undercut, conclusions about origins and identity in evolutionary genetics. In particular, I identify four types of reification: *population reification*, F_{ST} *reification*, *cluster reification*, and *statistic–parameter reification*.

As with any mathematical modeling methodology, there are both "bottom-up" practices of data abstraction and model generalization, and "top-down" model ontologizing, concretizing, interpretation, and application (Figure 5.2).[2] For instance, careful study of the assumptions and

---

[2] See chapter 3 in Winther (2020a).

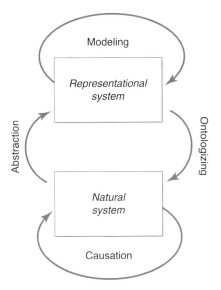

Figure 5.2. The mathematical modeling process
Mathematical modeling involves four rough sets of processes: (left) the abstraction of measures, units, and data from natural systems; (right) ontologizing representations (models, theories, and laws) for purposes of explanation, prediction, understanding, and intervention; (top) modeling and theorizing according to mathematical practices and heuristics; and (bottom) the natural system evolving according to causal regularities and processes. Concept by Rasmus Grønfeldt Winther, illustrated by Mats Wedin. © 2021 Rasmus Grønfeldt Winther.

methods of both variance partitioning and clustering analysis indicates that prior definitions based on, for example, linguistic and anthropological criteria of groups or populations influence modeling results and can result in pernicious reification. Therefore, some circularity seems inevitable in characterizing and proving the existence of groups. How seriously, then, should we take the genomic results of our identity and origins?

This chapter uses assumption archaeology to explain the basic logic and purpose of these two methodologies crucial to human evolutionary genomics, while identifying places where sensitivity to initial assumptions is especially relevant.[3] Specifically, what do the methodologies do, and what can we appropriately learn from them? And where can models get confused and conflated with the world, where can they go wrong, and where can they be reified?

---

[3] Winther (2020a, 2020b).

Understanding these two modeling strategies also sheds light on a distinct, individual-level inference procedure, namely *classification*, whereby individuals – including you and me – can be placed within our population(s) of origin and belonging. Classification can yield valuable information that is otherwise not easy to access. For example, by using a genetic ancestry kit, we can, fallibly and with some caution, find out our population(s) of origin (Figure 5.3). As ever, such information is not the end of our investigation – far from it – instead, it points toward new, potentially valuable lines of inquiry, including learning about and critically questioning the very nature of the data and the methodologies used by ancestry companies. This chapter (and this book) is a user manual for critical thinking about genomic modeling. Accordingly, it surveys the landscape of both population-level methodologies of variance partitioning and clustering analysis, and individual-level classification.

A Tale of Two Methodologies

The logic of variance partitioning and clustering analysis families of methodologies for the scientific identification of populations reminds us that there are complications and multiplicities in how we determine and define theoretical as well as natural populations. At the same time, the emergent patterns of the existence (or not) of human populations, and races, are somewhat robust, perhaps more so with variance partitioning than with clustering analysis.

Too briefly put, variance partitioning calculates the amount of total genetic variation, often – but not always – assessed by heterozygosity metrics and measures described in Chapter 4, distributed *within* and *among* groups or populations. Groups or populations can themselves be aggregated hierarchically: single populations versus large continental groups of populations, for instance. Variance partitioning methodology assumes the existence of populations, predetermined by geographic, linguistic, or anthropological criteria, and allows us to ask – and answer – the question of how much more genetically similar two randomly chosen individuals from the same group are, on average, compared to a randomly chosen individual from that group and an individual from another group (either from the same continental region or from a different one). If there is no variation among groups or populations, then two individuals from the same group will, on average, be as different as two individuals from different groups. If there is some variation among groups, then two

Central & South Asian
(Previously South Asian, no change)

100%

● Central & Southern Indian   41.4%

This genetic signature is more common in southern Indian states such as Tamil Nadu, Andhra Pradesh, Karnataka, Kerala, and Telangana. Do you have this ancestry? Tell us more about your family's heritage in the feedback section below to help us improve this report.

● Kannadiga, Tamil, Telugu & Sri Lankan   24.5%

Sri Lanka and the southern states of India harbor evidence of human activity going back thousands of years. Within the past two millennia, southern India saw great Hindu kingdoms come and go, including the Chola Dynasty and Vijayanagar – whose capital, Hampi, was among the largest cities of the medieval world. A genetic signature unique to South Asia reaches its highest levels in the region, and may be related to the spread of Dravidian languages, although the majority of Sri Lankans now speak Sinhalese, an Indo-European language.

● North Indian & Pakistani   10.3%

Today, around one-seventh of the world's people live along or near the basins of the Indus and Ganges Rivers. Most North Indian & Pakistani people speak Indo-European languages brought to the region from Central Asia around 4,000 years ago during the decline of the Indus Valley civilization. More recently, this region was home to a series of powerful empires, including the Gupta Empire (representing a golden age in Hindu art, literature, and science) and the Islamic Mughal Empire, which gave rise to many Indo-Persian architectural wonders like the Taj Mahal.

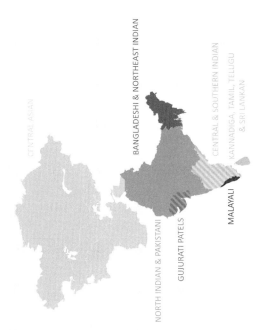

Figure 5.3. DNA ancestry report

Geographic elements of a DNA ancestry report from 23andMe. In 2019, the company introduced a finer-grained geographic analysis of individuals with Central and South Asian ancestry, applying genetic information from their customers to identify seven populations. This analysis helped redress the fact that, as the company's blog reports, "South Asian people represent around a quarter of the world's population but are severely underrepresented in global genetic studies." See https://blog.23andme.com/ancestry-reports/central-and-south-asia.

individuals from the same group will, on average, be more alike than two individuals from different groups.[4]

As we began to see in Chapter 4, one of the most famous examples of variance partitioning was the empirical result found first by Lewontin (1972), which showed that approximately 86% of all genetic variance is found within human populations (e.g., Han Chinese or Sami), 7% of the genetic variance is found across populations within continental regions, and 7% is found across continental regions. Although Lewontin's trailblazing study was methodologically unique, as I show below, downstream studies found overlapping but also somewhat distinct results of within populations, among populations but within continents, and among continents components of total genetic variance, to which I turn in Chapter 6. If it seems counterintuitive that most genetic variation in the human species is within populations, it is only because the standard anthropological, phenotypic paradigm, frequently touted by the political right,[5] overemphasizes the Galápagos-Writ-Large picture. In fact, variance partitioning such as that undertaken by Lewontin provides evidence for Planet Unity population similarity. Despite expedient political rhetoric, no serious scientist questions the admittedly somewhat broad family of results emerging from variance partitioning, nor should they.[6]

Clustering analysis aims to construct a particular number of robust groups or theoretical populations – more precisely, *clusters* – to which particular individuals are iteratively assigned. Clustering analysis finds the best arrangement of multilocus individual genotypes across clusters, where each cluster is a theoretical population, with a unique set of allele frequencies. (Individuals may also be assigned on a weighted basis to more than one cluster.) This procedure is subject to certain constraints. For example, each cluster should satisfy Hardy–Weinberg equilibrium expectations as much as possible. This means that inferred allele frequencies of each cluster should eventually satisfy, to the extent possible and among other constraints, the Hardy–Weinberg simple square expansion. For two alleles:

---

[4] See the thought experiment in Winther (2022a, pp. 22–23).

[5] See Murray (2020) and discussion in Chapter 9.

[6] Importantly, neither Edwards (2003), as reprinted in Winther (2018a) nor Smouse et al. (1982) dispute these results. Indeed, the latter article's central argument is "that if one utilizes a multiple-locus approach, one will discover that human subspecific taxonomy is quite efficacious, even with the sort of marker loci alluded to above [i.e., the results of Lewontin (1972)]" (Smouse et al., 1982, p. 445).

$$(p + q)^2 = 1 = p^2 + 2pq + q^2. \qquad \text{[Eq. 5.1]}$$

For three alleles:

$$(p + q + r)^2 = 1 = p^2 + q^2 + r^2 + 2pq + 2pr + 2qr, \qquad \text{[Eq. 5.2]}$$

where p, q, and r, and so forth, are allele frequencies.[7]

In clustering analysis, it is important to distinguish between the population-level activity of building clusters with distinct sets of individuals (*population*-level clustering analysis, as such), and the assignment, or *classification*, of a particular individual into their appropriate group(s) of origin and belonging (given the preexisting database collection of fine-grained groups). The latter, *individual*-level classification methodology is frequently used in biomedical or forensic contexts.[8] Genetic ancestry companies employ it. For instance, 23andMe classifies new individuals into the global clusters or theoretical or natural population(s) the company already has in its database.[9] This is so even if the company's programmers and proprietary algorithms often also simultaneously refine and redefine the allele frequencies of the clusters or theoretical or natural populations based on the stream of multilocus data of new individuals signing up for services.[10] Classification need not rely on the populations produced by clustering analysis. Rather, and as is the case for variance

---

[7] See Hartl and Clark (1989, pp. 39–43).

[8] Rosenberg (2018) distinguishes between *variance partitioning* and *classification*. I suspect that Rosenberg and I have gained from each other mutually due to multiple conversations.

[9] A brief thought experiment might help explain the logic behind classification. Consider two clusters or populations p_1 and p_2, with systematically different allele frequencies. For three biallelic loci, respective frequencies of the dominant alleles {A, B, C} are, for p_1, {0.8, 0.4, 0.49}, and, for p_2, {0.2, 0.6, and 0.5}. These are allele frequencies consistent with those found by Lewontin (1972), and others (see empirical pattern #4 in Chapter 6). We now have cross-loci information helping us predict and infer to which cluster(s) or population(s) an individual belongs. Think about it: If you were told that an individual's multilocus genotype is *AAbbcc*, what would you bet is her cluster membership? The answer is p_1. After all, allele A is fairly rare in p_2, and allele b is significantly more likely in p_1 than in p_2; admittedly, whether the individual has C or c provides very little information, even when combined with information about the first two loci. Notably, "multilocus *combinations* of allelic types differ dramatically in their probability of originating in different source populations" (Rosenberg, 2018, pp. 399–400). More generally, model-based statistical analysis (through either maximum likelihood or Bayesian statistical methods, e.g., Pritchard et al., 2000; Rosenberg et al., 2002, 2005) tells us with which (high) probability an individual belongs to p_1. The point is that if we know the allele frequencies of different clusters as well as the multilocus genotype of an individual, then we can safely identify the clusters (or populations) to which an individual belongs – that is, we can appropriately classify you (see also Smouse et al., 1982, pp. 448–450; Najmi, 2018).

[10] On proprietary algorithms, see, e.g., Janssens (2019).

partitioning, populations given by standard geographic or anthropo-logical criteria are more typically deployed. Interestingly and similarly to variance partitioning, classification must therefore be broadly under-stood (and practiced) in terms of the assumptions and sampling schemes deployed in establishing and calibrating the genomically based natural and theoretical populations on which its meanings depend.

Today, variance partitioning, clustering analysis, and classification rely on extensive genomic databases, as well as computational power and programs that have increased remarkably in terms of use and capacity, even over just the past 10 years.[11] That there are typically small allele frequency differences at many loci across historically and geographically distant human populations should be utterly unsurprising – evolution has after all occurred in *Homo sapiens*. These small differences at many loci provide the information – which is sampled, measured, modeled, and interpreted in different ways and to answer distinct questions – for variance partitioning, clustering analysis, and classification. Such human evolutionary genomic mathematical methodologies enable us to begin to answer our questions about individual and population origins with more specificity, and to reflect on how and to what extent you and I can reliably be assigned (or assign ourselves) to certain ethnic groups.

Variance Partitioning

Now, let's look more closely at what the methodologies can tell us about identifying and establishing populations, especially human populations such as racial groupings. Variance partitioning as typically practiced for *Homo sapiens* assesses how much of the total genetic variance can be apportioned at each of three levels: (1) within each local population, (2) among populations on a given continent, and (3) among continents. This framework reflects two levels of hierarchical population structure: the local base populations themselves and the continental aggregates of populations. By comparing within- and across-group genetic variation at different levels, we can potentially detect population structure that is relevant, for instance, to the existence of human races.

I explore key methodologies of variance partitioning: Lewontin's somewhat rudimentary 1972 analysis; Excoffier and colleagues' analysis of molecular variance (AMOVA); and Bruce Weir's and Rong-Cai

---

[11] See, e.g., Pritchard et al. (2000), Rosenberg et al. (2002), Edwards (2003), as reprinted in Winther (2018a), Novembre and Peter (2016), and Lawson et al. (2018).

Yang's fully hierarchical ANOVA interpretation of Wright's *F*-statistics. Both AMOVA and Weir's and Yang's interpretation are grounded in earlier seminal work by C. Clark Cockerham.

Recall that Lewontin defined and assessed genetic variation using a heterozygosity-like measure, the Shannon information measure, captured by Equation 4.5. Lewontin (1972) had varying amounts of background information on allele frequencies for each of 17 genes in different populations across the globe. (Most genes were biallelic, but a few had, or were assumed to have, three, four, or six alleles.) For each gene, he calculated three (heterozygosity-like)[12] Shannon information measures for the three different levels mentioned above: the diversity of each measured population, H_O; the racially averaged diversity, H_{race} (which he calculated from averaging allele frequencies across the measured populations of a given continent); and the species-averaged diversity, $H_{species}$ (which he calculated from averaging allele frequencies of all measured populations across the entire globe). From the various H_O values of the different populations classified as being of the same race, Lewontin calculated the average population diversity of a race, H_{pop}, which he then averaged across races, counting each population equally, to get the weighted average of every race's H_{pop} (i.e., $\overline{H}_{pop}$). Furthermore, again counting each population equally, he calculated the weighted average of every race's racially averaged diversity, H_{race} (i.e., $\overline{H}_{race}$).

Effectively, then, there are six measures in Lewontin (1972), the first three of which are primitive to a given level (population, continent, and species), while the other three are various kinds of averages of the first two. Interestingly, because of the Wahlund effect – or the "excess homozygosity" of populations relative to a total, aggregated population – a series of inequalities hold, such that variation at higher levels tends almost always to be higher than variation at lower levels. That is, $H_{race} \geq H_{pop}$, $H_{species} \geq \overline{H}_{race}$, and $\overline{H}_{race} \geq \overline{H}_{pop}$.[13]

What is the magic of Lewontin's variance partitioning strategy? With his framework of six measures in place, Lewontin partitioned total (heterozygosity-like) diversity at three distinct levels. His three explicit equations are as follows:[14]

---

[12] See Chapter 4.

[13] If this discussion seems cryptic or overly compressed, see Lewontin (1972) or, indeed, Winther (2022a), in which I make explicit these measures and the population genetic theory behind them, in addition to redoing all of Lewontin's calculations.

[14] Lewontin (1972, pp. 395–396); cf. Winther (2022a, p. 30).

- within populations : $\dfrac{\overline{H}_{pop}}{H_{species}}$, [Eq. 5.3]

- among populations but within races : $\dfrac{\overline{H}_{race} - \overline{H}_{pop}}{H_{species}}$, [Eq. 5.4]

- among races : $\dfrac{H_{species} - \overline{H}_{race}}{H_{species}}$. [Eq. 5.5]

Note that these three equations add up to a value of 1, since $\frac{H_{species}}{H_{species}} = 1$. Moreover, given the three inequalities, the denominator of each equation is almost always bigger than the numerator (except for rare cases when they are equal, which, in the case of Lewontin, 1972, almost occurred only for the within populations component of Xm), and for Equations 5.4 and 5.5 the left numerator value will almost always be greater than the right (except for rare cases when the values are equal, which, in the case of Lewontin, 1972, only happened for the among populations but within races component of Xm and almost occurred only for the among races components of Lewis and Xm, cf. Winther, 2021b).

As we shall see briefly, these equations resonate with downstream *F*-statistics and ANOVA methodology of variance decomposition, although not quite. For instance, Lewontin worried about using ANOVA methodology to decompose genetic variance into the three levels, and opted not to employ it.[15] Even so, Equations 5.4 and 5.5 are formally analogous to, and even equivalent to, (hierarchical) F_{ST} values, as given by Equation 4.3, at their respective levels: populations compared to races, and races compared to the entire species. This formal equivalence is especially true if we assume (little to) no inbreeding within populations, as is reasonable to assume for many human populations, such that each population consists of randomly mating individuals – F_{IS} is then (near) zero, and (almost) all hierarchical population structure, assuming Hardy–Weinberg equilibrium, is due to the Wahlund effect. Now, after calculating these three diversity apportionments per locus, Lewontin averaged them, respectively, across the 17 loci (with each locus counted equally) to come up with the grand, averaged apportionment: the famous (but not quite accurate) result of 85.4%, 8.3%, 6.3%.[16]

It is hard to overstate the importance of statistical geneticist C. Clark Cockerham to the emergence of the variance partitioning approach. In

---

[15] See Lewontin (1972, p. 386). [16] See Table 4 in Lewontin (1972, p. 396).

two rich articles, Cockerham connected Wright's F-statistics to an ANOVA framework in which he decomposed the total genetic variance at a biallelic locus into three hierarchical variance components: (1) for alleles within individuals, (2) for individuals within populations, and (3) for populations within a total population.[17] These three variance components add up to the total genetic variance, and the third was simply the total variance multiplied by F_{ST}. While these three levels may seem to be one level lower, as it were, than Wright's levels discussed in Chapter 4, there are explicit and robust conceptual and mathematical mappings between Wright's F-statistics and Cockerham's variance partitioning approach, even if variance partitioning methodology is interpretatively diverse, as I shall hint at.[18]

Excoffier et al. articulated an influential method for decomposing genetic variance into within populations, among populations but within continents (or regions), and among continents (or regions) components. Their AMOVA methodology "incorporated ... information on DNA haplotype divergence ... into an analysis of variance format, derived from a matrix of squared-distances among all pairs of haplotypes."[19] In fact, they drew on the Euclidean genetic distance metric we met as Equation 4.4 in Chapter 4 (see their equations 3a and 3b, pp. 480–481), not to assess population differentiation, but to measure "differences between single haplotypes,"[20] where a haplotype is, for example, mitochondrial DNA (though Excoffier and colleagues' method can be generalized to diploid chromosomes).

With this metric in hand, the idea of AMOVA is to compute the overall differences or deviations, at many loci, of many pairs of haplotypes sampled. The goal is to compare haplotype pair differences or deviations within populations (a) to haplotype differences across populations but within regions (continents) and (b) to haplotype differences among regions (continents). Again, these three sets of differences add up to 1. Excoffier and colleagues' Equations 8a–8c are part of the machinery for doing this and, importantly, correspond analogically to Equations 5.3–5.5 above (representing Lewontin's work), and are also each almost always greater than 0 due to the Wahlund effect.[21] That is, both

[17] Cockerham (1969, 1973).
[18] Readers interested in delving deeply into Cockerham's influential variance partitioning framework are invited to explore Hartl and Clark (1989, pp. 298–300) and Weir (1996, pp. 167–190), but, especially, Holsinger and Weir (2009) in its entirety.
[19] Excoffier et al. (1992, p. 479).   [20] Excoffier et al. (1992, p. 481).
[21] Excoffier et al. (1992, p. 483).

Lewontin and Excoffier and colleagues are taking the same kinds of averages of their respective measures (Shannon information or Euclidean genetic distances), at the three different levels (within populations, among populations but within regions, and among regions), and ending up with the same kinds of numerators (e.g., subtract an across-continents–averaged population-level average from an averaged continental-level average in Equation 5.4, above, and in Equation 8b from Excoffier et al., 1992).

Because Excoffier and colleagues' model is based on haplotypes, the authors effectively collapse together Cockerham's first two levels, alleles and individuals. They are thereby able, metaphorically speaking, to move up one level in Cockerham's three variance components (partitions) to redefine these in terms of, respectively, within populations, among populations but within regions (continents), and among regions (continents) levels. In short, inspired by Cockerham's ANOVA paradigm, as well as by the extension of this paradigm in a multiallelic direction by Long and colleagues,[22] Excoffier and colleagues formulate what they call *Φ-statistics*. They show how *Φ*-statistics are directly mappable onto, and analogous to, *F*-statistics, with its three variance components, as well as its three coefficients – F_{IS}, F_{ST}, and F_{IT}.[23] One example of an application of *Φ*-statistics is found in Rosenberg et al. (2002). Using AMOVA, they find that the total genetic variation of the 52 populations of their data set, which we shall meet again below, when divided into seven continental regions, partitions approximately thus: 94.1% within populations; 2.4% among populations but within regions; and 3.6% among regions.[24]

Biostatistician and statistical geneticist Bruce Weir, who was Cockerham's doctoral student and coauthor, provided the clearest explication of a fully hierarchical ANOVA interpretation of Wright's *F*-statistics.[25] Weir notes that individuals of a given species are often "sampled in a nested classification such as transects and sites within

---

[22] Long (1986) and Long et al. (1987); see Nei (1973) for an earlier attempt.

[23] See the precise definitions they give of the three equations (Equation 10a) in Excoffier et al. (1992, p. 482). As one measure of this paper's influence, consider that it has been cited 15,469 times as of August 1, 2021, according to Google Scholar. Equation 6.14 and Equation 6.15, each of which has three equations, of Hartl and Clark (1989, p. 299), capture the typical mapping of three variance components onto the three *F*-statistics coefficients, in both directions.

[24] Table 1 in Rosenberg et al. (2002, p. 2382).

[25] For instance, as of August 1, 2021, Weir and Cockerham (1984) has been cited 19,192 times, according to Google Scholar. For historical and philosophical background to this paper, see Weir (2012).

transects ... or even more complex structures such as drainages, rivers within drainages, and sites within rivers."[26] The first case corresponds exactly to the nested continental and population classification of *Homo sapiens* variance partitioning studies. It is in the context of this "further hierarchy of relationships when there are subpopulations nested within populations" that Weir developed explicit formal definitions of four variance components.[27] In these definitions, the bottom two remain the same as above: alleles within individuals, and individuals within populations. When summed together, these two correspond to the within populations variance, whereas the other two effectively partition the previous single level (and corresponding single component of genetic variance) into two new levels (and two corresponding variance components): first, among populations, within aggregates; and second, among aggregates (continents, regions, or transects). Correlatively, total, hierarchical F_{ST} apportions additively across these two above-population levels, analogously to Lewontin's or AMOVA's apportionment of among populations and among continents (regions) variance components.[28] As Yang showed, this fully hierarchical formulation of variance components can also be continued for "an arbitrary *s*-level hierarchy," with as many levels of population aggregation as desired.[29]

In sum, these are three different modeling strategies for variance partitioning: Lewontin's Shannon information heterozygosity-like measure and *F*-statistics-like Equations 5.3–5.5, Excoffier and colleagues' haploid Euclidean genetic difference measure and Φ-statistics, and Weir's (and Yang's) fully hierarchical ANOVA *F*-statistics, with associated and mappable variance components.[30] These three approaches apportion genetic variance at different levels of a total, aggregate population (or even of an entire species), with internal subdivision. In particular, the last two general approaches aim to produce descriptive statistics

---

[26] Weir (1996, p. 183).   [27] Weir and Hill (2002, p. 723).

[28] Consult the four variance component equations in Weir (1996, pp. 184–185). This approach constrains the relative values of F_{ST} (Weir, 1996, pp. 184–185), or of the variance components (Yang, 1998, p. 953; Goudet, 2005, p. 184), additively apportioned at different levels of the population hierarchy.

[29] Yang (1998, p. 950). Per Goudet (2005), a computer program, *Hierfstat*, "implements Yang's algorithm" (p. 184; the algorithm can be found in the appendix of Yang, 1998, p. 956) to compute and estimate the variance components for different cases of multiple hierarchical levels of population aggregation.

[30] Rosenberg (2011, p. 669) provides a narrative explication resonant with the framework first developed by Cockerham, clarified and clearly extended by Weir, and quantitatively captured by Equation 6.13 in Hartl and Clark (1989, p. 298).

estimating the true parameters of population structure. For cases with random mating within populations (i.e., no inbreeding), the F-statistics mathematics is greatly simplified, since F_{ST} becomes the main statistic and parameter of interest: $F_{IS} = 0$, $F_{ST} = F_{IT}$, and (hierarchical) F_{ST} fully describes population structure. These modeling strategies "do not need to make assumptions ... about the evolutionary processes that might have led to differences among [populations]."[31] Nevertheless, each is embedded in a rich mathematical and conceptual framework replete with assumptions. A deeper assumption archaeology could illuminate variance partitioning further.

The general result of fairly high within populations genetic variance apportionment has often been used as evidence that there is not much genetic differentiation among human populations – or between aggregates of human populations – defined with geographic, anthropological, or linguistic criteria. In fact, many observers use variance partitioning to reject racial categories as real or biologically relevant, as we shall revisit in Chapters 6 and 9. But there is no simple answer to questions such as "What is the relative overlap in genomic variation across Africa, East Asia, and Europe?" (see Figure 5.1).

Clustering Analysis

Whereas variance partitioning typically works with allele frequencies of populations predefined by geographic, anthropological, or linguistic criteria, clustering analysis builds its own populations, from the ground up, as it were. As a result, in the variance partitioning methodology, allele frequencies at different loci are measured by sampling preestablished, natural populations. In contrast, what is predetermined in clustering analysis is multilocus information about each individual as well as information about natural populations from which the individuals are sampled, though the latter information is not used directly in the modeling process. Allele frequencies of the emergent clusters are produced by the modeling process itself through the sculpting of the most robust clusters possible given the individual, multilocus genotypes available; the number of clusters, K, desired; and certain constraints, including assumptions that clusters meet Hardy–Weinberg expectations to the extent possible. We began exploring one clustering analysis algorithm,

---

[31] Holsinger and Weir (2009, p. 641).

Structure, in Chapter 3, where we looked at the dangers of conflating the theoretical populations calculated by *Structure* with natural populations of *Homo sapiens*, among other species. At this point, it seems appropriate to share a little more detail on the inner workings of this algorithm.

Here is a general description of the simplest algorithm of Pritchard, Stephens, and Donnelly, as built into *Structure*, the most influential computational model or algorithm for genetic clustering analysis:[32]

Step 1. Sample $P^{(m)}$ from $\Pr(P \,|\, X, Z^{(m-1)})$.
Step 2. Sample $Z^{(m)}$ from $\Pr(Z \,|\, X, P^{(m)})$.

Here, P is the vector of population (cluster) allele frequencies, Z is the vector of populations (clusters) of origin, X is the vector of individual multilocus genotypes, and m indicates the simple count iteration of the two steps, starting with $m = 1$.[33] Only X has been *measured* and, hence, is *known*. In contrast, P and Z are *inferred* using procedures such as Markov chain Monte Carlo (MCMC) methods, a standard tool in Bayesian analysis for estimating what model parameters might plausibly have given rise to the noisy data observed. Pritchard and colleagues describe the process as follows: "Informally, step 1 corresponds to estimating the allele frequencies for each population assuming that the population of origin of each individual is known; step 2 corresponds to estimating the population of origin of each individual, assuming that the population allele frequencies are known."[34]

The classical *Structure* algorithm is itself an instance of an MCMC algorithm type known as a *Gibbs sampler*.[35] It alternates sampling from

[32] Pritchard et al. (2000, p. 947). As of August 1, 2021, Pritchard et al. (2000) has been cited 31,784 times, according to Google Scholar.

[33] This is their first model, without admixture. Each of these three "multidimensional vectors" (Pritchard et al., 2000, p. 946) is a list, as it were, written in linear algebra form of, respectively, the allele frequencies at every locus for a given population (iterated for all populations), the (single) population of origin of every individual, and the two alleles of every locus for each individual (iterated for all individuals). See also Falush et al. (2003, 2007).

[34] Pritchard et al. (2000, p. 947). Furthermore, *Structure* starts with the prior assumption that in the first iteration, $m = 1$, "before observing the genotypes," any individual has an equal probability of being drawn from any cluster or population (Equation 3 in Pritchard et al., 2000, p. 947).

[35] Pritchard et al. (2000) puts it as follows: "Our main modeling assumptions are Hardy–Weinberg equilibrium within populations and complete linkage equilibrium between loci within populations ... Loosely speaking, the idea here is that the model accounts for the presence of Hardy–Weinberg or linkage disequilibrium by introducing population structure and attempts to find population groupings [clusterings] that (as far as possible) are not in disequilibrium ... *Under these assumptions each allele at each locus in each genotype is an independent draw from the appropriate frequency distribution, and this completely specifies the probability distribution* $\Pr(X \,|\, Z, P)$ ... [where] X

plausible distributions of allele frequencies given population member-ships (Step 1), and then of population memberships given allele frequen-cies (Step 2), until these distributions appear to have stabilized (subject to the constraints of Hardy–Weinberg expectations [per locus], and linkage equilibrium [across loci]).[36]

In a nutshell, clustering analysis infers the best sorting and placement of individual multilocus genotypes within and across clusters. At the con-clusion of the iterative clustering analysis inference process, each individ-ual is assigned to the cluster (or set of clusters, when admixture is assumed) that most closely matches what we could metaphorically call the individual *presence profile* of alleles. This means that the set of alleles an individual carries – that is, the individual's multilocus genotype – should fit comfortably within the large set of inferred and estimated allele frequencies of the cluster(s) best matching the individual. For example, if an individual is *AA* for a biallelic locus with alleles *A* and *a*, with respective allele frequencies p and q, we would, all other things being equal, intuitively prefer to assign the individual to a population or cluster with an *A* frequency of, say, 90% (i.e., $p = 0.9$), rather than to one with

---

denote the genotypes of the sampled individuals, Z denote the (unknown) populations [clusters] of origin of the individuals, and P denote the (unknown) allele frequencies in all populations [clusters]" (p. 946, emphasis mine; sentence order rearranged).

[36] To be more formally precise, this is a clear case of using Bayesian inference in statistical modeling, which is premised on Bayes's theorem: $P(A \mid B) = P(B \mid A) \star P(A)/P(B)$ [Eq. 5.1n]. Here we are interested in the probability of an event conditional on background or prior information of another condition that may be related to it (see Jeffreys, 1955, and Gelman et al., 2013). There are many applications of this theorem, including in drug or disease antibody tests. To take a reasonably simple example, the theorem was a hot topic of conversation during the early phases of the coronavirus pandemic. Some of us learned or applied concepts such as the *sensitivity* of an antibody test (i.e., $P[B \mid A]$, or the probability that your antibody test is positive [B] if – or given that – you have the virus [A]), the *prevalence rate* (i.e., $P[A]$, a "prior," or the probability of a random person having the coronavirus or, equivalently, the frequency or proportion of infected individuals in the population – in both formulations either at a point in time or over a period of time), and the *false positive rate* (i.e., $P[B \mid \neg A]$, or the probability that your antibody test is positive [B] if – or given that – you do *not* have the virus [$\neg A$]), and how to assess or calculate these in the context of Bayes's theorem. (See, e.g., Hagmann et al., 2020; Sarkar, 2020.) The case of *Structure* is a much more subtle case of iterative and formal Bayesian inference that is captured in computer code. Here we use known individual genotypes to infer and estimate unknown parameters: From which populations, with which allele frequencies, do the (known) individual genotypes come? Most abstractly: "$\Pr(Z, P \mid X) \propto \Pr(X \mid Z, P)\Pr(Z)\Pr(P)$" [Eq. 5.2n] (Pritchard et al., 2000, p. 947, where $\propto$ means "directly proportional to"; "Pr" and "P," whether italicized or not, are equivalent ways of writing "probability of"). Note that this is Bayes's theorem, ignoring only the denominator of Equation 5.1n, and conditional on two pieces of information, population(s) of origin and allele frequencies of populations. Ryan Giordano provided constructive feedback here.

$p = 0.5$. Furthermore, and turning to classification, as long as a sufficiently high number of genetic loci are used (roughly 20–50), individual humans can (eventually) be assigned to their appropriate single or multiple cluster(s) with extremely low probability of misclassification, even when the allele frequency differences across clusters for any one locus or gene are small. This is because information for cluster formation *or* individual classification accumulates for each additional locus included in the modeling methodology. In short, this is a "bottom-up" way of building theoretical populations from genomic data and is distinct in question and methodology from "top-down" variance partitioning.

A philosophical question remains: Which ontology is inferred from the facts that clusters can be robustly modeled, and individuals can be reliably classified and assigned to robust clusters? The outcomes of *Structure* could be taken as evidence for the strong reality of human groups, though few interlocutors have made this exact sort of statement, perhaps because of the potentially reactionary political repercussions such utterances may have or may imply.[37] A.W.F. Edwards, however, in discussing "Lewontin's fallacy," claimed as fallacious the argument "that the division of *Homo sapiens* into these [ethnic] groups is not justified by the genetic data" because it "ignores the fact that most of the information that distinguishes populations is hidden in the correlation structure of the data and not simply in the variation of the individual factors."[38] Edwards does not make ontologically strong pronouncements about the reality of human groups in this article. Even so, his arguments are consistent with a position stating that continental region classifications are real. The background argument here seems to be that the significant inferential reliability both of building clusters and of classifying individuals in clusters supports the reality of human group classifications. Although the results of variance partitioning and clustering analysis pull in opposite directions regarding the reality of human populations, their mathematics are mutually consistent.

---

[37] But see Wade (2014) and Murray (2020).

[38] A.W.F. Edwards (2003, p. 798), as reprinted in Winther (2018a, p. 250). See Najmi (2018) for a visual, geometric interpretation of why each further locus included diminishes the probability of misclassification. Rosenberg (2018) observes that in Lewontin's variance partitioning analysis "different loci amount to separate trials, each of which produces its own estimate of the desired variance components" (p. 399). This is also the case for Barbujani et al. (1997) and others using AMOVA. Variance components are often calculated solely from single-locus allele frequencies – individual multilocus genotypes are either unknown or effectively ignored or abstracted away from (see also Winther, 2022a). See also Chapter 6.

Mutual Methodological Consistency

Variance partitioning and clustering analysis are distinct ways of imagining, formalizing, and measuring the hierarchical structure of genomic variation in populations of single species, including *Homo sapiens.* Variance partitioning assesses the structure of genetic variance, at different hierarchical levels, with preestablished groups at each level. Clustering analysis assigns individuals to clusters or finds groups through modeling strategies, such as Bayesian modeling (or principal component analysis, also known as PCA).[39]

These methods are distinct, but they are mutually consistent. Roughly put, *the existence of populational difference (as measured, for instance, by heterozygosity in an F-statistics framework) between preestablished groups at different hierarchical levels is conceptually equivalent to the similarity of multilocus genotype individuals within populations (as clusters found, for instance, with the Bayesian clustering program* Structure*).* That is, the (often small) differences in allele frequencies at single loci among populations allow us, especially when using multilocus genotypes, to successfully cluster and classify individuals. Incidentally, there is no obvious modeling strategy for apportionment of multilocus F_{ST} variance components.

Neither methodology is "wrong" nor invalid, though each may be used inappropriately to answer a question it was not designed to answer.[40] Even though the vast majority of genetic variation for our species exists among individuals within populations, rather than among populations, clustering into groups can *still* be done, if we assume that a certain number of clusters exist, and we aggregate information across loci, as we can when we systematically examine individual genotypes at many loci. All we need is a little variation (read: excess homozygosity) among the allele frequencies in different populations, and on distinct continental regions. To take the argument to its extreme, if two groups or populations of individuals had identical frequencies at 9999 loci, but differed in frequency at just one locus, they could be different populations. Rosenberg et al. observed that "the challenge of genetic studies of

[39] While I cover neither PCA nor, indeed, factor analysis, these are also important for building and visualizing populations in genomic studies with large data sets. For general discussions, see Gould (1996 [1981], pp. 269–285) and Jolliffe and Cadima (2016); on genomic applications, see, for example, Novembre et al. (2008), Novembre and Peter (2016), and Li and Ralph (2019). More recently, network-based methods have been introduced as another methodology for finding population structure (e.g., Greenbaum et al., 2019). While promising, these latter methods have yet to gain traction.

[40] See Feldman and Lewontin (2008, pp. 89–90) for another way to make this point.

human history is to use the small amount of genetic differentiation among populations to infer the history of human migrations."[41] To this, I would at least add: *and to infer the group memberships of any particular individual (under the assumptions of the statistical model) – that is, to classify individuals.*

Ultimately, while our two families of mathematical methodologies are consistent, their aims, questions of interest, and basic assumptions differ significantly. Variance partitioning is particularly useful for evolutionary analyses of the opportunity for selection and random genetic drift in hierarchical populations. Clustering analysis and classification can be used for making medical predictions.[42] Whether race is taken to exist or not – and in which sense (e.g., biological or social, or both) – impacts the practices and commitments of these methodologies, evolutionary genetics in general, and neighboring disciplines and paradigms. Naturally, highly charged ethical, social, and political questions emerge, which we shall turn to in the following pages.

When Maps Become the World

The statistical methodologies described above are naked mathematics. Their logic is watertight – provided certain formal assumptions are met.[43] Modeling strategies can be powerful when applied to their particular contextual purposes, with appropriate data, but they can also be misleading when applied for inappropriate purposes, or when their particular assumptions are not met. As we started to see in Chapter 3, pernicious reification occurs in "conflating and confusing a single abstract representation with the world, thereby thinking of complex reality as *nothing but* the content of the representation."[44] Theoretical maps can come to stand in for the concrete world.

Identifying and familiarizing oneself with assumptions (assumption archaeology) and comparing and contrasting modeling strategies and overarching methodologies (integration platforms) can assist in blocking and overcoming pernicious reification. That is, suppositions of various sorts (methodological, ontological, data-analytical, and the like) need to be stated clearly and self-reflectively. Questions and aims must be explicitly articulated and understood, often in a comparative context in which

---

[41] Rosenberg et al. (2002, p. 2384).   [42] See Burchard et al. (2003) and Kumar et al. (2010).
[43] See also Gelman (2008).   [44] Winther (2020a, p. 94).

a plurality of modeling strategies and methodologies are considered. Both for epistemic and ethical reasons, critical care is required.

Here I identify four areas or places where pernicious reification can potentially occur with the two methodologies addressed in this chapter: *population reification, F_{ST} reification, cluster reification,* and *statistic–parameter reification.*

Population Reification

To understand the potential circularity of the origin of groups or populations and their dependency on various criteria, I characterize the potential reification of populations as *inappropriately conflating theoretical expectations stemming from social and human sciences such as anthropology and linguistics – or tacit expectations emerging from phenotypic biases – with genetic phenomena and the genetic world.*

In general, population reification can occur throughout the modeling process of either variance partitioning or clustering analysis. That is, it can unfold in both the abstraction, model-building phase and the ontologizing, interpretation phase, per Figure 5.2. But let's ask about the *data input* abstraction stage of modeling:

- Which presuppositions are made about the homogeneity and balance of sample sizes across groups and populations?
- How representative is the sampling of each group, and which populations are sampled? Consider the difference between sampling Han Chinese, with a population size of approximately 1.3 billion, and sampling the roughly 100,000 Samis of northern Scandinavia, or consider how many different populations from Africa should be sampled.[45]
- Are the sample data points homoscedastic and independent of one another? Is the variance of the residuals in data samples constant across loci and across populations?[46]
- Does the phenotypic appearance of potential donors bias – consciously or unconsciously – the researcher's choice of individuals whose blood

---

[45] With respect to clustering analysis, Nielsen (2022, p. 161) notes: "It is unclear if sample sizes, to be representative, should be based on sampling all individuals in the world living today with equal probability, should reflect local effective population sizes, or should be geographically representative. Different conclusions regarding categorizations of humans would be reached depending on which such sampling scheme was chosen."

[46] See Chapter 10.

and tissue will be sampled? If so, the collected data points are not independent, insofar as there is correlation between phenotypes and genotypes.

• Which group or population definitions should we start with, either in defining the groups whose respective genetic heterozygosity will be tested in variance partitioning or in framing the sampling scheme of individuals for our clustering analysis?[47]

These questions about the reliability of data input to our models pertain to both variance partitioning and clustering analysis methodologies. Let's explore each in turn, to see the potential dangers that arise when the assumptions about how populations are defined and characterized, and about how data are sampled and measured (i.e., data input), are not made explicit. Table 5.1, for example, presents the classifications that Lewontin used.

We can marvel at geneticists' level of detail, the many tens of thousands of kilometers they have traveled, the uninformed consent they probably pried from Indigenous peoples starting in the middle of the twentieth century, if not before, their likely dealings with government bureaucracies, and more.[48] The entries in Table 5.1 for Indigenous Americans and Oceanians, in particular, reflect the universalizing eye of the white man and generalized Western capitalism.

In the case of variance partitioning, the seeds of eventual population reification can be found at the earliest stages because the empirical starting point for variance partitioning is the set of natural populations preestablished by phenotypic, geographic, or cultural characteristics. Here there must be a set of properties that *gives* the classifications of races and populations against which genetic variation is compared. For instance, Lewontin uses races primarily meeting criteria and characteristics "external" to genetic data – "i.e., linguistic, historical, cultural, and morphological" and geographic – in seeking to answer the question

---

[47] On this last question, for example, Panofsky and Bliss (2017) investigate the multitude of populations adopted by over 600 research articles published in *Nature Genetics* in 1993, 2001, and 2009. They find ample "classificatory ambiguity." See also Winther (2022a) for discussion of classificatory confusion in Lewontin (1972).

[48] For instance, such concerns pertain to the data collection and management of many of the approximately 1700 research papers referred to by Mourant (1954) and the many hundreds consulted by Giblett (1969). Mourant and Giblett used these research papers as data sources for their rich data tables of allele frequencies. Lewontin (1972) drew upon the data tables presented in Mourant's and Giblett's books.

Table 5.1. *List of racial classifications and populations used in Lewontin (1972)*[49]

Europeans

Arabs, Armenians, Austrians, Basques, Belgians, Bulgarians, Czechs, Danes, Dutch, Egyptians, English, Estonians, Finns, French, Georgians, Germans, Greeks, Gypsies, Hungarians, Icelanders, Indians (Hindi speaking), Italians, Irani, Norwegians, Oriental Jews, Pakistani (Urdu-speakers), Poles, Portuguese, Russians, Spaniards, Swedes, Swiss, Syrians, Tristan da Cunhans, Welsh

Africans

Abyssinians (Amharas), Bantu, Barundi, Batutsi, Bushmen, Congolese, Ewe, Fulani, Gambians, Ghanaians, Habe, Hottentot, Hututu, Ibo, Iraqi, Kenyans, Kikuyu, Liberians, Luo, Madagascans, Mozambiquans, Msutu, Nigerians, Pygmies, Senegalese, Shona, Somalis, Sudanese, Tanganyikans, Tutsi, Ugandans, US Blacks, "West Africans," Xhosa, Zulu

East Asians

Ainu, Bhutanese, Bogobos, Bruneians, Buriats, Chinese, Dyaks, Filipinos, Ghashgai, Indonesians, Japanese, Javanese, Kirghiz, Koreans, Lapps, Malayans, Senoy, Siamese, Taiwanese, Tatars, Thais, Turks

South Asian Aborigines

Andamanese, Badagas, Chenchu, Irula, Marathas, Nairs, Oraons, Onge, Tamils, Todas

Indigenous Americans

Alacaluf, Aleuts, Apache, Atacameños, "Athabascans," Aymara, Bororo, Blackfeet, Bloods, "Brazilian Indians," Chippewa, Caingang, Choco, Coushatta, Cuna, Diegueños, Eskimo, Flathead, Huasteco, Huichol, Ica, Kwakiutl, Labradors, Lacandon, Mapuche, Maya, "Mexican Indians," Navaho, Nez Percé, Páez, Pehuenches, Pueblo, Quechua, Seminole, Shoshone, Toba, Utes, "Venezuelan Indians," Xavante, Yanomama

Oceanians

Admiralty Islanders, Caroline Islanders, Easter Islanders, Ellice Islanders, Fijians, Gilbertese, Guamanians, Hawaiians, Kapingas, Maori, Marshallese, Melanauans, "Melanesians," "Micronesians," New Britons, New Caledonians, New Hebrideans, Palauans, Papuans, "Polynesians," Saipanese, Samoans, Solomon Islanders, Tongans, Trukese, Yapese

Australian Aborigines

"How much of human diversity between populations is accounted for by more or less conventional racial classification?"[50]

---

[49] Lewontin's list (p. 387) contained a number of infelicities, which I have silently fixed in the table. Quotation marks are found in Lewontin's list itself.

[50] Lewontin (1972, p. 386); cf. Winther (2022a, pp. 19–21).

Similarly, Cavalli-Sforza and colleagues sought "a representative sample of the world aboriginal populations" that were as ancestral and isolated from neighbors, migrants, and invaders as possible.[51] Indeed, in an influential book, Cavalli-Sforza et al. were explicit about various aspects of their sampling protocol:

We have confined our analysis to aboriginal populations that were in their present location at the end of the fifteenth century when the great European migrations began. We have thus excluded Black Americans and all the recent colonizations of Caucasoid, Chinese, and Indian origins. We have also excluded all manifestly mixed populations; those stated to have 25% or more external admixture.[52]

Importantly, Cavalli-Sforza's research program laid the groundwork for the Human Genome Diversity Project, or HGDP, whose laboratory populations of human cell lines, derived from small blood samples and housed at the CEPH (Centre d'Étude du Polymorphisme Humain, today: Fondation Jean Dausset-CEPH) in Paris, scientists such as Noah Rosenberg and collaborators consult today.[53] The HGDP's geographic and cultural assumptions and informal measures are used not only to define *basic* populations (of which there are slightly over 50) but also to aggregate or pool populations. For instance, Cavalli-Sforza et al. argued for the following rank order of criteria in "pooling populations for generating higher categories": geographic, anthropological, and linguistic.[54] In a nutshell, basic as well as pooled populations are not only *natural* but also *theoretical* in the sense that they are individuated by theoretical assumptions or informal measures from social sciences such as anthropology, geography, and linguistics.

---

[51] Cavalli-Sforza et al. (1988, p. 6002).   [52] Cavalli-Sforza et al. (1994, p. 24).

[53] See Cavalli-Sforza (2005), M'charek (2005), Reardon (2005), and Griesemer (2020). Another genomic diversity research program is the International Genome Sample Resource (IGSR) (e.g., Clarke et al., 2017; Fairley et al., 2020), www.internationalgenome.org. The IGSR is built on the foundations of the 1000 Genomes Project of Auton et al. (1000 Genomes Project Consortium, 2015), which sampled 2504 individuals from 26 populations. Biddanda et al. (2020) deploy part of this database in a "variant-centric" approach, which we shall visit in Chapter 6. More recently, the Genome Aggregation Database (gnomAD) has been developed especially for medical genomic applications and for focusing on the functional aspects of genes and alleles, rather than evolutionary genetics applications per se (Karczewski et al., 2020).

[54] Cavalli-Sforza et al. (1994, pp. 20–24). On linguistic criteria, Marcus Feldman (personal communication, April 2014) recalled that *Ethnologue* (www.ethnologue.com) was a resource that he, Cavalli-Sforza, and their collaborators used.

In the case of clustering analysis, populations contain a social science–theoretical component in that they are built from data *themselves* employing cultural or geographic criteria.[55] Interestingly, different data input sampling designs (discussed above), as well as distinct genomic database sources (e.g., HGDP or IGSR) and computer programs can lead to importantly different high-level clustering results (Figure 5.4; Table 5.2). In particular, sampling and oversampling have an effect, as does whether data are drawn from admixed groups (e.g., Africa–America or America in 1000 Genomes; Figure 5.4) rather than Indigenous or Aboriginal groups as in Lewontin (1972), Rosenberg et al. (2002), and Tishkoff et al. (2009).[56] Which is correct? Researchers, scientists, and critics strive to answer the question of how to reach consilience among such a plurality of sampling schemes and modeling strategies.

The research programs summarized in Table 5.2 each define and classify populations differently, and different clusters are identified by each research program for $K = 5$. Only the results of Rosenberg et al. (2002) mirror the standard, colonialist, and Eurocentric racial classification. The other two research programs identify two (1000 Genomes Project Consortium, 2015) or three (Tishkoff et al., 2009) clusters within the African continent. And notice how *different* the clustering schemes are for these three state-of-the-art clusters (see also Figure 5.4). The differences illustrate the limitations of data, measurements, and analyses, underscoring the necessity of rigorously investigating the theoretical assumptions and interpretive contexts that give models meaning.

Acquiring a representative, independent, and random sample of each human population is a challenge. We have no single set of overarching cultural or geographic criteria for this purpose, and it is unclear whether we should only sample "aboriginal" populations. Moreover, unconscious

---

[55] Lewontin correctly reminded me that clustering generally requires sampling from geographically distinct populations and would be problematic if undertaken with historically "mixed" populations; after all, "human history has confounded the biological processes of differentiation" (personal communication, December 2013; Weiss and Fullerton, 2005, make a similar point).

[56] Whenever a continental region is sampled extensively (i.e., divided into many populations) while other regions are not, there is a tendency to find finer grains of difference – and more clusters – in the oversampled regions. Even so, despite the partial relativity of cluster identification based on sampling scheme – for example, Tishkoff et al. (2009) find three African clusters, and Rosenberg et al. (2002) find only one – certain populations will almost invariably cluster together under most reasonable sampling schemes, for instance, Icelanders and Norwegians versus Han (following Lewontin's categories). For overlapping concerns with clustering inconsistencies across studies, see Barbujani and Colonna (2010, pp. 288–289) and Nielsen (2022).

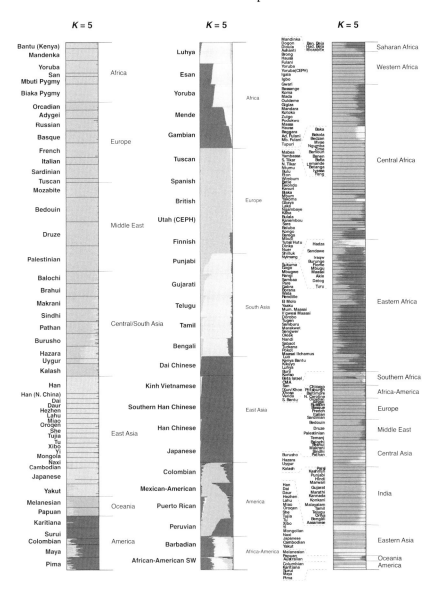

Figure 5.4. K = 5 clustering comparison
Clustering results for five clusters across three studies: (left) Rosenberg et al., 2002, p. 2382; (middle) 1000 Genomes Project Consortium, 2015, Extended Data Figure 5; and (right) Tishkoff et al., 2009, p. 1038. For each study, populations (left of the respective column) and large continental regions in which populations are classified (right of the respective column) are indicated. Each skinny horizontal line represents an individual; split lines with multiple grays (black and white image) or colors (color plate image) indicate the relative amount of that individual's

Table 5.2. *Clustering analysis results across three research programs*

	Rosenberg et al. (2002)	1000 Genomes Project Consortium (2015)	Tishkoff et al. (2009)
Data source	HGDP	(Became) IGSR	HGDP + African sampling[57]
Clustering analysis method	*Structure*	*ADMIXTURE*[58]	*Structure*
When K = 5, what are the clusters?	Africans; {Europeans, Middle Easterners, and Central/South Asians}; East Asians; Oceanians; Indigenous Americans	East Africans; West Africans; Europeans; South Asians; East Asians[59]	Eastern Africans; Western and Central Africans; {Baka, Bakola, Bedzan, Yakoma, Baluba, Hadza, and a few others}; {Europeans, Middle Easterners, Central Asians, and Indians}; {Eastern Asians, Oceanians, and Indigenous Americans}

biases (which likely guide the selection of individuals sampled) will probably remain unavoidable. Population reification is therefore always possible. In addition, ethical and political concerns must be addressed.

←——————————————————————————————

Caption for Figure 5.4. (cont.) genome fitting into different clusters. Respective copyright information: © 2002 The American Association for the Advancement of Science. All rights reserved. Reprinted with permission from AAAS. © 2015 The Author(s), under exclusive licence to Springer Nature Limited. Reprinted with permission. © 2009 The American Association for the Advancement of Science. All rights reserved. Reprinted with permission from AAAS. Adapted by Mats Wedin and Rasmus Grønfeldt Winther. (A black and white version of this figure will appear in some formats. For the color version, please refer to the plate section.)

[57] In addition to the HGDP, they also sampled "2432 Africans from 113 geographically diverse populations . . . 98 African Americans, and 21 Yemenites" (Tishkoff et al., 2009, p. 1035).
[58] Alexander et al. (2009) present this new program, which shares some features of *Structure*, but employs a maximum likelihood (ML) approach rather than a Bayesian approach.
[59] 1000 Genomes Project Consortium (2015, Extended Data Figure 5).

Also, owing to potential biomedical applications, there are strong profit motives for obtaining high-coverage, whole-genome sequence data from individuals in relatively well-off countries and populations. This means there are far more medically motivated sequenced genomes than anthropologically motivated ones, but the latter data are within our reach. In fact, as Bergström and colleagues assert, "producing high-coverage genome sequences for at least 10 individuals from each of the ~7000 human linguistic groups would now arguably not be an overly ambitious goal for the human genomics community. Such an achievement would represent a scientifically and culturally important step toward diversity and inclusion in human genomics research."[60] Clearly, a number of research projects focused on the sociology of science and the philosophy of science await.[61]

F_{ST} Reification

In this chapter and Chapter 4, we have extensively explored F-statistics, in particular F_{ST}. But how accurate and appropriate a measure of population structure in theoretical or natural populations is F_{ST}? Which assumptions and potential pitfalls should we be wary of with respect to F_{ST}? *Under which conditions might we run the risk of perniciously reifying our beloved statistic and parameter of population structure?*

Holsinger and Weir sing the praises of F_{ST} and identify some of its many uses:

The application of F-statistics to problems in population and evolutionary genetics often centres on estimates of F_{ST}. For example, when interpreting aspects of demographic history, such as sex-biased dispersal out of Africa in human populations, detecting regions of the genome that might have been subject to stabilizing or diversifying selection or correcting the probabilities of obtaining a match in a forensic application for genetic substructure within populations, estimates of F_{ST} often play a crucial part in interpretations of genetic data. Estimates of F_{IS} reveal important properties of the mating system within populations, but estimates of F_{ST} reveal properties of the evolutionary processes that lead to divergence among populations. Finally, in many populations of animals, and in human populations in particular, within-population departures from Hardy–Weinberg proportions are small [and, hence, F_{IS} is often near zero].[62]

---

[60] Bergström et al. (2020, p. 9). As justification of the statistic of approximately 7000 linguistic groups, they cite *Ethnologue*; cf. note 54.
[61] Cf. Weiss and Fullerton (2005) and Kaplan (2011).
[62] Holsinger and Weir (2009, p. 645), references omitted.

Indeed, F-statistics is a superbly powerful theoretical framework for human evolutionary genomics.

But there are also concerns with the interpretation and application of F_{ST}. That is, it can be perniciously reified as an indicator of population structure. First, once we have our F_{ST} measures and genetic variance components, how much difference must be present for us to be able to say that population structure is real? Is the total human F_{ST} of roughly 0.14, per Lewontin's distribution, sufficient for calling intracontinental and aggregate intercontinental populations distinct and attributing ontological status to them, or must we meet the typical boundary F_{ST} of between 0.25 and 0.30?[63] As we saw in Chapter 4, with respect to empirical interpretation, there is legitimate discussion about which F_{ST} values are sufficient for population identification, uniqueness, and structure.

More deeply perhaps, and turning to theoretical matters, F_{ST} should be interpreted with care. In fact, human population structure violates two assumptions necessary for F_{ST} to serve as a reliable index of population structure: "first, that expected genetic diversity is the same in every population; and second, that divergence between all pairs of populations is equal and independent."[64] Long adds that "it is odd that Lewontin felt that 15% [14%] of variation among groups is small and even odder that others have concurred."[65] At the very least, we can say that what counts as "high" or "low" F_{ST} is up for debate. Furthermore, Long writes that once we have abstracted and predicted F_{ST} from data and models, "we cannot be so naïve as to assume that we can turn the [modeling] process around and trust that we can translate that index [F_{ST}] back to the structure of a natural population" (p. 803; cf. Figure 5.2). In other words, some uses of F_{ST} may be cases of pernicious reification.

What is more, many estimates of F_{ST} are unusual or otherwise not indicative of the true population structure of our species – for instance, of populations across Africa, which generally have high diversity and heterozygosity.[66] The problem is that F_{ST} – even in the simplest case of two subpopulations, with one aggregate population – depends on several key mathematical factors, not just on empirical facts about natural

---

[63] Templeton (1999).

[64] Long (2009, p. 799); see also Long and Kittles (2003) and Long et al. (2009).

[65] Long (2009, p. 801). Long also reminds us that for Wright, such an F_{ST} "reflects moderately great differences," in part because random genetic drift can significantly change allele frequencies with such a mid-level F_{ST} value (pp. 801–802).

[66] See Jakobsson et al. (2013).

populations. First is the frequency of the most frequent allele across subpopulation pairs; when this frequency is small *or* large, F_{ST} is restricted to small values.[67] Second is the level of heterozygosity in populations; pairs of high-heterozygosity subpopulations are constrained toward low F_{ST}, compared to pairs of low-heterozygosity subpopulations.[68] Third is the type of genetic polymorphism studied; multiallelic microsatellites give lower F_{ST} values than do biallelic SNPs.[69]

To provide a better understanding of population structure in nature, Jakobsson, Edge, and Rosenberg counsel "developing an understanding of the ways in which possible statistics relate both to intuitive aspects of differentiation and to mathematical features of allele frequencies and genetic diversity."[70] I view this as a plea to develop an *integration platform* of statistics of biological populations and genetic variation, elements of which we have seen in this chapter and in Chapter 4. F_{ST} plays a particularly important role here because of its extensive history in evolutionary genomics and its role in framing and tying together a variety of models and methodologies, such as those we have explored in this chapter.[71]

Cluster Reification

We should neither confuse nor conflate clusters with evolving populations. That is, *we should not overinterpret the value of theoretical clusters as informative of ancestry, or even of actual population structure, and we should also be wary of believing that there is a natural number of clusters.* As we explored in Chapter 3, there are many cases in which *Structure*-based clusters, as such, are misleading when building hypotheses of ancestor–descendant relations of populations. In other words, we learned that theoretical populations emerging out of clustering analysis methodology cannot be automatically imported into the phylogenetic research context. Moreover, in the population reification section above, we saw how cluster identifications depend on particular population sampling schemes, and associated complexities of such schemes – hence, we must be wary of overinterpreting and reifying the clustering results of any particular research study or research program. Here, I briefly delve into another form of cluster reification: the potential danger of believing that there is a natural number of clusters.

---

[67] See, e.g., Figure 2 in Jakobsson et al. (2013, p. 519).
[68] See, e.g., Figure 9 in Jakobsson et al. (2013, p. 524).
[69] See Jakobsson et al. (2013, pp. 525–526); cf. Edge and Rosenberg (2014).
[70] Jakobsson et al. (2013, p. 526).   [71] Cf. Whitlock (2011) and Jakobsson et al. (2013, p. 526).

The ontologizing phase of Figure 5.2 can help demonstrate the potential dangers of reifying clusters as clearly and strongly segregated populations, and of believing that there is a natural number of clusters. This is the model *output* of *Structure* in our study of the worrisome potential of the pernicious reification of clusters. Given that "the problem of inferring the number of clusters, K, present in a data set is notoriously difficult," and because the "posterior distribution [with which we can "base inference for K"] can be peculiarly dependent on the modeling assumptions made," it is unclear how to interpret the reliability of the clustering for any particular K.[72] Indeed, a K of 3 or 4, or of 5,[73] is sometimes perceived to be a true and natural cut or division of human genomic variation. These divisions are taken to reflect continental regions, themselves cut in different ways. But this may itself be a reification. After all, "*Structure* identified multiple ways to divide the sampled individuals into K clusters when $K > 6$ (Rosenberg et al. 2002)."[74] Thus, the apparent naturalness of $K = 3$ or 4, or $K = 5$, is actually a conventional choice about how to interpret the robustness of modeling results, rather than a mirror of nature.[75] Cluster reification – in which we hold a particular value for K to be the right or natural number, capturing clusters clearly and strongly distinct from each other – is thus possible. The problem is worsened by the fact that for high K, there is rarely a single robust clustering outcome. How to interpret the model output depends on conventional judgments, such as choosing the appropriate K.

Statistic–Parameter Reification

A serious statistical pernicious reification – conflating and confusing sample statistics with population parameters – looms potentially large in human

---

[72] Pritchard et al. (2000, p. 949).

[73] For discussion, see, respectively, Bamshad et al. (2003) and Rosenberg et al. (2002).

[74] Bolnick (2008, p. 76).

[75] Kalinowski (2011) points to another related problem with *Structure*, which can also lead to reifications of clusters: "*Structure* is also frequently used to identify the main genetic clusters within species. In this second type of analysis, individuals are assigned to clusters … but K is deliberately set to be smaller than the actual number of populations … The mathematical model used by *Structure* was designed for clustering individuals into Hardy–Weinberg/linkage equilibrium populations. It was not designed for clustering individuals into groups of populations, and may not work as its users intend when this is done" (pp. 625–626). Kalinowski's simulations show that when too few K are chosen, *Structure* "will sometimes place individuals from unrelated populations into the same cluster" (p. 628). Again, this is a problem with neither the mathematics nor the computational algorithms; rather, it is a problem with the interpretation of nature that we impose on and from our modeling result.

evolutionary genomics. In Chapter 4, we began exploring the importance of distinguishing sample statistics, or the empirical information we have available to us for estimation, from population parameters, or the true value of whatever we are measuring (e.g., allele frequencies, F_{ST}). Clustering analysis in the sense of Pritchard and colleagues is explicit about this key statistical distinction. As we began to see in Chapter 3 and above, Pritchard et al. used MCMC methods and Bayesian analysis on multilocus genotypes for iteratively adjusting allele frequency *estimates* of the different clusters to eventually hit a stable distribution (across loci, across clusters) of inferred, allele frequency *parameters*.

In contrast, Lewontin's (1972) variance partitioning study conflated allele frequency estimates and parameters. Lewontin simply assumed that the allele frequencies in the data tables found in Giblett (1969) and Mourant (1954) were the true parameters and did not worry about allele frequency estimation procedures. This is a statistic–parameter conflation. Even so, because the sample sizes of Giblett (1969) and Mourant (1954) were rarely much below 100 individuals (typically between 100 and 500 individuals or so, and occasionally over 1000 individuals), the estimated allele frequencies were fairly accurate renditions of the true allele frequency parameters – sample sizes close to or above 100 individuals are sufficient for reliable allele frequency estimation.[76] Thus, Lewontin's statistic–parameter conflation does not lead to severe inferential errors, in this case.

Both AMOVA and the fully hierarchical approach of variance partitioning explicitly distinguish sample statistics from population parameters. Excoffier et al. understood what is at stake:

The point of the current exercise is neither to estimate unknown population parameters from our variance components nor to define exactly how or at what rate these population differences have developed. Our purpose here is to demonstrate how to delineate the extent of genetic differentiation within and among populations.[77]

---

[76] The language here is qualified because the exact reliability of the estimate depends not just on sample size, but also – in statistical argot – on the level of significance desired (*alpha*, which is the probability of a Type I error, i.e., of rejecting a null hypothesis that is actually true, and is frequently set at 0.01 or 0.05) and on the effect size. But sample sizes even just over 100 individuals give reliable allele frequency estimates under a fairly standard and rigorous statistical framework. See, e.g., Nei (1978), Hale et al. (2012), and Fung and Keenan (2014).

[77] Excoffier et al. (1992, p. 490). Similarly, in their investigations of how the F_{ST} parameter behaves, Jakobsson et al. (2013) observe: "We emphasize that in our formulation, F, H_T, and H_S are functions of the parametric allele frequencies, and our interest is in the properties of these functions and their relationships with the allele frequencies; we do not investigate their estimation

Weir also provides explicit estimation procedures for allele frequency and F-statistics parameters.[78] Indeed, Weir is a key player in making explicit useful estimation procedures for a variety of parameters fundamental to population genetic theory, including allele frequency and F_{ST}, and thereby in avoiding statistic–parameter reification.[79]

Because parameter estimation adds a significant new layer of conceptual and formal complexity, Chapter 4 set it aside, to an extent. But the distinction between statistic and parameter is close to the distinction between measure and metric, and even for cases when sample size is sufficiently large for reliable estimation, the distinction remains conceptually important. Furthermore, there are clear estimation protocols and equations for inferring fundamental parameters such as allele frequency, F_{ST}, effective population size, and mutation rate, parameters that play key roles in abstract evolutionary genetic theory. Thus, as a future project, revisiting genetic diversity, genetic heterozygosity, and genetic differentiation in terms of the dangers of statistic–parameter reification would provide fuller comprehension.

As discussed with the potentials of population reification, F_{ST} reification, cluster reification, and statistic–parameter reification, the pernicious reification of our biases and our theoretical representations can occur in various parts of the modeling process, described in Figure 5.2. The argument here is not that variance partitioning and clustering analysis lack merit or that the mathematics is wrong. But in certain parts of the modeling stream, it is hard to know how to interpret input and output, or how to apply the modeling machinery. Again, assumption archaeology and integration platforms are necessary.

Whither the Two Methodologies?

This chapter contributes to the ongoing search for conceptual and philosophical tools for identifying the promises and pitfalls of scientific abstractions in the study of populations. I wish to find criteria and norms that will allow us to differentiate between the generative and productive use of our scientific representations and abstractions and the pernicious

---

from data, nor do we consider how evolutionary models affect the underlying allele frequencies involved in their computation" (pp. 516–517).

[78] See Weir (1996), pp. 83–85 on allele frequency estimation, and pp. 161–190 on F-statistics estimation; see also Weir and Hill (2002) and Holsinger and Weir (2009).

[79] For discussion of fundamental parameter estimation procedures in state-of-the-art evolutionary genomics, see, e.g., Kim et al. (2011) and Lynch and Ho (2020).

reification – of which I distinguish four types – of modeling strategies. Two pertinent questions guide my argument: When is a particular population *grounded* in patterns of genetic variation, and when is it a biased and "viciously abstract" imposition on genetic data (to borrow a locution from William James)?[80]

The modeling machinery of each of the two methodologies for human evolutionary genomics investigated in this chapter may be well oiled. The mathematical variables, functions, and derivations of variance partitioning and clustering analysis are appropriately articulated and mutually consistent, as typically expected from formal systems. Reification of our expectations, biases, and preexisting theoretical maps (e.g., linguistic or geographic human populations) can occur. In other words, we can make concrete things out of our abstract representations in various ways, positing groups, natural kinds, and subdivisions where these do not exist (see Chapters 4 and 9). Moreover, what could be thought of as an ultimate statistical pernicious reification – conflating and confusing sample statistics with population parameters – looms behind much of the work discussed thus far (although Bruce Weir's efforts, for example, to formulate F_{ST} estimation methods clearly involve making the distinction between statistic and parameter; see also Chapter 10).

My arguments in this chapter are philosophical insofar as they consider the reality (or not, or both) of objects and processes of scientific inquiry, such as populations and groups, not to mention genes, and insofar as philosophy is *critique*, as we saw in Chapter 1, and as discussed in the final chapter of my book *When Maps Become the World*.[81] What is being critiqued in this chapter are the data and models used to answer empirical questions about whether our species is actually subdivided into groups or populations.

By applying a rigorous philosophical lens to our investigations into measures and methods, we gain insight into the contextual inputs informing empirical questions. But we also begin to see the reflections and refractions in the answers yielded by such questions. The political stakes are high because the mathematical, theoretical, and disciplinary origins of human populations apply to our notions of identity and belonging, freedom and potential.

---

[80] See Winther (2014) and references to philosophers William James and John Dewey therein.
[81] Winther (2020a).

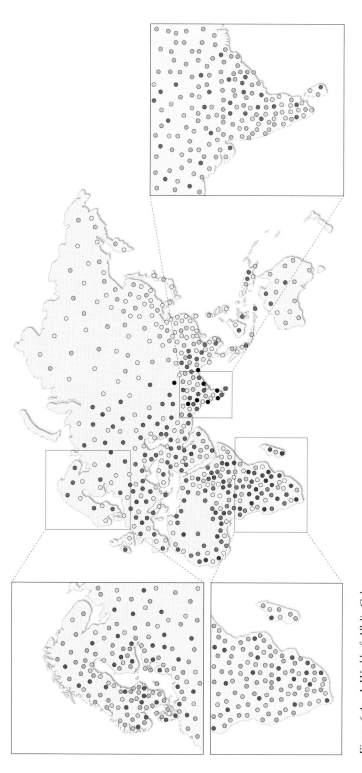

Figure 6.1. A World of Allelic Color

A geographic sketch of human allelic variation. Each circle represents a set of allele variants (i.e., a unique, partial genotype). Circle density hints at contemporary population density. The insets represent a kind of "base map," showing that many alleles are everywhere, in somewhat similar frequencies across geographies (e.g., Lewontin, 1972, Rosenberg, 2011, and Biddanda et al., 2020). There is also a tendency to lose allelic variation as we leave eastern/southern Africa. Both patterns – corresponding respectively to Planet Unity and Galápagos–Writ-Large – can be read from our genome. Concept by Rasmus Grønfeldt Winther, illustrated by Mats Wedin. © 2021 Rasmus Grønfeldt Winther. (A black and white version of this figure will appear in some formats. For the color version, please refer to the plate section.)

6 · *Six Patterns of Human Genomic Variation*

In pursuit of more nuanced answers about ourselves and each other, we must shift from metrics and models to interpretation and applications of genomic analyses, both in scientific and medical contexts, and in broader culture and politics. Accordingly, this chapter presents six general empirical patterns of human genomic variation. The patterns emerge from many data points, layered through measures and metrics such as heterozygosity and Euclidean genetic distance, statistical modeling machinery, such as variance partitioning protocols and Bayesian clustering algorithms, and theoretical concepts, such as population and genetic variation. Because the six patterns form an indelible, complex mix of genomic data, statistical methods, and evolutionary genetic theory, I call them neither *facts* nor *data*, nor *empirical results* nor *states of affairs*. They are also neither models nor methodologies. They are much too complex and unique for such simple and straightforward denominations.

Each pattern relies on the results of empirical protocols (e.g., gene sequencing and ethical sampling procedures), mathematical theory, and computational algorithms. But the rich data collected and managed – and the multidimensional, automated calculations – produce patterns irreducible to these data, theory, or algorithms, and these patterns are all the more cognitively interesting because of it. In other words, these patterns interweave metrics and measures, data, models and methodologies, concepts and theories, algorithms, and values, assumptions, and choices. They are large and equivocal results that require sustained and critical effort to understand and analyze. Patterns differ from models, in particular. Models are based on clear background theory, concepts, and assumptions, and involve derivation and calculation procedures. Patterns, though also informed by theory, represent an expansion, if not explosion, of basic theoretical machinery fed with complex data and then glossed through interpretive, hermeneutic frames. We will see this, for example, with Lewontin's 86%/7%/7% pattern of the global distribution of heterozygosity, and with the pattern of small, but often critical, interspecies differences.

This chapter explores six empirical patterns that shed light on our ongoing questions about origins and identity. In line with our preceding findings, the patterns illuminate a human genomic universe that is much more in line with Planet Unity than with Galápagos-Writ-Large, even if it possesses features of the latter (see Figure 4.1). The patterns also contain, in my view, the germ of understanding the vast connectedness of life and how much humans owe *all* that lives on land, in the sea, and in the air. As in other chapters, readers are invited to refer to the genomics glossary at the end of Chapter 2 for help in recalling basic definitions.

Six Empirical Patterns

The six empirical patterns of global human genomic variation revealed by our best contemporary science can be stated as follows:

1. There is low intraspecies genomic variation.
2. There are small, but often critical, interspecies differences.
3. Of all continental regions, Africa has the oldest – and thus the richest and most encompassing – human genomic variation.
4. Most genetic variation is among individuals within populations, not across populations within continental regions, nor across different continental regions (Lewontin's distribution).
5. Despite Lewontin's distribution, clustering populations and classifying individuals is possible.
6. Genomic heterozygosity of populations decreases with increasing distance from Africa, along human migration routes.

These basic quantitative features of human genetic variation can be understood without detailed reference to the depths of population genetic theory, which we plumbed especially in Chapters 4 and 5. However, they can also be further analyzed and comprehended with evolutionary genetic models that deploy theoretical parameters, such as mutation rate (μ), selection coefficient (s), and effective population size (N_e).[1] A more detailed understanding of the patterns could be

---

[1] Hartl and Clark (1989), Weir (1996), and Nielsen and Slatkin (2013).

provided regarding, for example, the effect of historical migrations on early human evolution, if we explored the logic and consequences of effective population sizes, or the potential future trajectories of the structure of human genomic variation, if we dove more deeply into mutation rates and selection coefficients. Indeed, the current depth and breadth of mathematical knowledge and modeling, as well as informed interpretation, paints a particular, reliable picture of human populations – and the individuals making up those populations – today. We can, however, understand this picture without diving – too much – into theory and parameters.

Let's take a look at the six empirical patterns in greater detail.

1. There Is Low Intraspecies Genomic Variation

Of species whose genomes have been mapped extensively, *Homo sapiens* has unusually low average nucleotide variation. Humans differ from each other on average at only 1 base pair out of 1000. Given our total, haploid genome size of just over 3 billion nucleotides, individuals will typically differ at just over 3 million nucleotides. This rough 0.1% average nucleotide difference holds across a variety of studies with different methods and conceptualizations of difference (e.g., average pairwise difference among sampled humans versus comparison of a typical, averaged human to a reference genome), although it is a little higher when looking at larger kinds of genetic polymorphisms, such as copy number variants (CNVs) and other structural variants.[2] For comparison: *Drosophila* fruit flies, the standard workhorse for genetic studies, as we learned in earlier

---

[2] Both Li and Sadler (1991) and Yu et al. (2002) calculated an average, species-wide nucleotide diversity of approximately 0.1%. The former article compared the coding and noncoding regions of 49 well-curated genes (at a total of over 75,000 base pairs). The latter "sequenced 50 noncoding DNA segments each ~500 bp long in 10 Africans, 10 Europeans, and 10 Asians" (p. 269), finding a total of 146 single-nucleotide polymorphisms and detecting more nucleotide diversity among Africans than among Europeans or Asians. The 1000 Genomes Project Consortium (1000 Genomes Project Consortium, 2015) sequenced the whole genomes of 2504 individuals from 26 populations, identifying SNPs at roughly 3% of the human genome. They found that the typical human differed from the standard human reference genome GRCh38 (https://www.ncbi .nlm.nih.gov/assembly/GCF_000001405.26/ or www.ensembl.org/Homo_sapiens/Info/Index) at approximately 3.9–5 million sites (see Figure 1b, p. 69). At a tiny fraction of these sites, however, structural variants of hundreds to many thousands of base pairs occur, meaning that the typical genome is more like 20 million base pairs different (in total) from the reference genome. This makes for a somewhat larger difference between a typical individual genome and the reference genome, closer to 0.6% (1000 Genomes Project Consortium, 2015, p. 68, Table 1, p. 71), assuming a total human genome size of 3.2 billion sites. Even so, Biddanda et al. (2020),

chapters, differ from each other by about 1% on average;[3] bonobos differ by 0.076%; chimpanzees by 0.134%; and gorillas by 0.158%.[4] Maize has even more nucleotide diversity than *Drosophila*, and soybeans have slightly more than humans.[5] Admittedly, *Homo sapiens* has more average heterozygosity than most big cats – according to one early meta-analysis, our average heterozygosity is roughly twice that of lions and leopards (unfortunately for their future prospects, cheetahs have near 0% genetic diversity).[6] What does this tell us? Again, wherever you may be from, you and I are genetically quite similar, located, figuratively anyway, somewhere on Planet Unity (see Figure 4.1). In fact, whatever the fine-grained structure of human genomic variation, which we will explore further below, there is, in line with our best genomic data, a *single overarching* human race. Interpretations and personal reactions to this reality are complex and multifarious, but there is a potent and relevant story for identity and belonging in human unity. We are, in a clear sense, all of the same family. We belong together.[7]

2. There Are Small, But Often Critical, Interspecies Differences

Comparing the broad genomic similarities between humans and other mammals often suggests that little separates us on the level of our DNA. For example, we see a 98.771% nucleotide match between humans and chimpanzees, 98.69% between humans and bonobos, and 98.36% between humans and gorillas.[8] Even human and mice genomes are roughly 85% identical.[9] However, this surface similarity belies a more subtle reality. Genes (or parts of genes) are rarely arranged in a linear and continuous fashion on the genome, and it is not just differences in individual genes that determine species, but also their order and structure. Meaningful interspecies comparisons therefore require more than straightforward matching of nucleotide sequences. Such comparisons

---

which based its analysis primarily on 1000 Genomes Project data, also put "human pairwise nucleotide diversity" at ~0.1% (p. 8). The front cover's mitochondrial sequence is from GRCh38.

[3] See Li and Sadler (1991, p. 513), who also define nucleotide diversity as "the number of nucleotide differences per site between two randomly chosen sequences from a population." Nucleotide diversity is a *further* measure of genetic diversity, and is itself interpreted in various ways.

[4] Table 1 in Yu et al. (2004, p. 1379).   [5] Brown et al. (2004, p. 15259).

[6] O'Brien et al. (1985).

[7] Consult the Galápagos-Writ-Large versus Planet Unity thought experiment from Chapter 2.

[8] Calculated from Table 3 in Yu et al. (2004, p. 1381).   [9] Batzoglou et al. (2000, p. 951).

should also consider structural variation in the genome,[10] especially when comparing humans to related great apes, such as bonobos, chimpanzees, gorillas, and orangutans (Figure 6.2; see also Figure 2.14).

In fact, recent research shows that earlier work, while underestimating interspecies nucleotide differences only slightly, was based on reconstructions of great ape genomes that were "humanized," such that these sequences "minimize[ed] potential structural and transcript differences observed between the species."[11] While our human nucleotide sequence differences with other great apes remain tiny (and nucleotide diversity within a great ape genetic pool, as it were, would be only slightly larger than nucleotide diversity within *Homo sapiens*), to the best of our genomic knowledge, structural variants unique to humans often occur in critical places, such as in regulatory regions of the genome, which contain genes that influence the expression of other genes.[12] For example, human-specific insertions, deletions, duplicates, or inversions in regulatory regions affect cell division and cell fate. Some structural variants, though seemingly small, result in key differences between humans and great apes. Human brain expansion stands as an important example: It is correlated with genomic structural variants – unique to the human lineage – that are involved in cortical neurogenesis.[13]

The example of brain expansion helps underscore some of the implicit weaknesses associated with simply comparing averaged and unstructured nucleotide sequences across species. To understand the origin of species, as well as the origin of adaptations, such as the greatly increased cognitive

---

[10] Structural variation includes translocations (relocation along the genome), as well as large insertions, deletions, duplications, and inversions (e.g., CNVs, *Alu* elements) of relatively large parts of the genome (say, over 100 or 1000 nucleotides); see Figure 1 in Pollex and Hegele (2007, p. 3131), and Ho et al. (2020). Such variation is not uncommon across the human genome, can itself comprise allele variation, and also makes us differ a little more from each other than simply comparing SNPs or indels, as we started seeing in the previous empirical pattern (1000 Genomes Project Consortium, 2015; Sudmant et al., 2015).

[11] Kronenberg et al. (2018, p. 1).

[12] For instance, see Figure 4D in Kronenberg et al. (2018, p. 7).

[13] On brain development, see Kronenberg et al. (2018) and Kanton et al. (2019). A plausible but not yet proven part of the story here is that the ancestral lineage leading to great apes may have had many copies or repeats of the same sequences, copies that were then lost or duplicated in different ways in the *Pongo*, *Gorilla*, *Pan*, and *Homo* lineages (Kronenberg et al., 2018, p. 10). Olson's (1999) "less-is-more" hypothesis postulating that gene loss is an "important engine of evolutionary change" (due to, for example, regulatory changes) might be useful in this context (pp. 18, 21). In all of this, the functionality of the structural variants must be assessed (Dicks and Savva, 2007; Gerstein et al., 2007).

Figure 6.2. Our cousins 2: Chimpanzees
Creative tool use in our primate cousins – a chimpanzee using a stick to fish for ants (cf. Goodall, 1964). Illustrated by Daphné Damoiseau-Malraux. © 2020 Rasmus Grønfeldt Winther.

power of *Homo sapiens*, we need to compare differences in general genetic architecture and genetic interactions. This is challenging because it involves not just cross-species genetic comparison but also clear knowledge of the anatomy, physiology, and development of each species, and, correlatively, the effects of genes on these complex processes of homeostasis and the emergence of organismic form and function. For this reason, even though it may seem far afield from the context of human evolutionary genomics, it behooves humans to rein in the destruction of the ecological habitats of natural populations of great apes, especially in

Africa and Southeast Asia. Why? Because we are, quite literally, destroying opportunities to keep alive our primate cousins, who could help shed light on the significance, possibilities, and potential pathologies and diseases of our own genes and genetic architecture.

3. Of All Continental Regions, Africa Has The Oldest – And Thus The Richest And Most Encompassing – Human Genomic Variation

Contemporary populations indigenous to Africa have the most direct genetic connection to our earliest ancestors of any humans on Earth. After all, humans first evolved in Africa, and the ancestors of today's Africans did not experience the same degree and number of genetic bottlenecks – caused by migrating long distances and crossing barriers such as mountains and oceans – as populations that left Africa, for other regions. All this helps make African genomic diversity the richest, most complex, and most encompassing on Earth, with most genomic variation found elsewhere also being found in Africa.[14] We can identify this pattern by exploring comparative genomic variation across continental regions as measured or assessed by (a) nucleotide diversity; (b) allele distribution of three kinds: (i) common and global, (ii) rare and private, and (iii) common and private;[15] and (c) genetic heterozygosity (Figures 6.1 [the chapter opener] and 6.3).

Nucleotide Diversity

African populations harbor significantly more nucleotide diversity than non-African populations. One study found that two people whose recent ancestors are of African origin differ on average by about 1 out of 870 nucleotides (0.115%), whereas two people whose recent ancestors are of European origin differ on average by only approximately 1 out of

---

[14] Long et al. (2009) write that "the diversity in non-Sub-Saharan African populations is essentially a subset of the diversity found in Sub-Saharan African populations" (p. 23).

[15] For the sake of simplicity, I address each of the three kinds of allele distribution using, respectively, Rosenberg (2011), Biddanda et al. (2020), and Bergström et al. (2020). Both Rosenberg (2011) and Bergström et al. (2020) are based on HGDP-CEPH samples of around 1000 individuals from slightly more than 50 global populations, addressed in Chapter 5. The former studied microsatellite markers, while the latter engaged in whole-genome, high-coverage sequencing to find approximately 67.3 million SNPs, 8.8 million indels, and 40,736 CNVs. These numbers are comparable to the deployment, by Biddanda et al. (2020), of the new high-coverage sequence data produced by the New York Genome Center of the original 2504 individuals sampled from 26 populations of the 1000 Genomes Project, also discussed in Chapter 5. The HGDP and 1000 Genomes Project only partly overlap in the tens of millions of SNPs they identify.

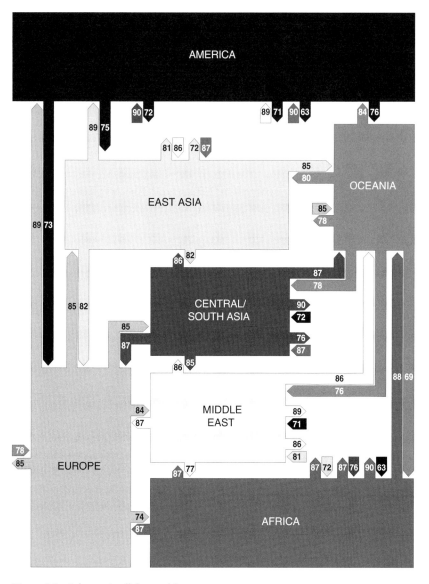

Figure 6.3. Schematic allele world map
"Schematic world map of the 'flow' of microsatellite alleles ... boxes represent
regions of the world, positioned geographically. Links entering into a geographic
region indicate the percentages of distinct alleles from the geographic region found
in other regions ... For example, averaging across loci, 87% of alleles observed in
Europe are also observed in Africa, whereas 74% of alleles observed in Africa are also
observed in Europe." From Rosenberg, 2011, p. 680. © 2011 Wayne State
University Press. Reprinted with permission. Adapted by Mats Wedin and Rasmus
Grønfeldt Winther.

1563 nucleotides (0.064%).[16] Another study found that – given one metric of nucleotide diversity – Mandinka peoples from Mali, Guinea, and the Ivory Coast and the San peoples of South Africa had nucleotide diversities of 0.12% and 0.126%, respectively, while Han Chinese and Basque peoples had nucleotide diversities of 0.081% and 0.087%, respectively.[17] In addition, Auton and colleagues provide a table with "median autosomal variant sites per genome" for all African and European individuals, finding that for single-nucleotide polymorphisms (or SNPs), Africa is 1.2 times more "diverse" than Europe (4.31 vs. 3.53 million sites), while for insertions or deletions – i.e., indels – Africa is 1.1 times more diverse (625,000 vs. 546,000 sites).[18] (Although this ratio is not strictly a nucleotide diversity measure, it is loosely analogous to one.)

Across these three studies, African nucleotide diversity is something like 1.1–1.8 times the genetic diversity of Europe. Comparing Africa to other geographic regions paints the same general picture, and relative diversity decreases the further we move from Africa along major migration routes. For instance, the visual summary that Bergström and colleagues provide indicates that, for total genetic variation, Africa is roughly (X) times more diverse than the Middle East (1.20), Central and South Asia (1.26), Europe (1.25), East Asia (1.33), Oceania (1.48), and the Americas (1.78).[19]

Allele Distribution

Common and global: *Most alleles that are common are also widely distributed around the world.* According to Rosenberg, at least 82% of common microsatellite alleles were found in Africa, followed by Central and South Asia and then, closely, the Middle East, each with at least 77% or so.[20] The order of prevalence of total, common alleles in a given region differs slightly from the diversity list above, though it's still congruent with the migration and local evolutionary effects of the Out of Africa hypothesis:[21] Africa, Central and South Asia, the Middle East, Europe, America, Oceania (with a minimum of at least 52.54%). Second, most alleles (87–90%) found in any single non-African geographic region were also identified in Africa, but not the reverse (Figure 6.3). Finally, approximately

---

[16] Percentages as found in Table 1 in Yu et al (2002, p. 271).

[17] Table 1 in Wall et al. (2008, p. 1355); cf. Campbell and Tishkoff (2008).

[18] Table 1 in 1000 Genomes Project Consortium (2015, p. 71).   [19] Bergström (2020).

[20] See Table 2 in Rosenberg (2011, p. 665); cf. Figure 6.3. A common allele in this study is one not restricted to only one person.

[21] Table 2 in Rosenberg (2011) is still missing 10.12% of the alleles. I was unable to secure the full table.

92% of common alleles are found in more than one region of the world. Roughly 82% of common alleles are found in three or more regions, and 46.60% of common alleles are found in *all* regions.[22]

Rare and private: *Almost all alleles that are rare are also private.* Biddanda et al. roughly corroborated the previous most-common-alleles-are-also-global picture using SNP variation data from the 1000 Genomes Project (see Figure 5.1, middle).[23] Biddanda et al. also found that "most variants are rare and geographically localized [i.e., private]."[24] To corroborate this visually, try clicking on "random" at the Geography of Genetic Variants Browser that John Novembre developed.[25] (See also Figure 5.1, bottom.) What do we make of all this rare *and* private allele variation? Biddanda et al. offer one plausible answer. Trawling the genome at high resolution significantly avoids "ascertainment bias."[26] Novembre and colleagues essentially found an enormous treasure trove of variants existing primarily due to young point mutations in one or very few (often related) individuals.[27] Such mutations are private to a single geographic region and have not been able to spread geographically, yet.[28]

---

[22] Figure 4 in Rosenberg (2011, p. 666).

[23] A caveat: Rosenberg looked only at nonsingleton alleles, whereas Novembre and colleagues in Figure 3, figure supplement 2, A (which provided the data for Figure 5.1, middle) included singletons, which are alleles found only in a single individual. However, the ranking and relative distribution across regions of common alleles is roughly the same whether singletons are included or not (e.g., cf. Biddanda et al., 2020, Figure 3, figure supplement 1, A vs. B).

[24] Biddanda et al. (2020, pp. 1ff.). For instance, when singletons are included, only a small fraction of the total variation is common in one or more regions (not much more than 6% or so), while 68% is rare and private to *exactly* one of the five regions sampled: Africa, Europe, South Asia, East Asia, and the Americas (see Figure 3B in Biddanda et al., 2020, p. 5). Excluding singletons nuances the picture, but common variation still only just exceeds approximately 17%, with much of the rest being rare and mostly found in only one or a small number of regions.

[25] See https://popgen.uchicago.edu/ggv.

[26] Whole-genome sequencing, such as that which Biddanda et al. (2020) drew on via the New York Genome Center, helps solve the "ascertainment biases" typical of many, limited SNP genotyping panels (e.g., Lachance and Tishkoff, 2013). Searching only for already known SNPs is reminiscent of the old yarn about the person looking for her keys under the lamp "because that's where the light is." Ascertainment bias includes missing rare SNPs that may be present in individuals not thus far sampled, and missing even common SNPs that may be present only or primarily in populations not yet sampled. The only way to avoid such bias fully is to map every individual in every population, but whole-genome sequencing on multiple individuals across many populations goes a long way. (Ascertainment bias is less of a concern for microsatellites, but caution is indicated; e.g., Eriksson and Manica, 2011; Fischer et al., 2017. Future work includes applying the approach of Biddanda et al., 2020 "to other classes of genetic variants such as insertions, deletions, microsatellites, and structural variants" p. 13.)

[27] See, e.g., Biddanda et al. (2020, p. 4); see also Figure 7.6.

[28] Extremely few alleles are rare and global (cf. Figure 3 in Biddanda et al., 2020, p. 5); I leave it as an exercise to the reader to hypothesize evolutionary reasons for why.

Common and private: *Only relatively few alleles that are common are also private.* Bergström et al. write:

Populations in central and southern Africa, the Americas, and Oceania each harbor tens to hundreds of thousands of private, common genetic variants. Most of these variants arose as new mutations rather than through archaic introgression, except in Oceanian populations, where many private variants derive from Denisovan admixture.[29]

Their article can give the impression that there are many common and private alleles.[30] But while their numbers may seem large, in the context of a total of 67.3 million SNPs, they are not, and indeed fall within the general picture of a relatively undifferentiated global human genome, with little common private variation. Even so, this class of variation clearly exists and is subject to the forces of random genetic drift and natural selection, as we shall further see in Chapter 7.

Genomic Heterozygosity
The metric and measure of heterozygosity is the final indicator of the empirical pattern of African genomic variation as the richest, most complex, and most encompassing on Earth. We turn to this in pattern #6.

This entire empirical pattern can be explained in terms of basic evolutionary principles and the Out of Africa migration of individuals of our species (see Figures 2.3, 2.4, 6.1, and 6.3). When a species migrates, relatively few individuals move. Migrating populations in nature tend to be small – and even smaller when barriers such as mountains, deserts, or oceans stand in the way. Two important consequences follow. First, because of random genetic drift, small populations tend to lose genetic variation over generations. Second, migrating natural populations also experience a series of genetic bottlenecks, a result of the "founder effect" whenever a small group colonizes new areas (see also empirical pattern #6).[31] The founder population consequently represents only a probabilistic sample of the full genetic variation of the parental population. Individuals reaching the Americas via the Bering Strait, for instance,

---

[29] Bergström et al. (2020, p. 1).

[30] For instance, "the highly diverged San of southern Africa harbour ~100,000 private variants at >30% frequency, ~1000 at >60%, and even ~20 that are fixed in our small sample of six individuals" (Bergström et al., 2020, p. 3, see their Figure 3A, p. 4). This adds new detail to the work of Rosenberg (2011) and Biddanda et al. (2020).

[31] See Ramachandran et al. (2005) and Handley et al. (2007).

experienced the most bottlenecks (though Oceanian populations experienced almost as many, some of them nonoverlapping with Indigenous Americans; see Figure 2.3). Over time, random genetic drift and genetic bottlenecks tend to lead to loss of (a) nucleotide diversity, (b) alleles of various kinds, and (c) genetic heterozygosity. It is therefore Africa, the second-largest continent on Earth, that retains the greater, or ancestral genetic variation, making it the putative capital of Planet Unity (see Figures 4.1, 6.1, and 6.3).

4. Most Genetic Variation Is Among Individuals Within Populations, Not Across Populations Within Continental Regions, Nor Across Different Continental Regions (Lewontin's Distribution)

What I have called *Lewontin's distribution* in Chapters 4 and 5 helps answer the question of how much less genetically different two randomly chosen individuals from the same population are, on average, than a randomly chosen individual from that population and a randomly chosen individual from *another* population – whether from the same continental region or from anywhere on the globe.

Lewontin deployed data from especially Giblett (1969) and Mourant (1954) in his influential 1972 study, and from Cavalli-Sforza and Bodmer (1971) in his 1974 book, wherein he partly corrected calculation errors from his earlier results.[32] Lewontin (1972) drew on reasonably extensive allele frequency data tables of serum proteins and red blood cell enzymes[33] and blood groups,[34] including two blood lipoproteins, *Duffy*, the Rh factor, and the ABO blood group complex. With varying degrees of breadth (numbers of populations sampled) and depth (numbers of individuals sampled in each population), Lewontin collected – and sometimes also had to finish calculating – global distributions of the allele frequencies of each of 17 genes or loci. He then used his particular form of the technical methodology of variance partitioning, which we reviewed in Chapter 5, on the allele variation for each of these genes or loci, doing so within populations, among populations on the same continent, and among continents.

---

[32] The earliest published data that explicitly give something close to this statistical result can be found in Table 12 in Cavalli-Sforza (1966, p. 367), which in column 4 indeed "showed that only about 15% of the overall gene-frequency variation occurred between populations compared with 85% within them" (Bodmer, 2018, p. 322).

[33] Giblett (1969).   [34] Mourant (1954).

Table 6.1. *Allele frequencies of three distinct genes across continental regions*[35]

Gene	Alleles	Africa	East Asia	Europe
Duffy	Fy	0.94	0.10	0.03
	Fy^a	0.06	0.90	0.42
	Fy^b	0.00	0.00	0.55
Auberger	Au^a	0.64		0.62
	Au	0.36		0.38
Xg	Xg^a	0.55	0.54	0.67
	Xg	0.45	0.46	0.33

Frequencies are rounded from four to two significant figures. Empty cells indicate lack of data. Following Lewontin (1974), the "a" and "b" superscripts differentiate alleles.

Table 6.1 focuses on three major continental regions to show three types of allele distributions, represented by three genes, that gave rise to the true 85.7%/7.1%/7.2% distribution of Lewontin's analysis. His results imply that at most variable loci, different human groups tend to have relatively similar allele frequencies. Thus, the *Duffy* gene is an atypical example, as it is more extremely diverged than average – 0.94% (Africa)/ 0.10% (East Asia)/0.03% (Europe) for the *Fy* "null" allele; based on similar Lewontin (1972) allele frequency data, *Duffy* has a true diversity apportionment of 63.6%/10.5%/25.9% (Table 6.2), which is a much higher among continents variance component than almost all other loci Lewontin studied in 1972 (think Galápagos-Writ-Large).[36] In contrast, *Auberger* indicates less variation across populations than the average human locus (think Planet Unity). The *Xg* gene is typical of the human

[35] From Cavalli-Sforza and Bodmer (1971), as presented in Table 33 in Lewontin (1974), a table with only these three regions represented (p. 153). As we saw in Chapter 5, Lewontin (1972) analyzed allele frequencies globally, including, at the highest continental regional level, South Asian Aborigines, Indigenous Americans, Oceanians, and Australian Aborigines, in addition to African, East Asian, and European populations (cf. Winther, 2022a). Depicting three major continental regions in this table is, however, sufficient for illustrating the range of allele frequency differences across continental regions – from extremely different (e.g., *Duffy*) to highly similar (e.g., *Auberger*). Figure 3 in Rosenberg (2011) selects three microsatellite loci with analogous levels of allele frequency geographic differentiation, with *Duffy* corresponding to D12S2070 (bottom row), *Auberger* to D6S474 (top row), and *Xg* to D10S1425 (middle row) (p. 664).

[36] See Table 1.6 in Winther (2022a, p. 36). The allele distribution for *Duffy*, and concomitant diversity apportionment, is unsurprising, as homozygous *Fy* individuals are resistant to the malarial parasite *Plasmodium vivax*, historically common in Africa and, to a lesser extent, in Southeast Asia (see, e.g., Szpak et al., 2019, pp. 1432–1435). This allele has thus been under selection, a topic to which we turn in Chapter 7.

Table 6.2. *Two genetic diversity apportionments from Lewontin (1972): Lewontin's and the true one*[37]

	Within populations		Among populations		Among races	
	Lewontin	True	Lewontin	True	Lewontin	True
Hp	0.893	0.893	0.051	0.050	0.056	0.057
Ag	0.834	0.835	–	0.003	–	0.162
Lp	0.939	0.942	–	0.025	–	0.033
Xm	0.997	0.997	–	0	–	0.003
APh	0.927	0.919	0.062	0.059	0.011	0.023
6PGD	0.875	0.877	0.058	0.055	0.067	0.068
PGM	0.942	0.942	0.033	0.032	0.025	0.026
Ak	0.848	0.740	0.021	0.154	0.131	0.105
Kidd	0.741	0.763	0.211	0.218	0.048	0.020
Duffy	0.636	0.636	0.105	0.105	0.259	0.259
Lewis	0.966	0.965	0.032	0.033	0.002	0.001
Kell	0.901	0.903	0.073	0.072	0.026	0.025
Lutheran	0.694	0.696	0.214	0.215	0.092	0.089
P	0.949	0.949	0.029	0.029	0.022	0.022
MNS	0.911	0.906	0.041	0.042	0.048	0.052
Rh	0.674	0.682	0.073	0.068	0.253	0.250
ABO	0.907	0.923	0.063	0.047	0.030	0.030
True mean	**0.857**		**0.071**		**0.072**	
Lewontin (1972) written mean	0.854		0.083		0.063	
Lewontin (1972) realculated mean (Table 4)	0.861		0.076		0.076	

[37] Lewontin's reported genetic diversity apportionment values, for each locus, from Table 4 in Lewontin (1972), as well as the true, correct apportionment of genetic diversity recalculated from Lewontin's Table 3, together with the true mean (for recalculations, see Winther 2022a). It remains unclear how Lewontin calculated his own written mean values, as they do not match a simple recalculation of his own Table 4. His overstatement of the among populations but within races diversity component at the expense of the among races diversity component is also somewhat curious. A note on nomenclature: while gene names and alleles today are typically italicized, as I have done in this book, also in order to distinguish them from their protein products (whose names are typically *not* italicized), Lewontin was not consistent – Lewontin (1974) italicizes genes and alleles (e.g., Table 33, p. 153) but Lewontin (1972) does not (e.g., Table 1, p. 384). Again, I italicize genes and alleles, except for Table 6.2 and Figures 6.4 and 6.5, which directly engage Lewontin (1972).

genome, showing some variation across continental regions. A key consequence of the hierarchical structure in genetic variation is that we can still use small allele differences globally to identify the population or cluster to which an individual belongs (cf. empirical pattern #5).

As we saw in pattern #3, we now know – thanks to the work of Noah Rosenberg, John Novembre, their respective colleagues, and many others – that most of the common alleles present in our species are globally distributed and that there are relatively few common, private alleles. Such a pattern was already presaged in Lewontin's meta-analysis, in which almost all the alleles of the 17 genes under study were found to be present in almost all populations.[38]

Interestingly, Lewontin's classic article is rather telegraphic in neither listing data sources nor explaining foundational evolutionary genetic theory, both of which I have now done in Winther (2022a), with data analysis available in Winther (2021b). I also found that Lewontin (1972) was replete with calculation errors for all of the 17 genes it covered, except one (P).[39] Importantly, Lewontin's calculation errors are not systematic (Figures 6.4 and 6.5; Table 6.2), although learning this required redoing all of his calculations and finding appropriate visualizations. However, the overstatement (by 0.7%) of the among populations diversity component and the understatement (by 1.3%) of the among races component, relative even just to Table 4 in Lewontin (1972), merits further discussion. Unfortunately, the erroneous distribution (even by his own data tables) of 85.4%/8.3%/6.3% is widely and influentially cited.[40] Now, Lewontin (1974) did correct some of these small calculation errors in its Table 34 (e.g., for the *Lewis* gene), for the same 17 genes as Lewontin (1972), and revised the overarching diversity apportionment to 84.9%/7.5%/7.5%, which is closer to his own data, even if still not quite consistent with it, and ends up rounding in the wrong direction. Even so, work focusing on the apportionment of human genetic variation that cites Lewontin (1972) rarely *also* cites Lewontin (1974).

There is reason for concern about Lewontin's own inappropriate reporting of his (true) results, and of the significantly greater influence of Lewontin (1972) compared to Lewontin (1974): When apportionments are commensurable among populations but within races, and

---

[38] See Table 3 in Lewontin (1972, pp. 390–394). [39] Winther (2022a).

[40] See, e.g., Latter (1980, p. 220), Barbujani et al. (1997, p. 4517), Brown and Armelagos (2001, p. 38), Long and Kittles (2003, p. 450), and Long et al. (2009, p. 23). None of these papers cites Lewontin (1974).

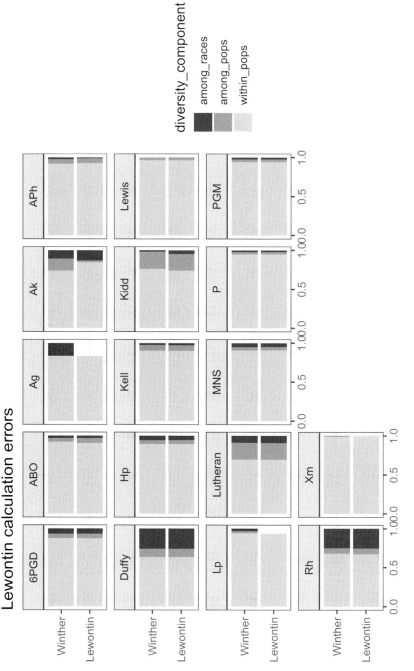

Figure 6.4. Lewontin–Winther bar plots

Bar plots of the three diversity components for each gene (on the same scale), as presented in Table 4 in Lewontin (1972), compared to the true recalculated values from Table 1.6 in Winther (2022a), juxtaposed here in Table 6.2. Lewontin did not calculate "among_pops" and "among_races" values for Ag, Lp, and Xm genes, hence there is some white empty space for those genes. Illustrated by Amir Najmi using ggplot2 in R. © 2021 Rasmus Grønfeldt Winther.

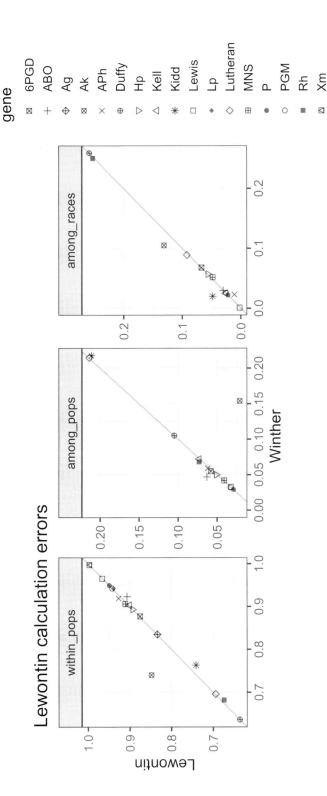

Figure 6.5. Lewontin–Winther scatter plots

Scatter plots of the differences between Lewontin's Table 4 values and the correct values from Table 1.6 of Winther (2022a). The diagonal line maps the identity of value between the two. Scales differ among the three plots because of the different magnitude ranges of the three diversity components (although the among populations and among races components are commensurable in magnitude). The "Winther" (x-axis) and "Lewontin" (y-axis) component values for each gene can be checked against Table 6.2. Note that while the "within_pops" scatter plot shows 17 genes, the "among_pops" and "among_races" scatter plots show only 14 genes because Lewontin had not calculated those respective values for Ag, Lp, and Xm genes. Illustrated by Amir Najmi using ggplot2 in R. © 2021 Rasmus Gronfeldt Winther (A black and white version of this figure will appear in some formats. For the color version, please refer to the plate section.)

among races, Lewontin's harsh and critical stance toward the genetic reality of races becomes incongruous with his permissive and even laudatory stance toward the genetic reality of populations, at least in this context.[41]

How has Lewontin's distribution fared since Lewontin's original study? From a bird's-eye view, it has withstood the test of time in a very general way: All subsequent studies show that most genetic variation is within populations. But zooming in on the details of these studies complicates the picture. We saw in Chapter 5 that F_{ST} depends on a variety of assumptions, and even on gene type studied (e.g., microsatellites versus SNPs). Recall also Long's statement about the oddity of Lewontin – and many others – feeling that F_{ST} on the order of 15% was small.[42] What counts as large or small in this context would seem to be a matter of judgment, as I argue through the conventionalist philosophical position defended in Chapters 4 and 9. And three points bear mentioning regarding Lewontin's actual percentages:[43]

• *Within populations variance component estimates for autosomal genes range from approximately 80% to 95%*: Six key studies indicate the following estimates of the within populations component, showing the maximal ranges, with different values resulting from differences in autosomal gene type, measure, and variance partitioning model used: 80.2–90.8%; 82.7–90.3%; 84.4%; 80.9–87.9%; 89.8–94.6%; and 88.9–94.0%.[44] This is a rather significant range of estimates, even if the average is only slightly higher than Lewontin's 86% within populations component estimate.

---

[41] Gannett (2022) and Winther (2022a) trace Dobzhansky's influence on his doctoral student, Lewontin, in emphasizing the ontological importance and uniqueness of populations.

[42] Long (2009, p. 801).

[43] For the purposes of simplicity, I here abstract away from indicating – for every case – the precise gene types, measures, and variance partitioning models used. The indicated ranges, or single values, are for the point estimates presented; I have not accounted for the standard errors that some publications give for the point estimates. Table 1 in Brown and Armelagos (2001, p. 38) provides a summary of nine different studies, including Lewontin (1972). Table 5 in Jorde et al. (2000, p. 983) summarizes the variance partitioning results for their article, which was "the first published comparison of within- and between-population genetic diversity in autosomal, mtDNA, and Y-chromosome loci in the same set of individuals" (p. 979). Table 4 in Rosenberg (2011, p. 670) presents variance components (and 95% confidence intervals) for each continental region singly, and for the entire globe, itself classified in different ways (i.e., 1, 5, or 7 regions). The partial lists could be much longer.

[44] Respectively, Table 4 in Latter (1980, p. 228), Table 1 in Ryman et al. (1983, p. 97), Tables 1 and 2 in Barbujani et al. (1997, pp. 4517, 4518), Table 5 in Jorde et al. (2000, p. 983), Table 1 in Rosenberg et al. (2002, p. 2382), and Li et al. (2008, p. 1102).

- *Among continents variance component estimates for autosomal genes are sometimes significant, easily exceeding 10% in a number of studies.* Six key studies represent the following among continents component estimates (and, for comparison in parentheses, the among populations but within continents component estimates): 2.8–14.0%; 7.2–15.4% (2.0–5.3%); 10.0–11.7% (3.9–5.5%); 10.4–17.4% (1.3–1.8%); 3.6–5.2% (2.4–5.0%); 3.7–9.0% (2.1–2.3%).[45] Again, these are maximal ranges for each study, with distinct estimates stemming from employing different (autosomal) gene types, measures, and models.

  A number of these studies present a high among continents ("races") component: up to 17.4%. Moreover, the sum of variance components above the population level is itself high, up to 19.8% (calculate this as 1 minus the within populations component). Averaging the range value extremes across all the studies, which itself is a strategy requiring discussion and qualification, produces a grand average of 9.2% for the among continents variance component, and of 3.2% for the among populations but within continents variance component. This is a somewhat higher among continents variance component than Lewontin's distribution, even if the sum of the two components is close to Lewontin's distribution 14% sum. Thus, averaging the studies seems to ground the relative stability of Lewontin's distribution, at least for autosomal genes. But the ranges merit discussion and the consistently higher among continents relative to among populations but within continents estimates belies the typical, easy, interpretation of Lewontin's distribution, calling into question its absolute nature, perhaps even its reproducibility. We might be better off learning the full range of estimates for each of the three variance components, and discussing each of their potential meanings (see Chapter 9).[46]

- *Mitochondrial DNA and Y chromosome among continents variance component estimates are as high as 25%.* The among continents variance component for both mitochondrial DNA (mtDNA) and Y chromosomes can be

---

[45] Identical sources to the previous note. Because Latter (1980) articulates a four-level hierarchy dividing continents into "regional subgroups" (e.g., Europe into "Northern, Central, Southern" and Oceania into "Australia, Melanesia, and Micronesia," p. 224), I here indicate only within populations and among continents estimates for his study. With the globe as one region, Rosenberg et al. (2002) estimated 94.6% within populations and 5.4% among populations.

[46] Furthermore, F_{ST} varies dramatically for different population pairs. Figure 5C in Bergström et al. (2020, p. 6) gives an immediate visual sense of the enormous variability in F_{ST} for all pairs possible of 54 HGDP-CEPH populations. Values can be as low as effectively 0 for some non-African population pairs and as high as roughly 0.28 for some African–non-African population pairs.

remarkably high (see Chapter 2). For mtDNA, key studies found maximal ranges of among continents component estimates of 15.73–21.99%; 12.5%; 22.0–24.9%; and 14.3%.[47] For the among races component of Y chromosomes, researchers estimated 52.7%, 7.8%, and 21.3%.[48] Wilder et al. argue that Seielstad et al.'s (1998) data sets "varied considerably with regard to sample sizes, populations represented and method used to assay genetic variation"; they found instead that "genetic differentiation between populations was similar for the Y chromosome and mtDNA at all geographic scales that we tested."[49] In a nutshell, the among continents variance component estimates for mtDNA and Y chromosomes are significantly higher than for autosomal genes. Furthermore, some studies find higher among continents, and among populations but within continents, genetic variance components for Y chromosomes than for mtDNA, while others find them to be somewhat comparable.[50]

Two patterns require some explication here: first, the significantly larger among continents component of mtDNA and Y chromosome genetic variance as opposed to nuclear, autosomal genes; and second, the differences (such as they are) between mtDNA and Y chromosome among continents, and among populations but within continents, variance components. Regarding the first, due to their haploidy and their uniparental inheritance, the customary, effective population size for mtDNA and Y chromosome loci is one-quarter that for autosomal loci, making genetic drift and consequential population differentiation much more significant, all other factors being equal.[51] Regarding the second, roughly 70% of human cultures and populations are

---

[47] Respectively, Table 3 in Excoffier et al. (1992, p. 486), Table 1 in Seielstad et al. (1998, p. 278), Table 5 in Jorde et al. (2000, p. 983), and Figure 2 in Lippold et al. (2014, p. 7) (this last estimate was not explicitly given in the article and had to be measured from the figure). Again, I abstract away from gene type, measure, and exact methodology.

[48] Respectively, Table 1 in Seielstad et al. (1998, p. 278), Table 5 in Jorde et al. (2000, p. 983), and Figure 2 in Lippold et al. (2014, p. 7) (this last estimate was not explicitly given in the article and had to be measured from the figure). Again, I abstract away from gene type, measure, and methodology.

[49] Wilder et al. (2004, p. 1122).

[50] See, e.g., Gunnarsdóttir et al. (2011). Wilkins and Marlowe (2006) discuss how any such data should be interpreted, why there might be conflicting results for the relative size of the among continents variance components for the Y chromosome and mtDNA, and "how the demographic shift associated with agriculture might affect genetic diversity over different spatial scales" (p. 290; see Chapter 2).

[51] See Hartl and Clark (1989, pp. 425–426), Jorde et al. (2000), and Storz et al. (2001). On effective population size, see Chapter 3.

patrilocal, whereby men tend to stay in their birthplace and women migrate to marry and form new families, at least at a fairly localized scale.[52] A standard anthropological genetics explanation is that "matrilocal groups have high within-group diversity for the Y chromosome and large between-group distances for mtDNA [because diverse men migrate from afar to relatively female-homogeneous villages], whereas patrilocal groups have high within-group diversity for mtDNA and large between-group distances for the Y chromosome [because diverse women migrate from afar to relatively male-homogeneous villages]."[53] Since more human cultures are patrilocal, estimates of the among continents and among populations variance components of Y chromosomes should be larger than the estimates of the same two variance components for mtDNA.[54] However, this might be especially true at more local scales, say, within subregions of continents, whereas at global scales there may be male-biased dispersal and migration patterns, leading to less difference in the Y chromosome and mtDNA among continents variance components.[55]

Lewontin (1972, 1974) did not include mtDNA or Y chromosome data. It is furthermore unclear how we should approach and weight these data in our overall understanding of human population structure. Data on haploid units uniparentally inherited edge us closer toward Galápagos-Writ-Large (see Figure 4.1).

5. Despite Lewontin's Distribution, Clustering Populations And Classifying Individuals Is Possible

Even if most variation is within populations, and Lewontin's distribution roughly and broadly holds, if we accumulate information across loci rather than averaging across loci, we can make somewhat reliable inferences about the contemporary populations that exist (clustering analysis) and about the population membership(s) of any particular individual (classification).[56] *Structure*, the clustering analysis computer program

---

[52] Arias et al. (2018, p. 2719).   [53] Oota et al. (2001, p. 21).   [54] Seielstad et al. (1998).

[55] See Hammer et al. (2001), Wilkins and Marlowe (2006), and Marks et al. (2012). Indeed, other explanatory factors are necessary for a fuller story. I have here side-stepped selection (but see Sayres, 2018, and Chapter 7). Moreover, Heyer et al. (2012) show how differences in effective population size between females and males is also important here.

[56] See Rosenberg et al. (2002), Edwards (2003), as reprinted in Winther (2018a), Witherspoon et al. (2007), Kaplan (2011), Tal (2012), Edge and Rosenberg (2015), Rosenberg (2018), and Winther (2018b). This is what, somewhat unfortunately, has been termed "Lewontin's fallacy" or, less frequently, "Lewontin's paradox." Even if most genetic variation is within populations, clustering

investigated in Chapter 5, was designed to do this.[57] Consider, for example, an extremely simple case of 100 loci spread across a haploid genome (meaning single, not paired, chromosomes). Each locus can have a *G* or an *H* allele. The frequency of *G* alleles is two-fifths in population 1 and three-fifths in population 2 – very much in line with Lewontin's distribution explained above, and analogous to tossing a biased coin (in statistical jargon, *binomial sampling*). Now, imagine that we genotype a person of unknown origin and find that she has a total of 40 *G* alleles across these 100 sites. Around 8% of people from population 1 will have this many *G* alleles, but only about 1 in 40,000 people from population 2 will have this many *G* alleles. We can thus infer with high confidence that the person is from population 1.

Thus, with even small differences in allele frequencies at each locus for distinct populations, we can, by examining enough different loci, become as confident about population membership(s) as we like.[58] Although human populations are often quite similar in allele frequencies at many loci, whenever we *accumulate* cross-loci information of many small differences across populations, we can reliably classify individuals. Knowing an individual's multilocus genotype permits near-certain classification. (Conversely, if we did not know the individual's multilocus genotype, but had to start afresh classifying that individual with every single locus, our classificatory error rate would indeed be high.)[59]

Because clustering analysis and classification are possible and reliable (as we examined in great theoretical detail in Chapter 5), we can use these methods to potentially trace subtle genetic differences between humans from distinct regions, even on Planet Unity (see Figure 4.1). This suggests that even with Lewontin's distribution, we can study many loci to construct population clusters fairly reliably and classify individuals accurately. Even so, as also addressed in Chapter 5, cluster reification is a serious concern.

---

and classification is still possible. In my view, there is no fallacy. As we are seeing here, and as we saw in Chapter 5, different questions are being asked, and different modeling strategies engaged in (e.g., variance partitioning and clustering analysis). See also Roseman (2021).

[57] Pritchard et al. (2000).

[58] See Figure 1 in Edwards (2003), as reprinted in Winther (2018a, p. 251); and Witherspoon et al. (2007).

[59] In Winther (2018b), I provide a thought experiment involving taking metallic objects out of two bags and using each object's weight, sound when struck, color, and so forth to classify from which bag it was taken (the averages and variances of each object property are known for each bag) (p. 5). Interestingly, Lewontin did not have access to multilocus data. The human genome projects emerging especially in the 1990s made oceans of multilocus data available.

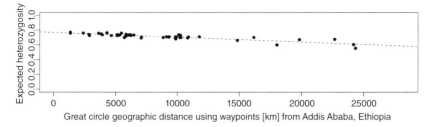

Figure 6.6 Heterozygosity to distance from eastern Africa global pattern
The amount of expected heterozygosity of each of approximately 40 worldwide populations diminishes as a function of their respective distance, along (approximate) migration routes, from Addis Ababa, Ethiopia ($R^2 = 0.7630$). Figure 4A from Ramachandran et al., 2005, p. 15946. © 2005 National Academy of Sciences, U.S.A. Reprinted with permission from *PNAS*.

6. Genomic Heterozygosity Of Populations Decreases With Increasing Distance From Africa, Along Human Migration Routes

Although genetic differences between any two populations tend to be relatively small, the genomic heterozygosity of populations correlates highly with the populations' distance from Africa, along human migration routes. This means one of two things. First, pairs of populations increasingly far from one another along standard human migration routes tend to have increasing levels of heterozygosity with respect to each other (read: the pairs have increasingly higher F_{ST} – they are increasingly genetically different). Moreover, calculating the expected heterozygosity of each single population (see Chapter 4) indicates that populations tend to have lower expected heterozygosities the farther they are from Africa, in part because their relatively small ancestral populations experienced more random genetic drift, which led to increased fixation and thus homozygosity at many loci (Figure 6.6).[60]

The concept of *migration* is particularly useful in this context. It refers to the human migratory routes that helped shape the human genome. But it also refers to *genetic* migration, or the process of gene exchange when populations mix. This empirical pattern illustrates the relationship between these two migratory concepts exceedingly well: In addition to random genetic drift and genetic bottlenecks occurring increasingly

---

[60] This has been confirmed across a variety of studies. Prugnolle et al. (2005), Ramachandran et al. (2005), and Henn et al. (2016) used HGDP-CEPH microsatellite data. So did Tishkoff et al. (2009), who supplemented HGDP data with sampling from, for example, 113 African populations, looking for microsatellites as well as genotyping African individuals for SNPs.

along human migration routes from Africa, genetic migration also tends to diminish along human migration routes as they veer from Africa. Indeed, taking Addis Ababa as the starting calibration point, Ramachandran et al. found a very strong positive correlation ($R^2 = 0.7835$) between the geographic distance (along human migration routes) of any two human populations and the relative population structure – the F_{ST} – of those two populations.[61]

The Out of Africa model, discussed in Chapter 2, nicely accommodates the global data of decline in genomic heterozygosity from our evolutionary cradle. After all, early humans had to pass through north-eastern Africa to leave it. And this empirical pattern clearly points to a cradle in Africa. What is less clear is what to make of the empirical pattern that emerges *in* Africa. After all, similarly strong global correlations are found for effectively *any* calibration point in Africa, a theme also explored in Chapter 2 under the Pan-African Cultural Patchwork hypothesis.[62] This suggests that genomic information is not sufficient for inferring precise regions of human origins, whether in eastern, southern, or, much less likely, western Africa. We also need archaeological and paleontological evidence that can deepen and broaden our analysis. Even so, genomic data alone provide strong evidence for a general Out of Africa scenario of human evolution. But, as with much else in human evolutionary genomics, interdisciplinarity and acceptance of a plurality of empirically adequate models and methodologies help nuance the full picture of our genomic universe and provide many empirical tools for further progress. Although the patterns indicate that Africa is indisputably the capital of our Planet Unity, we also recognize Galápagos-Writ-Large features as a consequence of human migration across the globe and local adaptation to distinct ecological environments that select for different suites of phenotypic characters and their associated (developmentally causal) alleles, as we shall see in Chapter 7.

---

[61] See Figure 1B in Ramachandran et al. (2005, p. 15943), which is a "different depictio[n] of the same phenomenon" (p. 15945) seen in Figure 6.6. See also Figure 7 in Rosenberg (2011), comparing the regression of expected heterozygosity to distance from (a) Nairobi, Kenya (a high correlation, or, technically, *coefficient of determination* [R^2], of 0.865, with heterozygosity declining with distance), (b) Vladivostok, Russia (effectively no correlation), and (c) Montevideo, Uruguay (a correlation of 0.662, with heterozygosity *increasing* with distance). Similar results are found and presented by Serre and Pääbo (2004) and Handley et al. (2007).

[62] Nairobi, Kenya, is used for comparing heterozygosity and distance in Rosenberg (2011). The absolutely strongest correlation in the data of Ramachandran et al. (2005) is found with (approximately) 4°N and 13°E as the starting point (p. 15946). This coordinate lies in Cameroon, in central/western rather than eastern Africa.

Pattern Matters

The six empirical patterns discussed in this chapter provide further means for interpreting the many and varied inputs informing our genomic understanding of ourselves and others, particularly in terms of the similarities and differences that define our identities and origins. How different are we from other species, in particular from our nearest great ape cousins? From where and how did humans emerge, and how do we know? To what extent can we classify individuals into their population(s) of origin? The patterns inform other questions and investigations, too, including those with political and moral weight. For example, are any of the small genomic differences among populations correlated with cognitive phenotypes? While it may be difficult to separate genomic-based findings from politically inflected rhetoric, we must work toward consolidating genomic discourse and attempt to keep it apart from our political lives: Our future as a more-or-less harmonious species may very well depend on such separation.

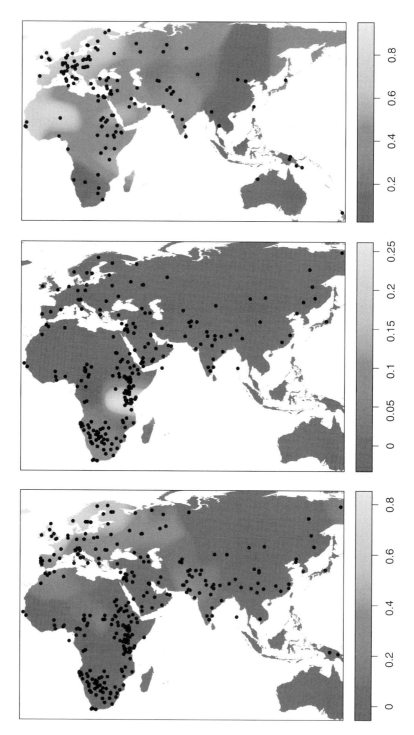

Figure 7.1. Old World distribution of lactase persistence

7 · *Natural Selection*

The six empirical patterns painted in Chapter 6 capture basic features of global human genomic variation. In truth, there are seven such patterns. However, the seventh pattern – natural selection – is so complex and interesting that it merits a longer investigation. In human evolutionary genomics, findings about natural selection emerge, as in the preceding six empirical patterns, primarily from analyses of contemporary, global, and populational genomic variation, even if the genomic investigation of remnants of ancient *Homo sapiens* and of archaic hominins is also increasingly relevant. Although all patterns provide evidence that our species is much closer to Planet Unity than to Galápagos-Writ-Large (see Figure 4.1), relative to the other six patterns, the signatures of natural selection provide far more, and perhaps clearer, evidence for some Galápagos-Writ-Large in our genomes. This is one reason why natural selection has served as the speculative vehicle for some interlocutors to rashly posit that unique selection regimes in local environments are responsible for racial differentiation, not to say racial adaptation.

This chapter frames the basic content of natural selection and its centrality in evolutionary theory. In so doing, natural selection is compared and contrasted with other complementary and contradictory

---

Caption for Figure 7.1. (*cont.*) (Top) Old World distribution of the lactase persistence (LP) phenotype, in percent equivalence (0–1 = 0–100%) of individuals with that genotype in contemporary populations native to the indicated geographic areas. Geographic distribution of −14010 (middle) and −13910 (bottom) alleles, with the color scale representing allele frequencies (note the differences among the three scales). While the functional role of the −13910 allele in causing LP has been plausibly established, there is less evidence for −14010 (Liebert et al., 2017). Contrast with Figure 7.4. Top image: Kindly prepared by and reprinted with permission of Dallas Swallow, UCL. Middle and bottom images: From Liebert et al., 2017, Supplementary Figure 2, without changes; published by Springer Nature under a CC BY 4.0: http://creativecommons.org/licenses/by/4.0/. Adapted by Mats Wedin and Rasmus Grønfeldt Winther. (A black and white version of this figure will appear in some formats. For the color version, please refer to the plate section.)

paradigms and explanations. The chapter then turns to key technical protocols and the current status of our knowledge about the signature of natural selection in the human genome. Finally, the chapter provides criteria for a powerful, integrated overarching evolutionary explanation that helps complement many alternative paradigms through a case study of Bajau "sea nomads" freedivers.

Natural Selection: Distinctions and Complementarity

Classically and simply, *natural selection* refers to individuals of a certain genotype leaving, on average, relatively more offspring in subsequent generations than individuals of another genotype. An individual competes with conspecifics for food, space, and other resources while surviving in the bare elements of its ecosystem. Yet survival alone is not sufficient in the evolutionary epic – reproduction must be secured, too. Here, sexual selection can become relevant, as, classically, males compete with other males for access to females, and females choose the best males. For evolution in a population to occur, there must also be *variation* in rates of survival and reproduction across genotypes, and the phenotypic characters mediating survival and reproduction must be *heritable*. Three conditions thus exist for evolution by natural selection: (1) differential success at survival and reproduction, (2) variation, and (3) heritability.

This calculus of evolution is admittedly harsh, but it does produce *adaptations* – phenotypes that match their environment well, such as camouflage or strong legs by which to hide from or run quickly away from predators. In this way, selection leaves its signature on the genome of all species.[1] Researchers have developed statistical methods to identify the historical action of natural selection by looking at nucleotide sequences, as well as at the relative frequencies of alleles, both across species and within species (including across populations of a species).[2]

A concrete example showing how natural selection has impacted human history might help. Consider the often-cited textbook case of the evolution of alleles for *lactase persistence* across the globe. The "Neolithic Revolution" some 12,000 years ago (e.g., Figure 2.2) included the domestication of animals whose milk can be consumed

---

[1] See Nielsen (2005), Vitti et al. (2013), Enard et al. (2014), Booker et al. (2017), and Stern and Nielsen (2019).

[2] See especially Vitti et al. (2013) and Stern and Nielsen (2019) for reviews covering various kinds of natural selection, including disruptive and balancing selection.

directly or used to make yogurt, cheese, and other products for down-stream consumption. The sugar known as lactose constitutes the main carbohydrate and energy source in such milk. Lactose is broken down by the enzyme lactase, which is produced in the small intestine of human infants, as in many young mammals. In approximately 70% of the human population, however, lactase stops being produced after 2–5 years of age – that is, after weaning. Nevertheless, in many cultures that were historically dependent on milk and dairy products, such as African nomadic pastoralists and Northern European peoples, lactase production persists throughout infancy and adulthood (i.e., lactase persistence). In fact, we know that some of the alleles involved in regulating the lactase gene are especially frequent precisely in dairy-centric agricultural or pastoral populations (Figure 7.1, chapter opener). In short, the cultural context – dairy production and consumption – created an environment that likely gave a selective advantage to individuals with alleles for lactase persistence, thereby spreading this trait in dairy-centric populations over many generations, but not in non-diary-centric populations. This is the standard *cultural evolution* or *niche construction* theory explaining lactase persistence.[3] Indeed, it is very difficult to comprehend evolution by natural selection in human populations without attention to culture and group structure and dynamics; without admitting that genes have multiple and sometimes indeterminate and unpredictable effects, and that their expression depends on environmental factors; without acquiescing to the fact that humans significantly change the selective environment acting on us; and without accepting that a decomposition and periodization of selection in an individual's lifetime cannot be performed as easily for a human as for a fruit fly (Figure 7.2). There are strong methodo-logical and epistemic limits on using natural selection to help us under-stand, explain, and certainly predict human evolution.

Even so, natural selection is foundational to evolutionary theory, and – while respecting explanatory caveats and limitations – can also potentially help to explain some key human traits, such as our relatively large cognitive capacity, opposable thumbs, and bipedalism, in addition to the textbook case of lactase persistence. But natural selection is also an

---

[3] See Cavalli-Sforza et al. (1994), Cavalli-Sforza and Feldman (2003), Nielsen (2009), Gerbault et al. (2011), Ranciaro et al. (2014), Swallow (2015), and Liebert et al. (2017). Admittedly, a *reverse-cause* theory, while unlikely, is possible, in which lactase persistence rose in frequency before the emergence and expansion of dairy production (perhaps through random genetic drift?), setting the stage for a robust culture of dairy production in agricultural and pastoral societies (e.g., Gerbault et al., 2011, p. 865). Niche construction could then kick in.

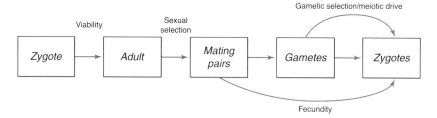

Figure 7.2. Selection during the life cycle
Stages of the life cycle of many animals during which selection may act in different ways (e.g., selection on viability or fecundity, sexual selection, gametic selection). Note resonances with aspects of Figure 3.2, especially the right-most column "survival responses." While fruit flies or flour beetles in experimental containers or in the field may be effectively modeled using such stages, the stages are inappropriate for modeling human evolution, where genetics and biology deeply and inextricably interact with environment and culture. From Hartl and Clark, 1989, p. 148. Reprinted with permission. Adapted by Mats Wedin and Rasmus Grønfeldt Winther.

interesting philosophical case study because it helps us see the role of a multiplicity of explanations, processes, and distinctions in human evolutionary genomics, and in science in general. In investigating complex systems, human evolutionary genomics relies on a plurality of paradigms and their components (explanations, assumptions, distinctions, and so forth). Such a plurality of explanatory and predictive representations and scientific practices requires detailed analysis and inspired integration.

In addition to *adaptationism* – which is the paradigm housing natural selection as the key explanatory factor in evolution – we should also consider alternative overarching evolutionary theory research paradigms such as mechanism and historicism. In *mechanism*, for instance, the composition and organization – that is, the machinery – of a system is disarticulated into parts, and the specific action of these parts in maintaining a functional and integrated whole is investigated systematically.[4] This is the biomedical side of human evolutionary genomics, and it helps us understand how our bodies (and minds) are put together (and how they fall apart). In *historicism*, researchers focus on the origin and history of populations and species, asking which populations gave rise to which other populations and how species have originated and mutually adapted

---

[4] For instance, the structure and mechanisms of hierarchical components of biological systems can be identified and modeled via material experimental protocols and computer simulations, as elucidated by Webster and Goodwin (1996), Craver (2007), Wimsatt (2007), and Winther (2011).

over time.[5] Historicism provides insight into the genealogical and demographic sides of human evolutionary genomics.

Here, as elsewhere, I counsel integration and complementarity. Adaptationism, mechanism, and historicism need not be seen as contradictory or as engaged in a zero-sum competition. Indeed, natural selection is a useful core evolutionary principle carrying enormous power and great influence. As such, it functions somewhat like the ubiquitous and authoritative Mercator projection in cartography and geographic information systems (or GIS; I have much to say about the Mercator and a plurality of other map projections in chapter 4 of *When Maps Become the World*). Analogous to the Mercator projection, natural selection must also be shown to have limits and must also be connected to alternative evolutionary forces, such as random genetic drift or migration, and to alternative overarching paradigms, such as mechanism and historicism. In part, natural selection gained its power via an adaptationist paradigm, a theoretical construct and set of research practices especially influential in the Anglo tradition since at least William Paley's 1802 *Natural Theology* (Paley saw the "fit" of organism to environment as indicative of God's design). Natural selection was institutionalized as the only naturalistic explanation of adaptation by Charles Darwin, most influentially in his 1859 *On the Origin of Species*.

To demonstrate the need for complementarity in a philosophical analysis of natural selection in the context of human evolutionary genomics, I frame this discussion with two devices. The first is a *metaphilosophy of distinctions*, and the second is an integration platform of *paradigm* – or *map* – *pluralism*.[6] These philosophical toolboxes are useful for investigating, in the case of the metaphilosophy of distinctions, the different possibilities available when two or more representations seemingly contradict one another, and, in the case of paradigm pluralism, the existing options for integrating the representations in practice so that they can be made to work together.

In the metaphilosophy of distinctions, *to cut* means "to understand." In fact, to slice reality is to act and even change it. Too often, we distinguish without noticing, but we carve beyond the limits of consciousness when we recognize colors or register different sound pitches. Social class,

---

[5] Felsenstein (2004) and Gaddis (2004) discuss methodologies for historical reconstruction and explanation in the fields of biology and (recent human) history, respectively. See also Crombie (1994), Frodeman (1995), Hacking (2002), and Winther (2012a).

[6] On integration platforms, see Winther (2020a, pp. 107ff.).

gender, and ethnicity are all intentional *cross-cuts* of the social Whole – these three distinction-wielders may produce classifications and groups that can intersect in surprising ways (i.e., intersectionality). Meanwhile, phylum, genus, species, and population are inclusive, *embedded* cuts of the biological One. Conscious or unconscious, individual or social, making and using distinctions is a basic tool for reasoning, for abstracting, and for world-making. Dichotomizing is therefore fundamental to our efforts at comprehending the world.

There are at least two ways to interpret the nature of a distinction: *dualism* or *interpenetration*. According to the first, sharply cut distinctions are *clean and robust differences*. In standard evolutionary theory, natural selection and random genetic drift, two forces we have already met, are often considered clearly distinct and dual with respect to each other. Ignoring other evolutionary forces in our mathematical models of theoretical populations results in clear and distinct equations for how each of these two forces operates in populations. We can thus model their relative effects, observing natural selection overpowering random genetic drift in and across populations as relative fitness differences among genotypes grow larger and, especially, as population sizes increase. Here, the two processes seem clear and distinct, and zero-sum. The force of one excludes the other.

However, according to interpenetration, *distinction poles are continuous and overlap*. Parts may indeed be cut, but they, as it were, meet and, therefore, require each other. In a concluding essay on "Dialectics," Levins and Lewontin expanded on the interpenetration of parts and wholes, of genetic and environmental causes, and of subjects and objects.[7] The poles of a distinction are now seen as mutually constitutive rather than mutually exclusive, and the distinction itself is interpreted as rich and non-dualistic. *Balance* is one turn of phrase sometimes used for this kind of cut. For example, consider the *levels of selection* debate, in which selection is taken to operate at various levels: Individuals compete with one another within populations to secure food and shelter, compete for mates, and leave relatively more offspring (measured in an averaged way by the relative number of alleles that different genotypes leave in the next generation – that is, relative fitness). But different populations or groups of a given species *also* compete by invading or destroying the living conditions for each other, or by sending out relatively more

---

[7] Levins and Lewontin (1985); Winther (2021c) analyzes Levins and Lewontin's concept of *dialectics*.

migrants. Darwin engaged with the family- or colony-level selection of social insects, where the queen is roughly analogous to the ovaries or eggs of an individual, and the worker wasps, ants, and bees are analogous to the body of an individual. In this way, colonies, also sometimes referred to as "superorganisms," compete with each other for food and space in the ecosystem.[8]

Random genetic drift and natural selection are often taken as opposing, dualistic poles. However, in the levels of selection debate, the distinction between individual- and group-level selection is often thought of in terms of interpenetration, because dynamics at the individual level cannot be understood without dynamics at the group level, and vice versa.

Although here we are only looking at the pairing, or yin-and-yang nature, of distinctions, we must also ask what, in this metaphilosophy of distinctions, we can or should do with these distinctions. While distinctions are in themselves useful, by shedding light on what exists, distinctions can also grow knowledge – can even lead us out of the distinction. How can we grow distinctions or develop and build a "thirdness" out of two poles of distinction? I suggest two ways: *neither–nor* and *synthesis*.

In the first case, we accept a distinction but move out of it by identifying another pole that falls outside the spectrum of the distinction. For instance, while many models in population genetic theory are built solely for exploring the consequences of either random genetic drift or natural selection (Figure 7.3) for theoretical populations,[9] other evolutionary forces exist and are also important. Mutation is key to new variation relevant for evolution by natural selection, and migration is important whenever there is some population structure of species, as well as flow of individuals among populations, both of which often exist. Thus, neither–nor thinking offers a key tool by which we might find other poles (e.g., mutation, migration) in the conceptual, multidimensional space of the distinction between natural selection and random genetic drift. In this chapter, I suggest we adopt such thinking. This will ensure that we are not caught in the prisons of single distinctions but, instead, work with, and through, a plurality of distinctions.

---

[8] For further details of levels of selection, see, e.g., Lloyd (1994), Sober and Wilson (1998), Winther (2005), and Winther et al. (2013).

[9] On random genetic drift, see, e.g., chapter 2 in Hartl and Clark (1989); on natural selection, see, e.g., chapter 4 in Hartl and Clark (1989).

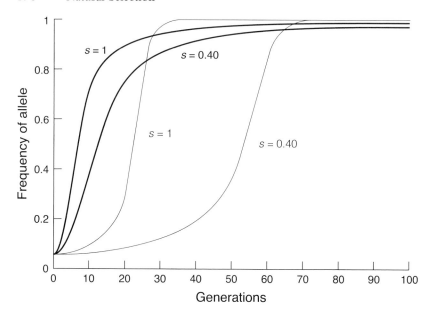

Figure 7.3. An allele under natural selection
A graph of allele frequencies for two different values of higher relative fitness, and when the fitter allele is dominant (thicker lines) or recessive (thinner lines), compared to the other, less fit, alternative allele. Assume that there are only two alleles at this locus. When the allele with the higher relative fitness is recessive, it increases slowly at first because most of the recessive alleles are hidden to selection in the heterozygote, which expresses the phenotype associated with the dominant, less fit allele (two right-most, thinner curves; with $s = 1$, where s is the selection coefficient, implying that the relative fitness is twice as large for the homozygote recessive as for the other two genotypes [i.e., homozygote dominant and heterozygote], and $s = 0.40$, implying that the relative fitness of the homozygote recessive is 1.4 times larger). In contrast, when dominant, the allele with the higher relative fitness increases rapidly at first because it always determines the phenotype, even in heterozygotes; its rate of increase slows down once the recessive, alternative allele is hidden in heterozygotes, allowing these less fit recessives to linger in the population (two left-most, thicker curves, with the same parameters of relative fitness). From Hartl and Clark, 1989, p. 156. Reprinted with permission. Adapted by Mats Wedin and Rasmus Grønfeldt Winther.

We can also move out of a distinction by coordinating, accepting, and synthesizing the features (potentially all the features) from each pole. According to the writings of nineteenth-century philosopher G.W.F. Hegel, we must synthesize thesis and antithesis. That is, both distinction poles function as parts of a larger whole. We see evidence for this in two relevant examples. First, Sewall Wright's shifting balance theory, discussed in Chapter 4, featured distinct phases for random genetic drift and natural selection. He incorporated both forces in his theory because, as he saw it, natural, theoretical, and laboratory populations were frequently within the right size ranges to be small enough for random genetic drift to operate but also large enough for natural selection (at both individual and group levels) to be relevant. Second, even in the levels of selection debate, the individual- and group-selection distinction can be synthesized: Not only do the poles interpenetrate, but they can be modeled simultaneously. In fact, evolutionary biologists can show that selection at the group level can frequently lead to *cooperation* at the individual level.[10] Consequently, while life is certainly competitive, it's not just competitive. It's also cooperative, even according to the adaptationist paradigm.

Of course, no single paradigm can capture the complexity of reality. Imagine a world map in front of you. Which projection does it have? The standard Mercator projection you almost certainly stared at in school, or the polar azimuthal projection found on the United Nations flag? Or perhaps you imagine the iconoclastic Gall–Peters projection, which the influential "dean" of cartography Arthur Robinson described thus: "the landmasses are somewhat reminiscent of wet, ragged, long, winter underwear hung out to dry on the Arctic Circle"?[11] Or perhaps you visualize a different map projection, such as the Mollweide, Peirce quincuncial,[12] or another projection (see, for example, Figure 2.3)? All of these projections are complete in and of themselves, but they are also incomplete representations of complex reality. Just as we cannot force Earth's curved surface onto a two-dimensional surface without discontinuity and distortion, we cannot expect any paradigm or explanation to capture the full complexity of existence.[13] Just as we instead require many map projections to provide a holistic image of our planet, we also require many theoretical approximations, models, and perspectives to understand the complex reality of the evolutionary and developmental processes associated with human evolutionary genomics.

The Genomic Signature of Natural Selection

A handful of key figures in the history of human evolutionary genomics who were specifically interested in the relation between natural selection

---

[10] See Sober and Wilson (1998) and Roughgarden (2009).   [11] Robinson (1985, p. 104).
[12] Winther (2020a, p. 45).   [13] See chapter 4 in Winther (2020a).

and random genetic drift serve as a prehistory and foundation of contemporary work and thinking on the genomic signature of natural selection. During the 1950s and 1960s, Theodosius Dobzhansky, Motoo Kimura, Hermann J. Muller, and Sewall Wright, among others, vigorously discussed the modes of action of natural selection on (human) genetic variation. These theorists disagreed strongly about (a) the relative importance of different kinds of evolutionary forces (e.g., random genetic drift versus natural selection); (b) the extent of genomic heterozygosity, both within and across populations (e.g., loci are mostly homozygous, with little standing variation within populations versus many individuals are heterozygotes, with many different alleles, at a large set of loci); and (c) the magnitude of differences in selection coefficients of alleles of a gene (e.g., small vs. large).[14] Due to limited data, their disagreements were primarily theoretical and from a bird's-eye view on the evolutionary landscape, but even so, there were ample appeals to experimental, field, and medical data and research. Today, these discussions continue, albeit in transformed ways, and with much more, diverse data and theory.[15] As is common in science, and certainly in evolutionary theory, a strong plurality of views exists even within a single paradigm, such as adaptationism.[16] Some theorists almost insist that natural selection is all-powerful, while others emphasize opposing distinction poles, such as random genetic drift or developmental mechanisms.

Today, natural selection – while identified with and in some ways invented by the indefatigable natural historian, breeder, and theorist Charles Darwin – is understood through molecular and genetic study. Perhaps no one was more influential in this arena than Richard Lewontin, whose name we have seen throughout these pages. More than almost any other figure, Lewontin moved the theory and

---

[14] Lewontin (1974); cf. Beatty (1987), the "Classical Versus Balance Hypotheses" section of Winther (2022a, pp. 10–12), and Winther (2021c). For a recent statement defending the ubiquity of various forms of natural selection in shaping the genome and the phenotype – a statement opposing Kimura's *neutral theory of evolution* – see Kern and Hahn (2018). For brief histories of the neutral theory, see Dietrich (1994) and Grodwohl (2018). I would add that the neutral theory – which broadly states that genetic variation within and among species is largely due to the combined forces of random genetic drift and mutation – continues to serve as a useful contrast to a theoretical framework emphasizing natural selection, and is often a starting point for null models in evolution. Moreover, and as outlined in Jensen et al. (2018), there is ample reason to believe that random genetic drift and mutation remain strong evolutionary forces, and that the neutral theory is hardly dead.

[15] See Jensen et al. (2018) and Kern and Hahn (2018).

[16] Cf. Beatty (1997) and Longino (2002, 2013).

experiment of evolutionary genetics toward the molecular and genomic level.[17] His work and influence make him a father, albeit a reluctant one, of theoretical evolutionary genomics. I say "reluctant" because Lewontin was a skeptic about applying the genetic theory, methods, and knowledge he used to investigate *Drosophila*, the fruit fly, to *Homo sapiens*.

Lewontin's aversion is understandable. Theorizing about the nature and consequences of selection in *Homo sapiens* has enormous political and ethical ramifications, to which I have repeatedly gestured, including the possibility of naturalizing ethnic groups and thus imputing hierarchies and power relationships among them; identifying genetic "internal" causes of cognitive differences or disease susceptibilities and thus potentially inviting invasive genetic engineering or even eugenic arguments and actions; or essentializing and reifying identity as foundationally and primarily genomic and biological. I want therefore to be as clear as possible below about how natural selection on genomes – including the human genome – is conceptualized and measured. My hope is to provide readers with both technical knowledge and critical context to start seeing the real, yet limited and not-to-be-overgeneralized role of natural selection on our genome. Many other evolutionary forces are at play, and the standard explanations from evolutionary theory are dramatically insufficient for explaining human evolution, whether in body or mind. Even so, natural selection is an important, partial, and perspectival explanatory force, which can be synthesized with other evolutionary forces and with which we can engage in neither–nor thinking. In addition, natural selection plays an important heuristic role in the many models and theories of evolutionary genetics.

Determining the Legible Signature of Selection

Subsequent to the theoretical disagreements among Dobzhansky, Kimura, Muller, and Wright, and to Lewontin's influential empirical push, statistical and molecular methods for detecting the effects of natural selection in the human genome improved in the early twenty-first century. Initial estimates of adaptive evolution in humans were quite low – even "nonsignificantly different from zero."[18] However, ongoing theoretical efforts have shown that there are statistical reasons for

---

[17] Cf. Winther (2021c). [18] Booker et al. (2017, p. 3), citing Eyre-Walker (2006).

underestimating the amount of adaptive evolution in the human genome.[19] While experimental and theoretical work on other species in this century has found a large molecular signature of natural selection, it is especially within the past decade that incontrovertible selective signatures in the human genome have been identified, such as the case of locating the typically dominant allele (SNP) variants associated with lactase persistence (Figure 7.4).[20]

Evidence suggests that interaction with viruses has been a key selective force in human genomic evolution. Kern and Hahn observe that "careful functional characterization of individual proteins has revealed that a large fraction of all genes which interact with pathogens show pervasive evidence for positive selection."[21] Enard and colleagues found that evolutionary arms races against viruses have driven approximately "30% of all adaptive amino acid changes," when comparing human proteins with the equivalent proteins found in other mammals.[22] Because of this, tracing pathogen transmission routes and understanding the local and global contexts of the emergence of epidemics and pandemics are critical for telling a broader evolutionary genomic story, about humans, about the viruses living in and attacking us, and about the degrading ecosystems in which we live (and yet continue destroying). For example, it seems likely that Neanderthals and modern humans exchanged viruses and that there was also relevant genetic exchange from the former to the latter for genes correlated with adaptive phenotypes involved in fighting viruses. Recall the discussion of Neanderthals from Chapter 2. These two human groups interbred at least when modern humans left Africa and found Neanderthals in Eurasia, where Neanderthals too were fighting possibly new and emergent viruses.[23] Given the often-severe mortality effects of viral epidemics and the repeated, nonstop emergence of such epidemics and pandemics in human evolutionary history, the strong selective signal of genes adapted to interaction with pathogens, especially viruses, is perhaps unsurprising.

This information is increasingly relevant today, when human encroachment on remaining wilderness and human-mediated habitat destruction increase human–animal interactions, providing ample and

---

[19] Messer and Petrov (2013a); on how the various concepts and parameters of effective population size, which we met in Chapter 3, fit into this thematic, see Messer and Petrov (2013b).

[20] Booker et al. (2017) and Kern and Hahn (2018).   [21] Kern and Hahn (2018, p. 1367).

[22] Enard et al. (2016, p. 1); see also Enard and Petrov (2018, p. 360) and Uricchio et al. (2019, p. 981).

[23] See, e.g., Enard and Petrov (2018); see also Figure 2.3.

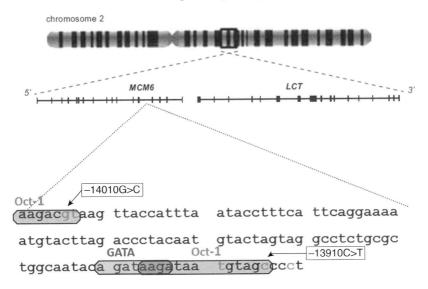

Figure 7.4. Genomic signature of lactase persistence
Positions on the human genome reference sequence, within the lactase gene
enhancer, of single-nucleotide changes −14010 (ancestral G to a derived C) and
−13910 (ancestral C to a derived T). The binding nucleotide sequences for Oct-1
and GATA transcription factors (proteins regulating gene transcription) are shown in
gray ovals. The numbers refer to the enhancer location (in nucleotide sites,
approximately 14 kilobases) upstream of the start codon of the lactase gene (*LCT*).
(The lactase gene enhancer is found within *MCM6*, a gene adjacent to *LCT*.)
The −13910 allele of the *LCT* enhancer has been experimentally established to
upregulate expression of the lactase gene and, thereby, the production of lactase in
individuals with that allele (e.g., Swallow, 2015). The three other lighter-shaded
nucleotides (i.e., the T at −14009, the T at −13915, and the C at −13907) refer
to other LP-associated alleles not discussed here. From Swallow, 2015, p. 21.
© Portland Press. Reprinted with permission. Adapted by Mats Wedin and
Rasmus Grønfeldt Winther.

repeated evolutionary opportunities for pathogens to jump from animal
hosts to humans and evolve human-to-human transmission.[24] In fact,
today, through genetic and molecular engineering, we are finding ways
to design vaccines that will produce planned immune responses. While
salutary from a public health standpoint, we are thereby injecting human

---

[24] Consult Wolfe et al. (2007) and Parvez and Parveen (2017). Already in 2007, Wolfe et al.
proposed "a global early warning system to monitor pathogens infecting individuals exposed to
wild animals" (p. 279).

intentionality into the process of adaptation and selection.[25] Given the increased likelihood and frequency of pathogenic release as humans mercilessly and unwisely hunt down and destroy the last wild habitats on Earth, there is a strengthening relationship between the selection of virus-fighting genes, the deepening of biotechnology, and the evolution, if not bare survival, of modern humans. Genetic and molecular engineering is indeed a likely, partial future of biomedicine.[26]

The signature of selection to lactase persistence, or to viruses, over the course of human evolutionary history is captured as changes in nucleotide sequences, the paradigm, as we have seen, of heritable change. A codon is the three-letter nucleotide sequence corresponding to a particular amino acid. Amino acids are coded for by a variable number of codons – as many as six codons for amino acids leucine and serine, and as few as one for methionine and tryptophan – although four or two is the typical number. The 64 codons are represented in Figure 7.5. (While the content of this figure is frequently called the "universal" genetic code, there are exceptions. Mitochondria and various species of bacteria and yeast have variant genetic codes. Thus, a plurality of code maps is, again, required: Here I have indicated only the most frequent and dominant map.) More specifically, there is a standard biochemical machinery whereby the cell reads parts of the genome. This gene expression system converts nucleotide sequences – which can be thought of as sentences written, as it were, in codon words – via RNA transcription and subsequent translation, to amino acid sequences. The amino

---

[25] Certainly, the coronavirus pandemic that continues wreaking havoc worldwide at the time of this book's production exemplifies what can happen when appropriate, adaptive responses to a virus (in this case, SARS-CoV-2) are absent from the human genome. Callaway (2020) shows how two vaccine development strategies are premised on directly reading the coronavirus RNA genome, which has been tabulated, along with the virus's protein suite, by Corum and Zimmer (2020). In a *viral-vector vaccine*, relevant parts of the SARS-CoV-2 genome are engineered into another, modified virus such as a chimpanzee or human common cold virus, so that the new common cold virus/coronavirus hybrid can be injected into the body and trigger an appropriate immune response that is effective against coronavirus (e.g., the AstraZeneca vaccine uses a chimpanzee common cold viral vector and the Janssen vaccine is based on a human common cold viral vector). In a *nucleic acid (RNA) vaccine*, parts of the coronavirus genome are packaged in a lipid envelope and directly inserted into the body – after the RNA is absorbed by human cells, they produce SARS-CoV-2 proteins, triggering an immune response (e.g., Pfizer-BioNTech and Moderna vaccines). Two other vaccine forms inactivate or weaken *entire* SARS-CoV-2 themselves, or their *protein parts*, respectively. Through genetic, molecular engineering we are becoming ever more effective at guiding our own evolution – and that of our mischievous microbial mates.

[26] As just one example beyond pathogens and disease, see Greely (2016) on the potential future landscape of sex and reproduction in *Homo sapiens*.

Figure 7.5. Universal Genetic Code Mandala
An illustration of the mapping from the 64 three-letter nucleotide codons to the
20 amino acids to which each given codon corresponds. Just as messenger RNA is read
as three-letter codons, from the 5′ to the 3′ end of the nucleotide string, so the
figure should be read from the center out. For instance, codons UUU and UUC
correspond to phenylalanine, UUA and UUG to leucine, and so forth. Standard
three-letter and one-letter abbreviations of each amino acid are presented in the
Mandala. Color coding of the four nucleotides follows the primary colors plus green,
whereas amino acid coloring follows the delicate color scheme humorously provided by
Taylor.[27] Both nucleotide and amino acid color schemes are called "Taylor" in the

---

[27] The Taylor (1997) system is also visually cunning: "hydrophobic amino acids are green ... ; amino
acids found in loops are red and orange ... ; large polar acids are purple and blue" (p. 744). Full
names of amino acids, single letters, and corresponding colors are, clockwise from the Mandala
top middle, with repeats when necessary: phenylalanine (F) (emerald), leucine (L) (grass), serine

acid sequence itself determines the folding – and thereby structure and function – of the protein composed of that amino acid sequence. And, of course, proteins in the form of enzymes regulate the entire set of biochemical reactions of our bodies.

One way to measure the signature of selection involves using gene-based statistical methods for comparing rates – both across and within species – of synonymous to nonsynonymous nucleotide substitutions. In order to understand this distinction, we must turn to point mutations, where just *one* nucleotide in the DNA sequence is replaced (a uracil becomes a guanine, for example). Point mutations are often the outcome of cellular errors in copying DNA, more technically known as *DNA replication*. Such errors can have surprising effects.[28] Consider the example of a point mutation that changes just the first letter of a codon. Although this almost always causes a change in the corresponding amino acid, there are two exceptions, arginine and leucine (Figure 7.5). For instance, both CUG and UUG code for the amino acid leucine, so when a point mutation ends up changing CUG to UUG, the amino acid remains leucine. This is an example of a *synonymous mutation*. The nucleotides change (from C to U) *without* changes to the amino acid, or – all else equal – to the protein amino acid sequence.[29] An analogy for a

---

Caption for Figure 7.5. (cont.) popular MegAlign Pro program used to align two or more sequences of amino acids or nucleic acids: https://www.dnastar.com/manuals/MegAlignPro/17.0/en/topic/selecting-color-schemes. There are three stop codons, indicated by black dots, and one start codon, methionine, indicated by the "play" icon. Concept by Rasmus Grønfeldt Winther, illustrated by Larisa DePalma. © 2021 Rasmus Grønfeldt Winther. (A black and white version of this figure will appear in some formats. For the color version, please refer to the plate section.)

(S) (scarlet), tyrosine (Y) (turquoise), cysteine (C) (yellow), tryptophan (W) (cyan), leucine (L) (grass), proline (P) (tangerine), histidine (H) (peacock), glutamine (Q) (magenta), arginine (R) (blue), isoleucine (I) (lime), methionine (M) (green), threonine (T) (vermillion), asparagine (N) (purple), lysine (K) (indigo), serine (S) (scarlet), arginine (R) (blue), valine (V) (lemon-lime), alanine (A) (lemon), aspartic acid (D) (red), glutamic acid (E) (violet), glycine (G) (orange).

[28] Yet another reminder on terms: Within a species, a point mutation that is sufficiently frequent (e.g., affecting more than 1% of a given population) tends to be called a *SNP* or *variant*, rather than a *(point) mutation* (Karki et al., 2015; cf. the glossary at the end of Chapter 2).

[29] The genetic code is typically written in the language of RNA, which has uracil instead of DNA's thymine. Thus, the mRNA codon UUU (not TTT) was originally read from the complementary DNA's AAA codon. Now, from the perspective of mRNA, this is the "backward" *transcription* direction. But in the genetic code we care about the "forward" *translation* direction, that is, how the mRNA's codons map onto amino acids (e.g., UUU ➔ F/Phe/Phenylalanine; see Figure 7.5).

synonymous mutation is represented in Figure 7.6 – replacing "The" with "Thy" in the sentence does not (really) change the meaning.

If we contrast this with a similar mutation for the other 18 amino acids (i.e., those other than for arginine and leucine), by which a point mutation changes the codon's first letter, the difference is stark. For instance, a mutation from cytosine to guanine makes a difference when we keep the second codon letter U – a valine amino acid is produced rather than a leucine, an example of a *nonsynonymous mutation* (Figure 7.5). In the analogy in Figure 7.6, a nonsynonymous mutation corresponds to the change from "Rat" to "Bat," thereby radically changing the meaning of the sentence.[30] Because changing single amino acids in the amino acid chain often changes the way proteins fold, and because protein structure largely determines biochemical function, a nonsynonymous mutation of a single amino acid can change the sequence's function – in most cases, for the worse. Interestingly, changes to just the *second* codon letter (always) result in nonsynonymous mutations, while any kind of change to just the *third* codon letter always results in synonymous mutations for five amino acids (P, T, V, A, G) and will always result in synonymous mutations for exactly one of the two possible first two codon letter orders for, respectively, three amino acids (S, L, R), but not for the other 12 amino acids (Figure 7.5).

Some statistical tests of selection rely on the distinction between synonymous (i.e., neutral) and nonsynonymous (i.e., nonneutral) mutations, identifying, counting, and comparing these two mutually exclusive and collectively exhaustive types, both for allele substitutions across species and for different alleles (point mutations) within a species (i.e., genetic polymorphisms). Even if somewhat complex, such tests use many of the same basic statistical concepts and methods we have already met, including frequencies, population samples, maximum likelihood analyses, and significance tests. They also permit us to detect the signature of natural selection.[31]

Let's imagine comparing the genomes of average and idealized mice and humans to find allele substitutions across these *two* species.[32] Say we

---

[30] Let's set aside the fact that some changes in amino acids are far worse than others for the ensuing proteins and their effects on fitness. That is, some amino acids are fairly similar in size and in biochemistry, and hence may not change the protein's structure and function too much, making them almost equivalent to synonymous mutations.

[31] See Nielsen (2005), Vitti et al. (2013), and Stern and Nielsen (2019).

[32] Don't worry about intraspecies genomic variation for the calculation of substitutions across species. There are statistical, molecular, and conceptual ways to handle such variation. For starters,

THE CAT ATE THE RAT

Point mutations: one base is substituted for another

TH<u>Y</u> CAT ATE THE RAT	THE CAT ATE THE <u>B</u>AT	THE CAT ATE THE<u>.</u>
- The mutation may not affect the expression or function of a protein	- The mutation may stop the protein from working properly	- The mutation does not code for an amino acid
- These types of missense mutations are *unlikely* to cause disease	- These types of missense mutations *may* lead to disease	- These types of nonsense mutations cause the cell to stop making the protein; therefore, the effects can be *severe*
- In this exanple, the **E** of **THE** has been replaced by a <u>Y</u>; however, we can still understand the meaning of the sentence	- In this example, the **R** of **RAT** has been replaced by a <u>B</u>; the change has completely altered the meaning of the sentence	- In this example, the **R** of **RAT** has been replaced by a <u>FULL STOP</u>; the mutation has ended the sentence prematurely

Frameshift mutation: bases are either inserted or deleted

THE <u>B</u>CA TAT ETH ERA T	THE ATA TET HER AT	CAT ATE THE RAT
- An insertion causes a frameshift; the bases are +1 from where they should be	- A deletion causes a frameshift; the bases are –1 from where they should be	- A deletion causes an in-frame deletion mutation; all the bases are –3 from where they should be
- The effects of these mutations are usually *very severe*	- The effects of these mutations are usually *very severe*	- The effects of these mutations *may* or *may not* be serious
- In this example, a <u>B</u> has been inserted into the sentence between **THE** and **CAT**	- In this example, the **C** of **CAT** has been deleted from the sentence	- In this example, **THE** has been deleted
- The sentence no longer makes any sense!	- The sentence no longer makes any sense!	- As the words are in-frame, we can still understand the sentence

Figure 7.6. Two types of single-site DNA mutations
Of the many kinds of DNA mutations that exist, two occur at a single site: point mutations, where one base is changed to another (e.g., SNPs); and frameshift mutations, where, in the cases shown, either one or three bases are added or removed (a type of indel, which are insertions or deletions of one to many thousands of nucleotide base pairs). © MRC Mitochondrial Biology Unit, University of Cambridge. Published with permission. Adapted by Mats Wedin and Rasmus Grønfeldt Winther.

are only comparing genes for point mutations, and we ignore any frame-shift mutations, which occur when one or more nucleotides are either deleted or added (bottom half of Figure 7.6), thereby completely changing the downstream codons (e.g., "ate" becomes "tet" and "the" becomes "her" in the bottom, middle box of Figure 7.6) and making the meaning of the two sequences quite different indeed. For point mutations at a given locus (gene), calculating the temporal rate of synonymous substitutions requires, in part, assessing the ratio of the number of synonymous differences actually found *to* the total number of nucleotide sites (and ways) where synonymous point differences could potentially happen, across mice and humans – and ditto for nonsynonymous differences.[33] One interesting general pattern is that the rates of the synonymous substitutions of different genes, across species, are roughly similar, while "the rate of nonsynonymous substitution is extremely variable among different kinds of genes."[34] Basically, selection acts on the latter in different degrees and directions, but it does not act on the former. Would we expect *higher rates* of synonymous or nonsynonymous substitutions between mice and humans, both for particular genes and across genes? When I first thought about this question some years ago, I expected that there would be higher rates of synonymous substitutions, because most nonsynonymous mutations are deleterious and would be quickly eliminated by negative or purifying natural selection. After all, I had learned in my first evolution course that most mutations "visible" to selection (i.e., nonsynonymous) were deleterious.[35] All of this is

---

intraspecies variation is already quite low for many of these studied genes (e.g., empirical pattern #1 in Chapter 6), and there are statistical methods for averaging inter-individual variation at particular loci.

[33] Exactly how to count and assess the total number of sites, including how to avoid histories of multiple substitutions at single sites is tricky. See, e.g., Yang and Bielawski (2000), Yang and Nielsen (2000), and Nielsen and Slatkin (2013).

[34] Li et al. (1985, p. 160). Li and colleagues's well-designed work accounted for historical divergence times among mammalian orders (e.g., primates, especially humans; artiodactyls, in particular cows, goats, and pigs; and rodents, specifically mice, rats, and Chinese hamsters) and also averaged out somewhat different within species nucleotide sequences, and their careful findings reflect that (Li et al., 1985).

[35] The McDonald–Kreitman test is one key evolutionary genomic test for detecting the signature of selection, and I can only partially explain it here. The McDonald–Kreitman test simply tests the null hypothesis of *equal* nonsynonymous to synonymous substitution ratios within and between species. If all mutations are neutral, then the null hypothesis should hold; if we find we must reject the null for a given locus, then (positive or negative) selection has likely been detected. This statistical test is a specific way to use nonsynonymous and synonymous mutations that does not per se involve directly calculating rates of substitution. The McDonald–Kreitman test is one way – among many – of using nonsynonymous and synonymous mutations to test the neutral

Table 7.1. *Comparison of substitution rates between humans versus rodents or artiodactyls (cows, goats, and pigs)*

Gene	Synonymous substitution rate[a]	Nonsynonymous substitution rate[a]
Growth hormone	4.37	0.95
Prolactin	5.59	1.29
α-hemoglobin	3.94	0.56
Albumin	6.72	0.92
Average[b]	4.65	0.88

The data here help to illustrate points discussed in the text.
[a] Substitution rates are in units of substitutions per site per billion years.
[b] Average of substitution rates for all 40 genes in Table 2 of Li et al. (1985).
Data from Table 2 in Li et al. (1985, pp. 158–159) and Table 8.2 in Nielsen and Slatkin (2013, p. 165).

indeed the case for mice and humans (Table 7.1), even as synonymous, neutral mutations accumulate in our genomes, and even if positive selection is also possible and statistically detectable.[36]

My peers and I also learned that, all else equal, alleles that quickly increase in frequency, or that are found at significantly different frequencies in different populations of a *single* species (say, at 10% in one population and at 80% in another), could be under selection. While it is true that nonsynonymous (nonneutral) mutations are often eliminated, the story is a bit more complicated and nuanced. For one, other evolutionary forces, including random genetic drift, can also change allele

---

null hypothesis, and thereby identify the signature of selection. See McDonald and Kreitman 1991, who worked on *Drosophila*; explanation in Nielsen and Slatkin (2013, pp. 183–184), Vitti et al. (2013, pp. 100–103), Booker et al. (2017, pp. 1–3), and Stern and Nielsen (2019, pp. 406–407). Rasmus Nielsen provided constructive feedback here.

[36] As Vitti et al. (2013) explain, all other things equal, a "relative excess" of synonymous to nonsynonymous substitution rates for a locus (gene) indicates negative or purifying selection (across species, selection preserves, to the extent possible, protein structure and, relatedly, as we saw above, amino acid composition and order; equivalently, nonsynonymous mutations tend to be deleterious), while a relative excess of nonsynonymous substitution rates for a locus indicates positive selection "favoring novel protein structures" (Vitti et al., 2013, pp. 100–102; cf. Nielsen and Slatkin, 2013, pp. 180–181). Vitti et al. (2013) also confirm that "synonymous substitutions . . . tell us about the background rate of evolution" (p. 100); synonymous substitution rates roughly serve as one kind of molecular clock (cf. Chapter 2).

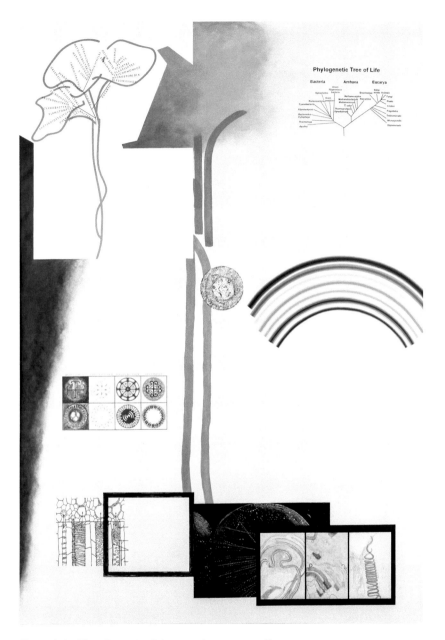

Figure 1.1. Plant Perception 2 (see caption on page 0)

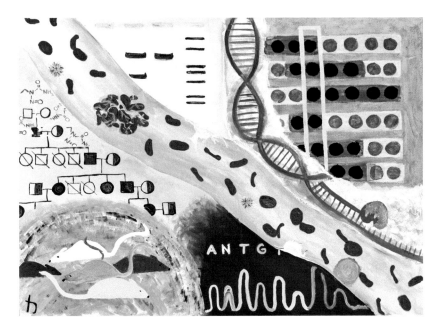

Figure 1.2. Acrylic Genetics (see caption on page 9)

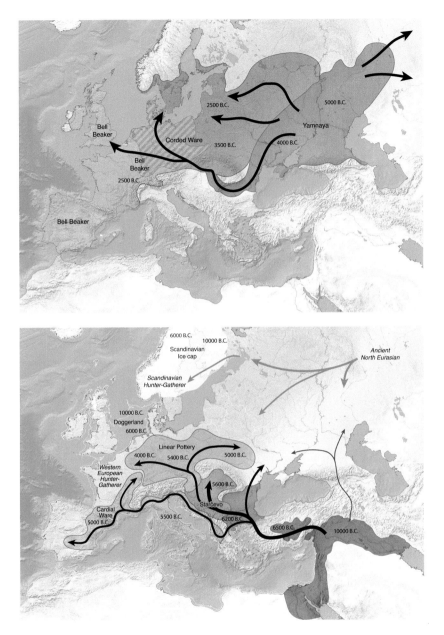

Figure 2.2. Making Europe (see caption on page 36)

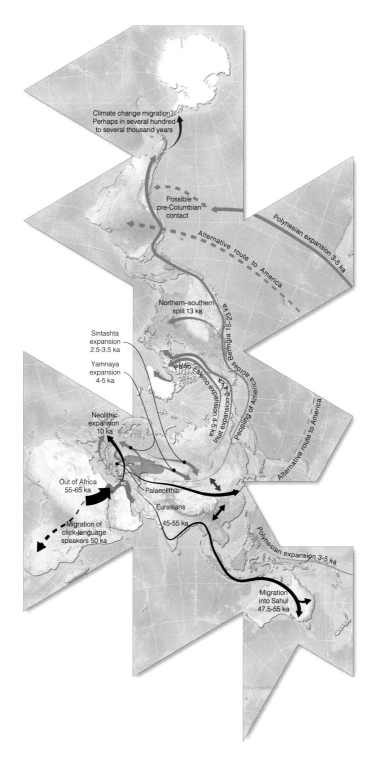

Figure 2.3. Out of Africa global migrations (see caption on page 39)

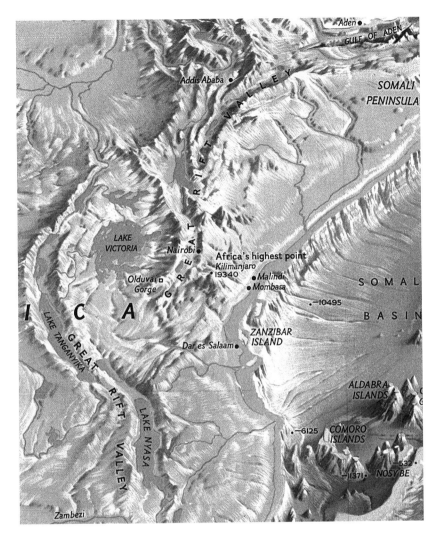

Figure 2.4. The East African Rift System (see caption on page 41)

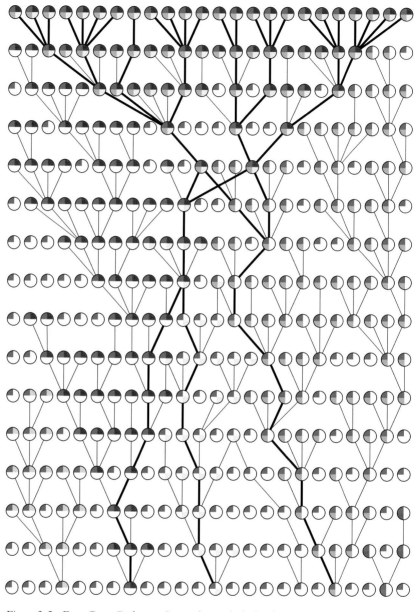

Figure 2.5. *Four Gene Coalescent (Incomplete and Idealized)* (see caption on page 45)

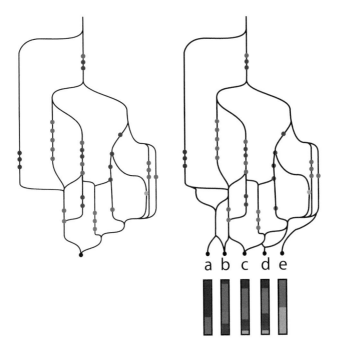

Figure 2.7. Ancestral recombination graph (see caption on page 51)

Figure 2.8. Hominin time spiral (see caption on page 53)

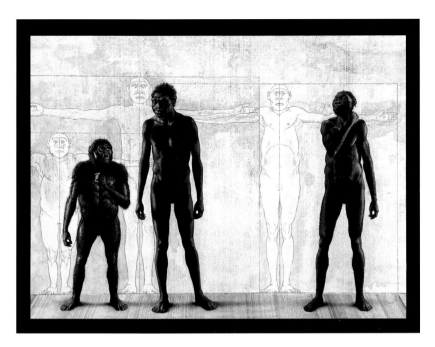

Figure 2.15. Vitruvian Hominins (see caption on page 66)

Figure 3.1. A population of *Australopithecus (Praeanthropus) afarensis* (see caption on page 72)

Central & South Asian
(Previously South Asian, no change)   100%

- Central & Southern Indian   41.4%

  This genetic signature is more common in southern Indian states such as Tamil
  Nadu, Andhra Pradesh, Karnataka, Kerala, and Telangana. Do you have this
  ancestry? Tell us more about your family's heritage in the feedback section below
  to help us improve this report.

- Kannadiga, Tamil, Telugu & Sri Lankan   24.5%

  Sri Lanka and the southern states of India harbor evidence of human activity
  going back thousands of years. Within the past two millennia, southern India saw
  great Hindu kingdoms come and go, including the Chola Dynasty and
  Vijayanagar – whose capital, Hampi, was among the largest cities of the
  medieval world. A genetic signature unique to South Asia reaches its highest
  levels in the region, and may be related to the spread of Dravidian languages,
  although the majority of Sri Lankans now speak Sinhalese, an Indo-European
  language.

- North Indian & Pakistani   10.3%

  Today, around one-seventh of the world's people live along or near the basins of
  the Indus and Ganges Rivers. Most North Indian & Pakistani people speak Indo-
  European languages brought to the region from Central Asia around 4,000 years
  ago during the decline of the Indus Valley civilization. More recently, this region
  was home to a series of powerful empires, including the Gupta Empire
  (representing a golden age in Hindu art, literature, and science) and the Islamic
  Mughal Empire, which gave rise to many Indo-Persian architectural wonders like
  the Taj Mahal.

CENTRAL ASIAN

BANGLADESHI & NORTHEAST INDIAN

CENTRAL & SOUTHERN INDIAN

KANNADIGA, TAMIL, TELUGU
& SRI LANKAN

NORTH INDIAN & PAKISTANI

GUJURATI PATELS

MALAYALI

Figure 5.3. DNA ancestry report (see caption on page 135)

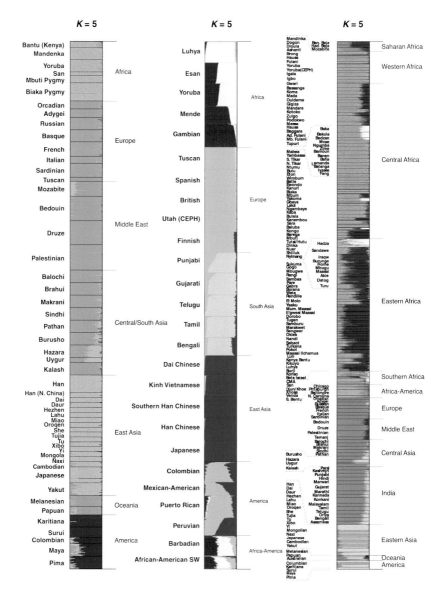

Figure 5.4. K = 5 clustering comparison (see caption on page 155)

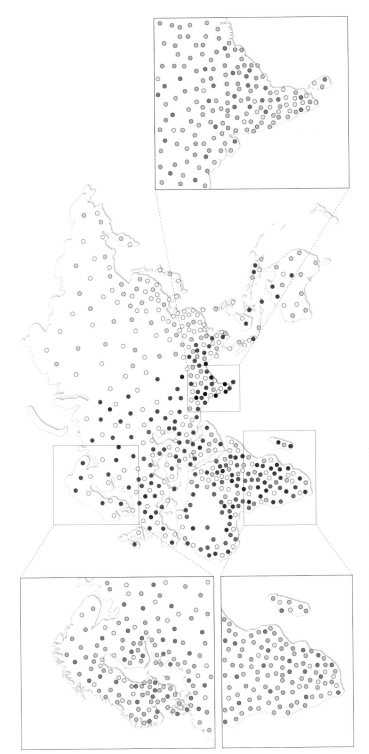

Figure 6.1. A World of Allelic Color (see caption on page 164)

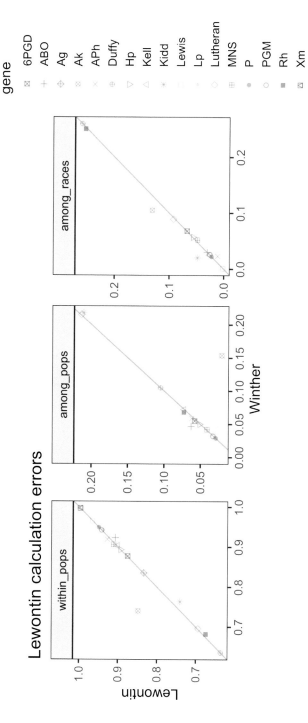

Figure 6.5. Lewontin–Winther scatter plots (see caption on page 181)

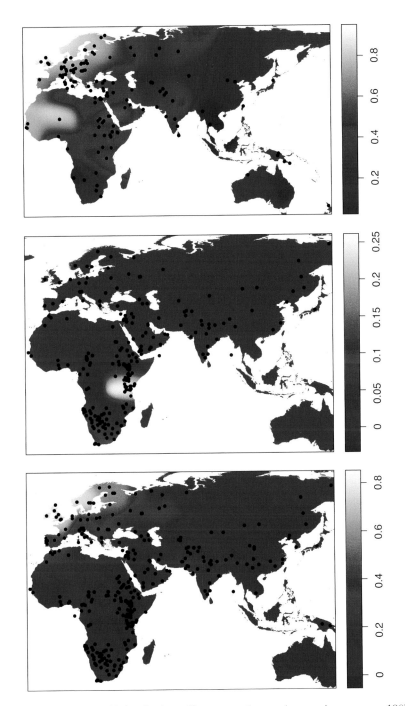

Figure 7.1. Old World distribution of lactase persistence (see caption on page 190)

Figure 7.5. Universal Genetic Code Mandala (see caption on page 205)

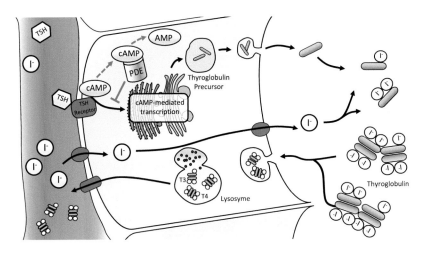

Figure 7.9. Mechanisms of PDE10A action (see caption on page 220)

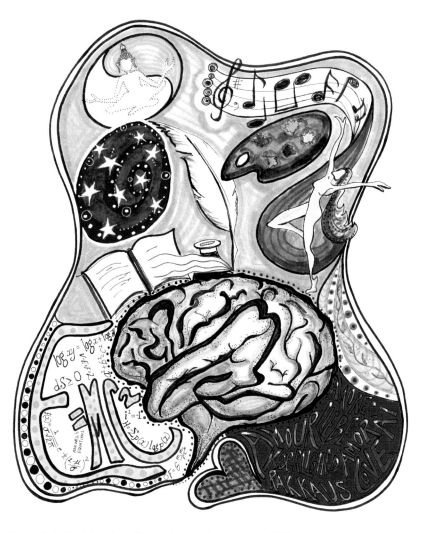

Figure 8.4. Varieties of Intelligence (see caption on page 239)

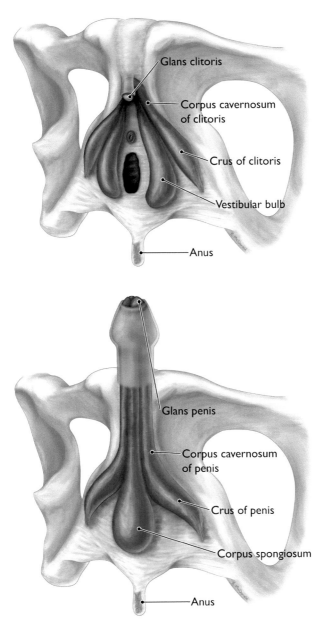

Figure 8.5. Adult human genitalia (see caption on page 250)

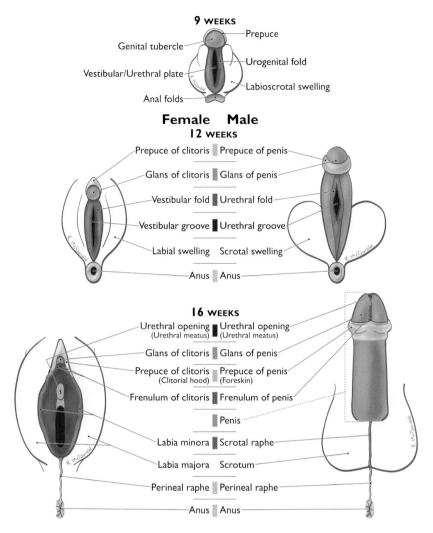

9 WEEKS

Genital tubercle —
Vestibular/Urethral plate —
Anal folds —

— Prepuce
— Urogenital fold
— Labioscrotal swelling

Female   Male
12 WEEKS

Prepuce of clitoris | Prepuce of penis
Glans of clitoris | Glans of penis
Vestibular fold | Urethral fold
Vestibular groove | Urethral groove
Labial swelling | Scrotal swelling
Anus | Anus

16 WEEKS

Urethral opening | Urethral opening
(Urethral meatus) | (Urethral meatus)
Glans of clitoris | Glans of penis
Prepuce of clitoris | Prepuce of penis
(Clitorial hood) | (Foreskin)
Frenulum of clitoris | Frenulum of penis
| Penis
Labia minora | Scrotal raphe
Labia majora | Scrotum
Perineal raphe | Perineal raphe
Anus | Anus

Figure 8.6. Development of human genitalia (see caption on page 252)

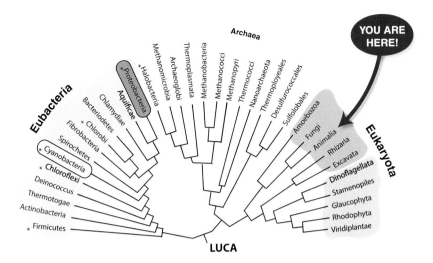

Figure 10.4. Branching life (see caption on page 310)

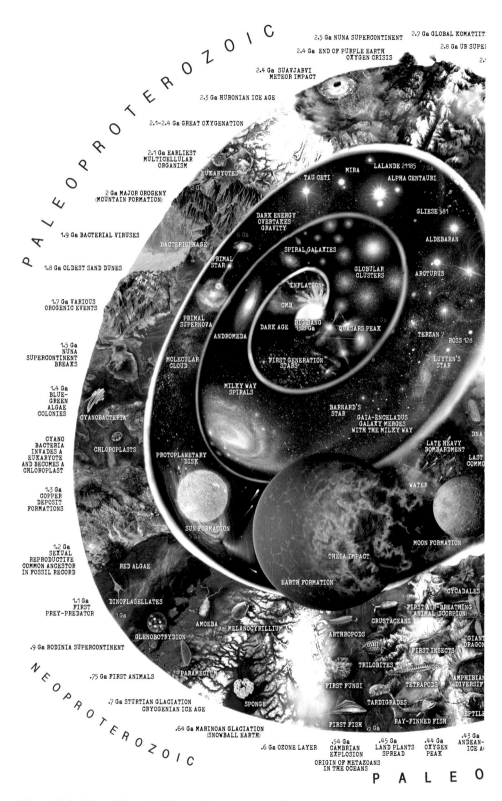

Figure 10.2. Nature Timespiral (see caption on page 290)

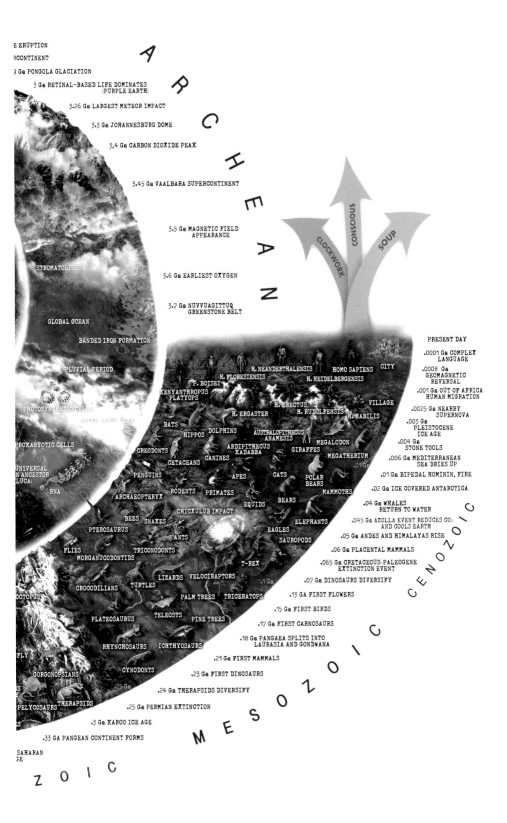

E ERUPTION

RCONTINENT

3 Ga PONGOLA GLACIATION

3 Ga RETINAL-BASED LIFE DOMINATES
(PURPLE EARTH)

3.26 Ga LARGEST METEOR IMPACT

3.3 Ga JOHANNESBURG DOME

3.4 Ga CARBON DIOXIDE PEAK

3.45 Ga VAALBARA SUPERCONTINENT

3.5 Ga MAGNETIC FIELD
APPEARANCE

3.6 Ga EARLIEST OXYGEN

3.7 Ga NUVVUAGITTUQ
GREENSTONE BELT

A R C H E A N

STROMATOLITES

GLOBAL OCEAN

BANDED IRON FORMATION

PLUVIAL PERIOD

PHOTOSYNTHETIC CELLS

PROKARYOTIC CELLS

UNIVERSAL
N ANCESTOR
(LUCA)

RNA

ARCHAEOPTERYX

BEES
PTEROSAURUS   SNAKES

ANTS

FLIES
MORGANUCODONTIDS   TRICONODONTS

CROCODILIANS   TURTLES   LIZARDS   VELOCIRAPTORS

OCTOPUS

PALM TREES   TRICERATOPS

PLATEOSAURUS   TELEOSTS   PINE TREES

RHYNCHOSAURS   ICHTHYOSAURS

FLY

GORGONOPSIANS   CYNODONTS

PELYCOSAURS   THERAPSIDS

SAHARAN
GE

Z O I C

CONSCIOUS

CLOCKWORK   SOUP

PRESENT DAY

.0001 Ga COMPLEX
LANGUAGE

.0008 Ga
GEOMAGNETIC
REVERSAL

.001 Ga OUT OF AFRICA
HUMAN MIGRATION

.0025 Ga NEARBY
SUPERNOVA

.003 Ga
PLEISTOCENE
ICE AGE

.004 Ga
STONE TOOLS

.006 Ga MEDITERRANEAN
SEA DRIES UP

.01 Ga BIPEDAL HOMININ, FIRE

.02 Ga ICE COVERED ANTARCTICA

.04 Ga WHALES
RETURN TO WATER

.045 Ga AZOLLA EVENT REDUCES CO$_2$
AND COOLS EARTH

.05 Ga ANDES AND HIMALAYAS RISE

.06 Ga PLACENTAL MAMMALS

.065 Ga CRETACEOUS-PALEOGENE
EXTINCTION EVENT

.07 Ga DINOSAURS DIVERSIFY

.13 Ga FIRST FLOWERS

.15 Ga FIRST BIRDS

.17 Ga FIRST CARNOSAURS

.18 Ga PANGAEA SPLITS INTO
LAURASIA AND GONDWANA

.21 Ga FIRST MAMMALS

.23 Ga FIRST DINOSAURS

.24 Ga THERAPSIDS DIVERSIFY

.25 Ga PERMIAN EXTINCTION

.3 Ga KAROO ICE AGE

.33 Ga PANGEAN CONTINENT FORMS

CITY

H. NEANDERTHALENSIS   HOMO SAPIENS
H. FLORESIENSIS   H. HEIDELBERGENSIS
P. BOISEI
KENYANTHROPUS
PLATYOPS   H. ERECTUS   VILLAGE
H. ERGASTER   H. RUDOLFENSIS
H. HABILIS
BATS
HIPPOS   DOLPHINS   AUSTRALOPITHECUS
ANAMESIS
CREODONTS   ARDIPITHECUS   MEGALODON
KADABBA   GIRAFFES
CETACEANS   CANINES   MEGATHERIUM
PENGUINS   APES   CATS   POLAR
BEARS
RODENTS   PRIMATES   MAMMOTHS
EQUIDS   BEARS
CHICXULUB IMPACT   ELEPHANTS
EAGLES
SAUROPODS
T-REX

MESOZOIC   CENOZOIC

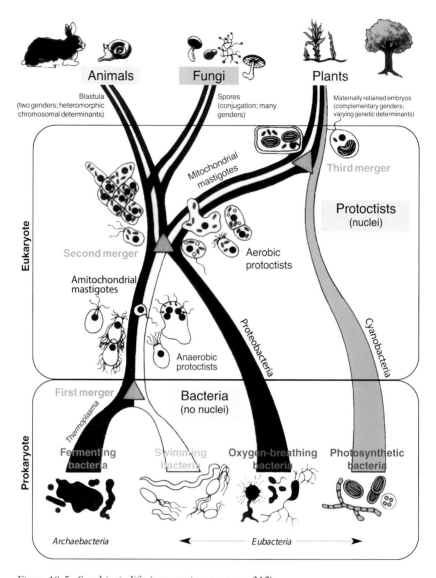

Figure 10.5. Symbiotic life (see caption on page 312)

frequencies. Thus, statistical methods have been developed to detect when selection is the only, or the main, force at play. Some of these methods examine patterns of allele frequency distributions at *neighboring* loci, which may be "hitchhiking" with the locus under putative, positive selection. Such loci tend to be co-inherited together, exhibiting *linkage disequilibrium*.[37] Loci hitchhiking with a locus under positive selection undergo a "selective sweep," which can be identified – they will have reliably different patterns of allele frequencies than loci that are simply changing allele frequencies due to mutation and random genetic drift, and these patterns can be detected statistically.[38] In summary, there are different models and methodologies for detecting selection across species or within a single species.

In my introductory explanation of the importance of natural selection as an empirical pattern (and of its imprint on the human genome), I suggested that findings associated with natural selection are often co-opted to serve political ends. Genes associated with skin color variation serve as an apt example. The gene *SLC24A5*, which influences skin color variation, has been identified as potentially under strong selection.[39] The ancestral allele associated with dark skin is practically fixed – that is, present in almost all individuals – in most African (and most East Asian, Southeast Asian, and Indigenous American) populations. The derived allele, highly correlated with light skin, is effectively fixed in most Middle Eastern and European (and many Central Asian) populations. Thus, this is a case in which the allele frequencies vary wildly among populations, thereby increasing F_{ST}, which is also a measure used in some of the statistical methods for detecting selection.[40] Importantly, *SLC24A5* was studied for its unusually divergent geographic distribution and its link to known molecular mechanisms of

---

[37] Cf. Clark et al. (2003) and Vitti et al. (2013); on linkage disequilibrium more generally, see Slatkin (2008).

[38] For discussion, see Nielsen and Slatkin (2013, pp. 166–171), Vitti et al. (2013, pp. 104–108), and Stern and Nielsen (2019, pp. 399–404). Stern and Nielsen (2019) list three features of selective sweeps that can be identified with appropriate statistics and mathematical evolutionary genomic theory and are evidence for selection if found: "decreased variability (e.g., the number of segregating sites), increased identity by descent (IBD; i.e., DNA sequence identity due to recent common ancestry), and increased haplotype homozygosity" (pp. 401–402), reference omitted.

[39] Sabeti et al. (2007) and Table 2 in Vitti et al. (2013, p. 113).

[40] Recall *Duffy* from Chapter 6, which is also under strong positive selection (malaria) and therefore exhibits unusually large variance components among populations but within continents and among continents.

melanin production.[41] The full explanation for selection on skin color is not yet completely clear.[42]

Another epicenter of interest regarding the impact of natural selection on the human genome concerns genes for human cognitive abilities and behavioral features. Despite journalistic yarn-spinning[43] – which occasionally has been informed by genes such as *FOXP2*, the so-called language gene – our knowledge about causal links between genes and actual behavioral features remains scant.[44] We do not yet know if there are certain genes (rather: alleles) for cognitive abilities or behavioral traits "hidden" in the relatively small fraction of private common alleles or in the among continents variance component, which we explored in detail in Chapter 6. As with other questions about natural selection, this is clearly an area of immense political consequence for our best genomic biology, to which we shall return in Chapters 8 and 9. (In contrast, information about genes associated with physiological or morphological variation is often less fraught.[45])

Regardless of the genes at the center of the work, investigating natural selection is complex, and interpreting our findings is always charged and limited. In fact, even when we identify a gene that seems to have been the target of selection for a certain phenotypic trait, we cannot rule out that our target allele might be pleiotropically causative of one or more other phenotypic characters (themselves under selection). Hence, our target allele may have increased in frequency *not* because it caused (alternatively, was associated with) the candidate phenotypic trait, but because it caused *another* trait that was under selection. We must therefore have a full picture of all selection on all traits (a perhaps impossible achievement) to gain this kind of insight. To take a different set of phenomena, it is also possible that, for a given polygenic character involving many genes, genes other than the one we study are more

---

[41] On the first, see Sabeti et al. (2006, p. 1619); regarding mechanisms, see Lamason et al. (2005), who "present evidence that the human ortholog of a gene associated with a pigment mutation [the so-called "zebrafish *golden* phenotype"] in zebrafish, *SLC24A5*, plays a role in human skin pigmentation" (p. 1782). Cf. www.ncbi.nlm.nih.gov/gene/283652.

[42] Wilde et al. (2014) and Szpak et al. (2019, p. 1435).

[43] Wade (2014); effectively reviewed by Orr (2014).

[44] Enard et al. (2002) and Table 2 in Vitti et al. (2013, p. 113); see also Table 1 in Szpak et al. (2019, p. 1434) for partial lists of other genes under selection, including *EDAR*, for which "the derived allele was shown to associate with increased scalp hair thickness . . . and incisor tooth shoveling . . . in multiple East Asian populations" (Szpak et al., 2019, p. 1436), references omitted.

[45] For instance, altitude adaptation in Tibetans, perhaps caused by gene mixing with archaic hominins (Huerta-Sánchez et al., 2014); see also Sabeti et al. (2006).

causally explanatory of the trait itself, or interact epistatically with the gene under study.[46] The finger of selection is complex and evasive, and deducing its action from the fingerprints it leaves on the genome is a humbling task.

Even so, clear evidence and theoretical methods are available that indicate natural selection has acted on the human genome.[47] Philosophical and critical work can contribute to placing natural selection in its appropriate theoretical and methodological context, in evolutionary theory in general, and in human evolutionary genomics in particular. We must keep in mind that natural selection has its limits, and other complementary, standard evolutionary forces, such as random genetic drift and migration, are also powerful (as recognized by both Kimura's neutral theory and Wright's shifting balance theory). Moreover, mechanistic and developmental features must be integrated with natural selection in order to understand, predict, and explain human evolution. It is to this topic that I now turn.

---

[46] Statistical methods for modeling genetic and phenotypic correlations exist, although the idea goes back at least to Darwin's notion of *correlation of growth* (Darwin, 1964 [1859], pp. 143ff.; see also Winther, 2000). One can build a *variance–covariance matrix* for genotypes as well as phenotypes, indicating, for example, the amount of genetic correlation between phenotypic traits. To the extent that there is shared genetic variance across traits – a population-level signal of genetic pleiotropy – selection on one trait will indirectly lead to population distribution changes on other traits (e.g., Lande and Arnold, 1983; Goswami et al., 2014; in their chapter 19 on "Correlated Characters," Falconer and Mackay, 1996 deploy concepts such as *cross-variance* for genetic correlations, and *coheritability* for phenotypic correlations in the context of direct and indirect [i.e., correlated] responses to selection; see also chapter 21 of Lynch and Walsh, 1998). A caveat: Genetic correlations are generally theorized as occurring among loci that each have an additive, independent effect on the phenotype. But this picture is complicated by *epistasis*, where gene action is intrinsically interactive and mutually dependent. The concept of epistasis has been used in multiple ways (e.g., Cordell, 2002; Wade, 1992, 2002, 2016; and Wade et al., 2001), for instance, to describe when the effects of (variance in) one genetic locus on the phenotype or fitness depends on (variance in) one or more other genetic loci. Recent work on detecting and disentangling selection on polygenic traits includes Racimo et al. (2018), Edge and Coop (2019), and Stern et al. (2021).

[47] Again, this is neither to deny the neutral theory, either in content or as a null model (e.g., Jensen et al., 2018), nor to defend methodological or empirical adaptationism in the sense of Godfrey-Smith (2001), following Gould and Lewontin (1979). As I understand Godfrey-Smith's (2001) categories, a *methodological* adaptationist commits to the assumption that biological research is best carried out by understanding organisms or cells, say, as bundles of adaptations (even if they are not, in all respects), whereas an *empirical* adaptationist holds that most phenotypes, or most parts of phenotypes, are adaptive, and that natural selection is, to a large extent, omnipresent and omnipotent.

Case Study: Freediving Physiology

Tibetans living on the roof of the world with rarefied low-oxygen air. Inuit braving the coldest Arctic, thriving on a diet rich in fats. The Bajau "sea nomads" freediving in the tropics with ease. Each of these peoples have lived for many generations in extreme environments. As a result, they sport adaptations distinguishable from those of neighboring or historically related populations inhabiting less ruthless environments. These adaptations have also likely left a genetic signature in the form of alleles mechanistically and statistically associated with the adaptive phenotype. Scientists have found that *Homo sapiens* populations inhabiting extreme environments – for even just hundreds or thousands of years – are ideal places to search for the action of natural selection on our bodies.

Indeed, the evidence has flooded out. Investigating the potential genomic signature of natural selection in these populations can provide insight into specific phenotypic targets of selection in human evolution and, in so doing, indicate which other components of an evolutionary explanation are necessary.

One of the clearest examples of natural selection as a shaping influence can be found among the Bajau peoples of Southeast Asia. This case involves a physiological set of phenotypic adaptations of the spleen, with strong genomic correlates, specific to the mode of life of the Bajau population and understandable as an adaptation to extreme environments. In my view, it is possibly the best and most paradigmatic case – in its singularity, uniqueness, and completeness – of human evolutionary genomics to date. This case satisfies what I take to be six crucial elements or components of an adaptationist explanation in human evolutionary genomics. Taken together, these six components form a rubric for adaptationist total explanation: There is convincing evidence for a clear and distinct *phenotypic trait*, strongly associated with a small set of individuated *candidate genes*.[48] The *mechanisms* involved in the production of the phenotype from genes – and in the action of the gene product protein or phenotype – can be traced developmentally and physiologically, and natural selection has been plausibly shown to have operated on this phenotypic trait. In other words, a clear *adaptive scenario* has been

---

[48] For instance, we have database information for the gene *SLC24A5*, which is also mechanistically associated with skin pigmentation in humans (Lamason et al., 2005; www.ncbi.nlm.nih.gov/gene/283652).

articulated, and there is *evidence of natural selection*.[49] Comparative *historical evidence* for populations lacking both the trait and its correlated genes has also been provided, further confirming the selective scenario of that phenotype and associated genes.[50]

The Bajau or Sama-Bajau or Sama Dilaut is a nomadic and seafaring ethnic group living along the coasts of parts of the Philippines, Malaysia, and Indonesia. Many Bajau individuals can freedive down to depths of 60–70 m on a single breath for up to almost a quarter hour, walk calmly on the sea floor, and spear fish with perfect aim. Wooden googles or masks, and sometimes weight belts, are their only attire – no fins, neoprene suits, or other modern diving technologies. Swim with them or watch a video and be impressed. As an avid scuba diver, I can attest to their impressive prowess, equaled only by a small number of other individuals globally.

Their exact geographic origin in this region is controversial, but the Bajau probably emerged as a people only some 1000–1500 years ago (Figure 7.7). They are known as "sea nomads" because they are so closely tied to the oceans, having traditionally lived on houseboats with stilts or even on boats, and eating a diet rich in fish and other sea life acquired by the community's many freedivers. However, the Bajau traditional way of life has been increasingly lost over the past few decades. While varying according to national and local contexts, resettling and migration to cities leads to attrition. Reasons for such shifts include heavy-handed government action; an ongoing desire, especially on the part of young individuals, to avoid stigmatization and oppression; and general globalizing and

---

[49] Evidence for selection in broader morphological studies – not per se human evolutionary genomics – can include using functional morphological design principles for bone arrangements in vertebrates. See, e.g., Winther (2006b).

[50] Another way of carving up a total adaptationist explanation can be found in Sinervo and Basolo (1996, p. 150) and is used by Lloyd (2005a, pp. 4–7). Sinervo and Basolo argue that "variation in a trait must have a genetic basis" and that the trait under investigation "must influence fitness or a component of fitness"; moreover, there must be "mechanistic links . . . between the trait and the fitness" in natural populations. This seems fine, but for a variety of reasons I develop a different account or rubric of the explanation of adaptive genetic traits. For instance, Sinervo and Basolo also argue that field experimental manipulations are essential (cf. Chapter 3). Given obvious ethical limits, we cannot do these in human evolutionary genomics, and many paradigmatic cases in our discipline (e.g., lactase persistence, sickle-cell anemia, Bajau freediving adaptations) suggest that field experiments would not even be helpful in growing our knowledge. Lloyd's (2005a) carving of total genetic adaptive explanation (pp. 47–48) resonates better with my analysis into trait individuation, gene individuation, mechanisms, adaptive scenario, evidence of selection, and evolutionary history; I shall also return to my rubric, for the case of the evolution of female orgasm, in Chapter 8.

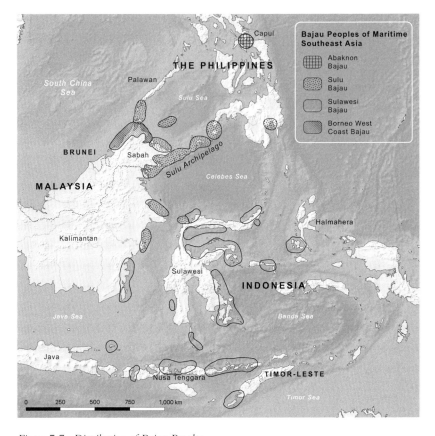

Figure 7.7. Distribution of Bajau Peoples
Geographic distribution of four groups of the maritime Bajau peoples of Southeast Asia. From Nagatsu, 2017, p 43. © National Museum of Ethnology, Osaka. Reprinted with permission. Adapted by Mats Wedin and Rasmus Grønfeldt Winther.

urbanizing mass media. Most Bajau are stateless, making education, employment, and assimilation to their broader societies challenging. Although there is much more to discuss about the Bajau culturally,[51] the focus now is on the genomic and evolutionary details of their free-diving physiology adaptations.

Much of our information comes from Melissa Ilardo, who lived in a Bajau village on the coast of Sulawesi, Indonesia, as part of her doctoral

---

[51] As a sample of relevant literature, see Sather (1997), Field et al. (2009), Acciaioli et al. (2017), and Arunotai (2017).

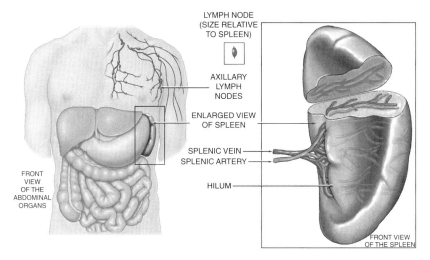

Figure 7.8. Spleen
The spleen – a large lymph node that synthesizes antibodies, breaks down old blood cells, and holds a reserve store of blood – is enlarged in Bajau peoples. © Nucleus Medical Media Inc / Alamy Stock Photo. Reprinted with permission.

dissertation, which she completed in 2018. She was formally hosted by Eske Willerslev at the GeoGenetics Center of the University of Copenhagen.[52] She dove with Bajau freedivers, secured saliva DNA samples with informed consent from all participants, and measured spleen size using a portable ultrasound instrument, also with consent.[53] What she and her collaborators discovered grounds a strong narrative of recent selection in extreme environments.

Trait Individuation

Spleen size is one plausible, potential trait of interest because larger spleens can store more blood and, thus, more oxygenated red blood cells (Figure 7.8). When the body is "induced by apnea and cold-water facial immersion," the spleen contracts, as part of the mammalian diving reflex.[54] Contracting a large spleen can boost oxygen supply up to 10%. Both diving and nondiving Bajau have unusually large spleens compared

---

[52] Ilardo (2018).
[53] It seems evident that Ilardo and collaborators satisfied interpersonal village norms and fulfilled Indonesian and Danish government protocols appropriately. See Rochmyaningsih (2018).
[54] Ilardo et al. (2018, p. 569).

to their land-based neighbors, the Saluan. This was shown with a linear regression model, where population of origin was just one factor, and gender, age, weight, height, and other potential confounding factors or covariates of spleen size were included. This synoptic, statistical model drove the conclusion "that Bajau have significantly larger spleens than the Saluan, even when correcting for several potentially confounding factors."[55] Put differently, and following earlier themes, variance in population of origin is significantly explanatory of variance of spleen size. Although Ilardo investigated size, we can easily imagine future work investigating other features or phenotypes of the organ as possible traits under selection. For example: the amount of blood a given spleen emits when contracted; its contraction rate during one dive or many, or an average of all possible dives, and for one individual or for a class of individuals, and so forth. Ultimately, many descriptions of the parts of an organism and their relevant characters or traits are possible.[56] It remains somewhat unclear which exact phenotype natural selection has here acted upon.

Gene Individuation

After identifying a potential, plausible trait, Ilardo homed in on gene individuation. Using statistical significance tests relevant to natural selection, as partially reviewed above, Ilardo and coworkers explored the entire genome and identified a variant (a SNP) "significantly associated with spleen size" and unusually frequent in the Bajau population, compared to Saluan and Han Chinese populations, and compared with allele distributions elsewhere in the genomes of Bajau individuals.[57] The variant was found within the gene *PDE10A* and hence is an allele of that gene.[58] Although the gene has been identified, there is still work to do in individuating the exact allelic "causal variant" of the *PDE10A*

---

[55] Ilardo et al. (2018, p. 570). Relevant comparisons here are between related populations, with plausibly similar morphologies, physiologies, and sizes. Interestingly, Bajau spleens and the spleens of "exclusively white" German volunteers in a 2016 study (Chow et al., 2016, p. 309) overlap in mean and range of spleen volume, as can be seen by comparing Figure 1 in Ilardo et al. (2018, p. 570) and Figure 2 in Chow et al. (2016, p. 310). That spleen size "matters" for freediving performance, see Schagatay et al. (2012).

[56] See, e.g., Winther (2006b, 2009a, 2011).   [57] Ilardo et al. (2018, pp. 571–574).

[58] As database sources for the *PDE10A* gene, see, e.g., www.genecards.org/cgi-bin/carddisp.pl?gene=PDE10A and www.omim.org/entry/610652.

gene.[59] Recall that development of any one trait almost invariably involves many genes, and a gene can have effects on multiple traits. The causal genotype–phenotype map is complex, and although we have found here one gene associated with one trait, the full story requires future elucidation. Indeed, a fuller picture of other allele variants, their nucleotide sequences, and their relative frequencies in the Bajau population and beyond would also be useful information.

Mechanisms

Even so, according to Ilardo's work, *PDE10A* is expressed in tissues such as neurons and thyroid cells.[60] The gene codes for a phosphodiesterase protein that hydrolyzes or breaks down cyclic nucleotides. In this case, the protein hydrolyzes the cyclic nucleotide cAMP, thereby regulating the production of two thyroid hormones, T3 and T4 (Figure 7.9). Although the PDE protein generally downregulates the production of T3 and T4 by inactivating cAMP in the complex pathway of T3 and T4 production (Figure 7.9), the allele prevalent among the Bajau is different: It is statistically significantly "associated with elevated T4 circulating [blood] plasma concentrations."[61] The variant under study in the Bajau thus appears to *upregulate* the production of T4. Potentially relevant, T4 triggers the development and increased size of spleens, at least in mice.[62] Ilardo and coworkers were the first researchers to integrate, and further discover, the strong evidence linking *PDE10A* to spleen development in humans.

Of course, we have next to no information about how the allele prevalent among the Bajau works in apparently upregulating T4 production. For example, it is possible that this allele leads to a form of hyperthyroidism. Also, it is possible that the allele interacts epistatically with alleles in other genes coding for proteins of biochemical and physiological pathways not yet understood. Further, although we know that T4 increases spleen size in mice, there are attendant concerns with using model systems as stand-ins for humans. Finally, we must consider the regression model employed to show that the Bajau and Saluan really do differ in spleen size (due to population-level genetic differences).

---

[59] Ilardo et al. (2018, pp. 576–577).
[60] Interestingly, blocking the action of PDE10A apparently has potent antipsychotic effects in the treatment of schizophrenia (e.g., Suzuki et al., 2016).
[61] Ilardo et al. (2018, p. 576).   [62] Ilardo et al. (2018, p. 575).

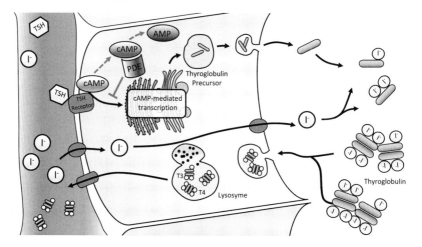

Figure 7.9. Mechanisms of PDE10A action
"The thyroid hormone synthesis pathway. The binding of the Thyroid Stimulating Hormone (TSH) triggers synthesis of thyroglobulin, a thyroid hormone precursor, through a cAMP-dependent pathway. This pathway can be inactivated by phosphodiesterases (PDEs) such as PDE10A. The thyroglobulin precursor is discharged into the follicle lumen [light gray; gray]. Enzymes attach iodine to tyrosines (a part of the thyroglobulin molecule), and the iodinated tyrosines are joined together to form T3 and T4. These molecules are endocytosed into the follicle [middle gray; blue] where lysosomal enzymes cleave the thyroid hormones T4 and T3 from thyrogolobulin, and the hormones are released into the blood stream [dark gray; red]. Decreased expression of PDEs increases the synthesis of thyroglobulin, thus increasing thyroid hormone production." From Ilardo and Nielsen, 2018, p. 80. © 2018 Elsevier Ltd. Reprinted with permission. (A black and white version of this figure will appear in some formats. For the color version, please refer to the plate section.)

Despite the important fact that Ilardo and collaborators found the association between *PDE10A* and spleen size to also be true in a European cohort,[63] environmental factors could still, in principle, explain some of the variance in spleen size. That is, perhaps other causal covariates, such as diet or even trace elements or environmental contaminants associated with certain geographic regions,[64] might *also* help explain spleen size. We must also investigate the potential importance of interaction among covariates. The aim of Ilardo and colleagues' regression was to find and support differences in trait distributions across populations, in order to individuate a clear phenotypic trait. They were not attempting to build a

---

[63] Ilardo et al. (2018, p. 576).   [64] See, e.g., Rembert et al. (2017).

causal model of spleen size, in order to identify genetic mechanisms. However, their regression model *could* be retrofitted in a causal or mechanistic direction (see also Chapter 10), with potential insight into the relative causal powers of environmental factors and of genotype-by-environment interactions.[65]

Adaptive Scenario

Larger spleens are adaptive because they are one clear way of improving freediving physiology. By providing more oxygen to active body tissues, larger spleens permit divers to spend more time submerged and go deeper.[66] However, this is emphatically not the only adaptation to freediving, and spleen contraction is not the only feature of the diving reflex – a slower heart rate and the constriction of peripheral blood vessels are also adaptive features of the diving reflex.[67] There are many physiological processes and anatomical structures involved in most behavioral processes, including freediving, and thus there are many possible adaptations. To expand our knowledge, future research can inquire into the exact fitness consequences of different features of the adaptation – of spleen size variation (does fitness vary linearly with size?), of amount of blood, and of contraction rate, for example. Put differently, which parts or properties of the spleen are under selection? Might other parts of the body's morphology or physiology be correlated and interact with the spleen in adaptively relevant ways? Perhaps there is a wider adaptive suite of traits, which includes aspects of the spleen.

Evidence of Selection

Ilardo's work used statistical techniques to identify the signature of selection and found that *PDE10A* was subject to natural selection. These techniques, again partially reviewed above, included looking for unusually high allele frequencies in the population under study, and looking for genes physically close on the genome that may have "hitch-hiked" with the selected allele during selective sweeps.[68] Interestingly,

---

[65] Melissa Ilardo provided constructive feedback here. For further details about the population structure of Bajau populations across Indonesia, see Kusuma et al. (2017).

[66] See, e.g., Baković et al. (2005).   [67] Ilardo et al. (2018, pp. 569–570).

[68] I sidestep here *BDKRB2*, a gene associated with increased constriction of peripheral blood vessels and thus relevant to the diving reflex, which also has a strong and relevant selection signature in

such techniques have also identified strong selection to extreme environments in analogous studies. Ilardo et al. declaim: "selection scans with similar designs and similarly small sample sizes have been successful in identifying variants underlying important physiological adaptations in other human populations, such as the Inuit (Fumagalli et al., 2015) and Tibetans (Yi et al., 2010)."[69]

Consequently, although the frequency distribution for this allele was unusual, and there were selective sweep statistical patterns indicating the role of natural selection, it would be interesting to gain a fuller picture of the list of alleles of *PDE10A*, and their relative frequencies, in the Bajau population and elsewhere. We will likely never know the exact selective history, or the exact distribution of selective effects across different individuals in Bajau society, many generations back into the deep past. For instance, was selection correlated with sex or occupation or age, or any combination of the above? Such demographic matters are important for understanding the exact action of natural selection in extreme environments. Regarding the possibility of genetic correlations, perhaps *PDE10A* is causative of other phenotypes under selection – therefore, spleen size might not be under *direct* selection. Further studies on genes potentially correlated with *PDE10A*, and a fuller narrative of which other phenotypic parts and features could be under selection, will be useful.

Evolutionary History

The case study of the Bajau populations helps show that we must build out our historical knowledge in order to better understand which alleles, allele frequencies, and phenotypic traits might be unique to, or even just somewhat characteristic of, a given population compared to other historically related populations. In the example of the Bajau, recall that the mean and variance of spleen size were compared with those of the Saluan. The Saluan and Bajau share common ancestry, having diverged some 16,000 years ago.[70] The best historical reconstruction made by Ilardo et al.'s mathematical model suggests low migration from the Saluan to the Bajau population, but high migration from the Bajau to

the Bajau, and, while explored by Ilardo et al. (2018), requires significant further investigation. See Baranova et al. (2017).
[69] Ilardo et al. (2018, p. 573).   [70] Ilardo et al. (2018, pp. 571, 576).

the Saluan.[71] Statistical methods comparing populations and identifying the signature of selection in the genome show that the identified SNP allele variant is not private to the Bajau; it is present elsewhere.[72] Even so, and because Ilardo et al. also used data from archaic human populations, including Denisovans, the allele was found to likely not "originate from adaptive introgression . . . from Denisovans,"[73] who, after all, also lived across Southeast Asia and did introgress some genes into the modern human lineage.[74]

To provide an even fuller picture of the local genomic adaptation of peoples of extreme environments and of their relationship to other peoples of less extreme environments to which they are historically connected, we can gather more information regarding migration patterns. Bajau from elsewhere might have immigrated into the village studied, for example, and other peoples might have immigrated as well. Perhaps rough data about relative census (not to say effective population) sizes in different periods of history could be amassed. Nongenomic sources for such information could include archaeological and linguistic evidence, government records (including from colonial governments), and even accounts by "explorers" and travelers.

Taking a Step Back

To sum up, Ilardo and colleagues' study of Bajau freedivers stands as an impressive study of human evolutionary genomics. Indeed, the Bajau, and other peoples living in extreme environments, are excellent cases for thinking about niche construction, and how genes and culture evolve in response to each other.[75] Cultural practices over many generations have relied on the maintenance of specific, local ecological environments to create a given peoples' life-space. In this unique ecology, diet, work, and

---

[71] Ilardo et al. (2018, p. 571).   [72] On private alleles, recall empirical pattern #3 of Chapter 6.
[73] Ilardo et al. (2018, p. 577).
[74] See Bergström et al. (2020); in both senses of "adaptive introgression" from Jagoda et al. (2018), that is, "immediate adaptive introgression" and "standing introgressed variation."
[75] Genetic and cultural transmission are interacting processes of intergenerational heredity, not always easily differentiated, and certainly replete with feedback effects. Indeed, coevolution happens, in part, through niche construction. Classic statements of gene–culture coevolution and niche construction are, respectively, Cavalli-Sforza and Feldman (1981) and Odling-Smee et al. (2003). Ilardo et al. (2018, p. 577) briefly mention gene–culture coevolution. Various essays of Levins and Lewontin (1985) hinted at niche construction, and Lewontin influenced Odling-Smee and collaborators.

other socially sanctioned behaviors and habits could have strong selective effects on the physiological, morphological, and behavioral variation that exists. Insofar as there was a genetic basis for such variation, adaptive biological evolution could then occur. In other words, culture constructs a lived ecological niche, some features of which then operate selectively on the population.[76] Here, culture leads and selection follows.

Moreover, the work of Ilardo (and others) has clear consequences: "The variants that allow the Bajau to undergo repeated bouts of acute hypoxia with minimal physiological stress might affect an individual's ability to withstand acute hypoxia during a medical crisis such as traumatic brain injury."[77] Medical implications of the investigations of Tibetan and Inuit adaptations to cold environments and to, respectively, high altitude and low oxygen, and fatty diets, are also promising. That is, by exploring potential genetic mechanisms underlying, e.g., gas exchange in the blood, and fatty metabolism, possible genetic engineering interventions and medical treatments for physiological disease conditions could be hypothesized and tested.

Ultimately, the Bajau case study is not closed, and important future work remains. The Bajau peoples and ongoing research on them can tell us one narrative about the origins of a particular human population, and about belonging and identity. It is also an ideal example to showcase the value of human evolutionary genomic population- and subpopulation-specific investigation. Further theoretical work on evolution in small populations and future ethical and medical work could expand our knowledge of who we are as populations and peoples. To ensure that such work is meaningful, we must use a reflective and conceptual approach to frame and point the way toward the future, engaging in productive work that is empirically honest, theoretically sophisticated, ethically grounded, and pragmatically oriented toward consequences.

Taking a step back, this case study also shows the importance of paradigm pluralism. To provide complete explanations and narratives, even of just one episode of natural selection in human populations, we need to distinguish, supplement, and *integrate* a broad panoply of

[76] This is not necessarily the "Baldwin effect" or West-Eberhardt's (2003) "genetic accommodation," which seem to be much more specialized evolutionary developmental processes, not particularly pertinent to cases of human adaptation to extreme environments. These cases can be accommodated in human evolutionary genomics theory.
[77] Ilardo and Nielsen (2018, p. 81).

paradigms and explanations – here, adaptationism, mechanism, and historicism. To gain a view of the whole of human evolutionary genomics, and to fully understand the action, implementation, and limits of natural selection, even from the technical, statistical, genomic point of view, we must delve into, and integrate, alternative perspectives.

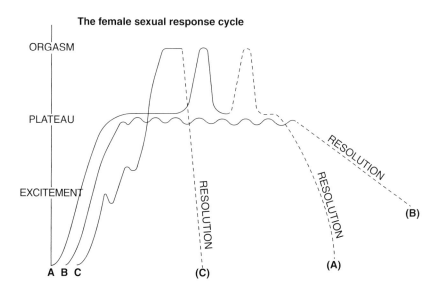

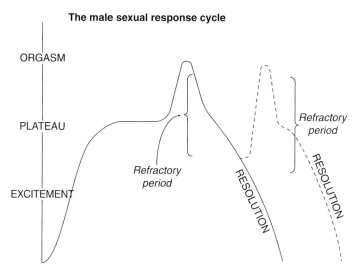

Figure 8.1. Female and male orgasm cycles
The female and male sexual response "cycles" according to Masters and Johnson
(1966), with three distinct patterns shown for women. From Masters and Johnson,
1966, p. 5. Adapted by Mats Wedin and Rasmus Grønfeldt Winther.

8 · *Intelligence, Female Orgasm, and Future Discovery*

In Chapter 7, we explored dichotomies between natural selection and various forces, explanations, and paradigms complementary to natural selection, such as random genetic drift and developmental mechanisms. Both poles of each of these multiple binaries need to be respected and, ultimately, integrated. As we have discovered throughout *Our Genes*, the dichotomy between *individual* and *population* calls for continuous measurement as well as careful comparison and conceptualization. In light of populational statistical analysis, genomics shows us that individuals differ but that individuals are also, in effect, almost the same, and certainly more similar than what social discourse typically suggests.

This chapter continues and extends the concerns discussed against the backdrop of natural selection. It also allows me to apply my perhaps idiosyncratic obsession with trying to understand characteristics central to human experience that help illuminate the relationship between individuals (*Who am I?*) and populations (*Who are you? Who are we?*). Accordingly, this chapter makes use of the tension between *gene* and *environment* that always, as we now know, informs our understanding of who we are, for better and worse, and consciously and unconsciously. This tension is built into the very history and foundations of statistics, via Galton's concept of correlation and Fisher's ANOVA methodology, for example, and it is neither simple nor singular. The gene/environment binary is related to the *nature/nurture* binary, but it is also distinct: The latter is more metaphorical and abstract, less scientific and measurable. This chapter perhaps overly reifies the distinction between gene and environment, but it does so in order to deploy it as a window for case studies allowing for more rigorous investigation of the interplay between gene and environment in disease, intelligence, and female orgasm.

Gene versus Environment in Disease

As complex, somewhat adaptive, and evolutionarily fine-tuned creatures, humans have phenotypic traits that typically emerge through interacting causes, with multiple genes directly or indirectly involved, and with a slew of environmental factors required – water, appropriate temperatures, the chemistry of nourishment, familial and social interaction, and so forth. Philosophers and other scholars have long debated the content and meaning of the *genetic information* encoded in our long chains of DNA nucleotides, and the consequences of postulating a given factor as either genetic or environmental.[1]

We can occasionally characterize a trait as (at least) somewhat genetic in cases where a particular single mutation makes a significant difference to the trait's presence. In contrast, if a particular environmental intervention (such as heat shock during *Drosophila* development) causes a series of phenotypic effects, or a significant difference in the presence of one trait, we might be justified in characterizing those effects as (at least) somewhat environmental. However, because there are so many environmental (nongenetic?) and genetic (nonenvironmental?) levers or settings that can be adjusted, as it were, by nature or humans – and because distinctions are often blurred by interactions (e.g., environmental hazards can cause mutations) – every trait is both genetic and environmental. Even so, we can usefully make the distinction between gene and environment when variation in a given genetic (or environmental) factor makes a quantitatively large difference to the phenotype, holding other factors roughly equal. Under such contextual conditions, the trait is *largely* genetic (or environmental). No doubt, philosophers and scholars will find many weaknesses with my characterization, but I wish here to

---

[1] *Genetic information* is a philosophically challenging and rich concept. Some interlocutors hold that genetic information is a pernicious reification – it does not exist in the world and there can be no appropriate scientific conceptualization or theorizing about it (e.g., chapter 3 in Levins and Lewontin, 1985; Winther, 1996; and Oyama, 2000a, 2000b). Others argue that the notion plays a useful role in genetics, development, and evolution. Under the latter family of views, genetic information is characterized in a variety of ways – as *a semiotic system* (Emmeche, 1990); as *transmission information*, in the Claude Shannon sense of information (Bergstrom and Rosvall, 2011); as *inherited representations* (Shea, 2013); or as *proximate* or *ultimate information*, depending on whether it plays a role in development or in the evolution of species (Griffiths, 2016). Walsh (2015) moves "the units of phenotypic control" from genes to the "entire gene/genome/ organism/environment system" but finds the critiques of genetic information as a reification "inconclusive" (pp. xi, 132). The analysis in Noble (2006) resonates with Walsh's. Griffiths and Stoltz (2013) provide a synoptic overview of genetic information. I have gone from critic of (Winther, 1996) to pluralist about (Winther, 2020a) genetic information.

simply tie the concepts of genetic and environmental traits to variation in (or cause of) traits.

I thus offer a statistical analysis of variance understanding of the dichotomy between gene and environment to better explore extreme cases of genetic and environmental disease. By placing these cases in the context of two distinct overarching modern medical paradigms – imagining disease as primarily genetic (as in precision medicine) or as environmental (as in population health) – I seek to further our comprehension of ourselves and of each other and help determine where interventional opportunities may lie. Deep knowledge of genetic and environmental categories not only points our way to therapeutic options but also sheds light on future work in precision medicine and population health.

In terms of the public's exposure to DNA and genomics in the context of biomedicine, the first publication of an annotated nucleotide map of *Homo sapiens* – which appeared in a 2001 article in *Science* – stands as a watershed moment.[2] This instant classic, which included a foldable gene map poster, featured a New England Biolabs advertisement on the page opposite the article's first page. Cultural resonances can be drawn out of this advertisement: In trying to sell custom restriction enzymes for the amplification of gene sequences, it featured a map of Africa draped and superimposed over a woman's body, inviting us to visualize a sail and a "cutter ship" pushing us forward into unexplored territories of genetic engineering and biomedicine. Exuding scientific authority, the ad promised new frontiers of genetic engineering, gene therapy, and corporate profits.[3]

Genomic technologies, such as those advertised 20 years ago by New England Biolabs, and genome projects, such as the Human Genome Project, promised that they would reveal essential truths about humans as individuals and collectives. The hype and hope associated with these technologies at the dawn of a new millennial era inspired the expectation that genomics would help eradicate disease. The era's baptism through sexualized (woman; Eve) and racialized (Out of Africa) advertisement imagery, which likely energized and titillated the mostly (white) male consumer base, also helps us track ethics and power.

[2] Venter et al. (2001); see Dupré (2004) and Gannett (2008).
[3] In *When Maps Become the World* (Winther 2020a), I cite the Biolabs advertisement (p. 223) as an example of a phenomenon I call *cartopower.* "the representation's intrinsic plea, seduction, or argument, via the operation of the ontological assumptions of the ontological layer, to interpret the world one certain way, and only that way" (p. 129).

The heady atmosphere of excitement heralded by this 2001 article and its accompanying advertisement continues today. DNA has become the embodiment of genomic knowledge, regardless of its limitations or nuance. Consider consumer DNA companies. When you send your saliva to AncestryDNA, 23andMe, and other such companies, they try to extract your health information and predict some of your future ailments and diseases. Your genome is sequenced to assess whether you might have particular functional alleles known from biomedicine to be associated with particular medical conditions, such as breast cancer, Huntington's disease, or Tay–Sachs disease. But where and how did this modern genetic scouring of the human body, and its weaknesses and predispositions to certain diseases and ailments, start?

Consumer DNA companies and many research agendas in contemporary biomedicine depend heavily on the possibility that certain traits or conditions are largely – not to say straightforwardly – genetic, as opposed to environmental. But, even if we know enough to know that human evolutionary genomics holds at least some key answers to our questions, we also know that data-mining the human genome throws us endless curveball surprises. The genomic signature of lactase persistence, which we examined in Chapter 7, stands as an apt example. Here, genomic research yields the ongoing discovery of new alleles, their particular geographic and populational distribution, and their mechanistic mode of action.[4] This research also has potential biomedical impact, since the etiology of diseases such as inflammatory bowel disease and various types of cancers (e.g., ovarian and breast) may sometimes be correlated to the lactase persistence phenotype, and perhaps even to its genetic basis and to the dairy consumption it permits.[5]

Returning to the big picture, researchers, doctors, and other stakeholders comb the genome as though it were a book, trying to find nucleotide sequences – and, more generally, genes and their alleles – that might be indicative of medical conditions. Consider the case of a strongly genetic disease like Huntington's disease. A late-onset neurodegenerative disease, Huntington's is strongly, if not uniquely, associated with a single, dominant allele in affected individuals – a pattern confirmed by a genealogical analysis of more than 3000 Venezuelans in the early 1980s (Figure 8.2). The gene (allele) responsible for the condition

---

[4] Recently, Charati et al. (2019) pinpointed the ethnic distribution of certain lactase persistence alleles in Iranian populations.

[5] See, e.g., Szilagyi (2015).

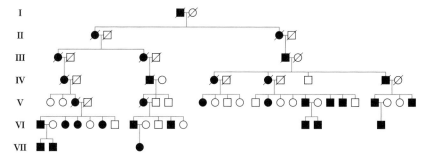

Figure 8.2. Original Huntington's disease pedigree
"Pedigree of the Venezuelan Huntington's disease family. The pedigree represents a small part of a much larger pedigree … DNA prepared from the lymphoblastoid lines [was] used to determine the phenotype of each individual at the G8 locus." Roman numerals represent generations; circles = females; squares = males; a black symbol = individuals with Huntington's disease; a slashed symbol = deceased individuals. From Gusella et al. 1983, p. 236. © 1983 Nature Publishing Group. Reprinted with permission. Adapted by Mats Wedin and Rasmus Grønfeldt Winther.

was mapped to a location on the fourth human chromosome, thanks to a molecular study using DNA from lymph cell lines taken from this study's Venezuelan population.[6] Although researchers had never before specified the etiology of a primarily genetic disease in this manner, Huntington's disease still has no cure.[7] Lacking clear treatments, medical personnel often prescribe antidepressants and antipsychotics.[8]

The genetic basis of Huntington's disease, which fortunately manifests very rarely at a global level, provides a glimpse of the importance of learning about family history (Figure 8.2). Throughout *Our Genes*, we have discussed the questions associated with identity that motivate investigation into human evolutionary genomics. However, we have mostly set aside those "subpopulations" typically most accessible to us: our families. Indeed, consumer DNA companies like 23andMe provide a

[6] Gusella et al. (1983). On gene mapping in general, see chapter 8 of Winther (2020a).

[7] Interestingly, Huntington's disease affects people of European descent almost exclusively. (The cases studied in Venezuela can likely be traced to a single European ancestor.) There are molecular genetic reasons for this pattern. A slightly old but still relevant discussion with a useful bibliography can be found in Liou (2010). For a more contemporary perspective, see Dayalu and Albin (2015). For a personal account by the sister of one of the co-authors of the original study, see Wexler (2010).

[8] See www.mayoclinic.org/diseases-conditions/huntingtons-disease/diagnosis-treatment/drc-20356122.

bridge back, in a manner of speaking, from the insights gained through genomic analysis of populations to the genomic analysis of the individual (see Figure 5.3). Statisticians and geneticists of these companies construct this bridge by comparing your or my allelic genomic profile to the genomic profiles of the many customers in the company's genomic database, including your or my nearest family members. For instance, evaluating across many genes and alleles, a very close match between two customers indicates that they are especially close relatives. Family history inferences are constrained by the number of individuals the company has sequenced, though by now it is on the order of many millions (e.g., Ancestry claims to have "over 20 million people in the world's largest consumer DNA network").[9] This matters for relatively rare diseases because you are more likely to be correctly identified as a carrier of, say, rare, recessive genetic diseases (e.g., albinism) the more total individuals from many backgrounds have been tested. In addition, at least some individuals with the rare disease need to be in the company's database for it to be informative for tracking the likelihood and distribution of the disease across all individuals.[10]

The potential of genomic findings is applicable to contemporary *precision medicine*, "an emerging approach for disease treatment and prevention that takes into account individual variability in genes, environment, and lifestyle for each person."[11] We could even say that precision medicine is personalized medicine. Treatments are developed in accordance with a person's DNA, as well as their environment and lifestyle. To gain an idea of how precision medicine informs our understanding of disease, consider cancer. Many cancers may include an environmental component, but Francis Collins and Harold Varmus argue that "inherited genetic variations contribute to cancer risk, sometimes profoundly." This genetic component suggests that the "individualized, molecular approach" available in precision medicine could be particularly

---

[9] See www.ancestry.com/corporate/about-ancestry/company-facts.

[10] The power of consumer genomics reaches beyond biomedicine. In the United States, populations can be combed for identificatory genetic similarities and differences in criminal cases, where the FBI CODIS (Combined DNA Index System) can be consulted. Alternatively, using DNA from the crime scene, the FBI can request a family history study directly from consumer DNA companies. (On CODIS, see www.fbi.gov/services/laboratory/biometric-analysis/codis/codis-and-ndis-fact-sheet; on the close relation between law enforcement and genetic companies, see Regalado, 2019.) With just a few dozen markers, relatives of a perpetrator, or even the perpetrator's identity, can be identified (Edge and Coop, 2018; but see Rohlfs et al., 2012). On privacy concerns, see Kulynych and Greely (2017).

[11] See https://ghr.nlm.nih.gov/primer/precisionmedicine/definition.

effective in the fight against "inherited genetic disorders and infectious diseases."[12] Their pronouncements – Collins headed the National Institutes of Health (NIH), and Varmus is a Nobel Laureate scientist who has also headed the NIH as well as the National Cancer Institute – powerfully shape the field of biomedicine.

Although precision medicine need not be solely practiced at the individual level, or primarily focus on genetic factors, it often is and does.[13] This is evident in the recent interest in polygenic trait risk scores, representing and summarizing the aggregate effect on particular phenotypic traits of the many genes – and their associated alleles – likely causative of that trait, within and across different populations. Such scores are a potentially significant boost for personalized medicine and could lead to identification of alleles associated with various cancers. Historically, the scores are built on the results of genome-wide association studies (GWAS), which search as much of the genome as is possible for a given study, looking for single-nucleotide polymorphisms (SNPs) that reliably differ between the group of people who have the disease and the group who do not.[14] GWAS, and the polygenic risk scores emerging from these studies, could promise to be useful in examinations of disease etiology. This is especially the case for complex diseases such as cancers, rather than, say, Huntington's disease, which is more genetically simple, as variation in just a single gene significantly explains disease expression. The genomic focus of precision medicine could push technologies forward (e.g., CRISPR), and precision tools developed for individuals could lead to important diagnostic and treatment modalities for families, groups, and larger populations.

Diseases with more environmental etiology suggest a different, if complementary, primacy: The individual is individual only relative to the population – and the environment – that produced it. Consider lung cancer. Two important studies on the etiology of lung cancer were Doll and Hill's preliminary report on the effect of smoking and Cornfield and

---

[12] Collins and Varmus (2015, p. 794); cf. Berger and Mardis (2018).

[13] For contrasts between uses of the concepts of *precision* and *personalized* medicine, see Bresnick (2018). For personalized medicine and human genetic diversity, see Lu et al. (2014).

[14] Warren (2018). Each GWA study involves many individuals, often in the many thousands. For early, cautiously optimistic reviews of GWAS methodology from the heartland of human evolutionary genomics, see, e.g., Rosenberg et al. (2010) and Visscher et al. (2012). For a contemporary view, see Visscher et al. (2017), Visscher and Goddard (2019), and the public database www.gwascentral.org. See also www.genome.gov/genetics-glossary/Genome-Wide-Association-Studies.

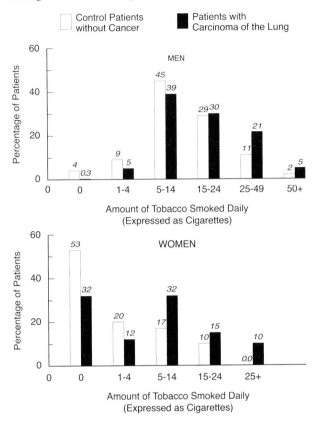

Figure 8.3. Original lung cancer and smoking graph
"Percentage of patients smoking different amounts of tobacco daily." From Doll and Hill, 1950, p. 742. © 1950 BMJ Publishing Group Ltd. Reprinted with permission. Adapted by Mats Wedin and Rasmus Grønfeldt Winther.

colleagues' study of relative risks (Figure 8.3).[15] Adducing copious survey and medical data, these studies showed that lung carcinomas were much more common among smokers than nonsmokers, and that lung cancer incidence had strongly increased in parallel with an increase in smoking behavior in the first half of the twentieth century.[16] Associations between

[15] See Doll and Hill (1950), Doll (2002), and Cornfield et al. (2009 [1959]).

[16] Not only statistical evidence supported their conclusion. For instance, Cornfield et al. (2009) cite "additional confirmations ... on the induction of cancer of the skin in mice painted with tobacco-smoke condensates" (p. 1176). As many readers know, these studies were gaslit by the tobacco industry. But they were also challenged by at least one accomplished statistician: R.A. Fisher suggested the possibility of a spurious correlation between smoking and lung cancer, in case

environmental effects and diseases such as cancer have been increasingly found: ultraviolet light exposure and skin cancer, radon and lung cancer, chemicals such as vinyl chloride and leukemia and liver cancer, and even parasites such as human papillomavirus (or HPV) and cervical cancer. This is of interest because it shows that independently of genotype, factors external to the individual and operating during an individual's lifetime – from radioactive substances to noxious chemicals to dangerous bacteria and viruses – are correlated to, and causative of, disease.

The existence of environmental disease, of environmental factors in disease etiology, and of the systemic, unequal distribution of such factors across different populations, living in different (urban) environments, is of particular interest to a second contemporary medical paradigm: *population health*. According to the Center for Population Health Sciences at Stanford University, "population health is the study of relationships between many health determinants and/or health outcomes in large populations."[17] It includes identifying and evaluating lifestyle choices, infectious agents, and local environmental factors, both for individuals and for entire populations. Dunn and Hayes assert that "a population health perspective is fundamentally concerned with the social nature of health influences."[18] Population health can thus also involve "redistribut[ing] the power and authority concentrated in the healthcare sector."[19] Diagnostic opportunities and therapeutic interventions should be made available to populations that are less well-off and more prone to environmentally mediated diseases. For instance, "African Americans in the US still have the highest death rate and lowest survival rate of any racial or ethnic group for most cancers."[20] Genomics provides little insight into the social influences and economic factors of healthcare. For example, genomics cannot trace the rich spatial and temporal structure of familial, local, and global environmental factors, nor can it shed light on the effects of systemic racism and class injustice.[21] Rather, when it comes to environmental and complex diseases, such as lung cancer or

some third factor – possibly a genetic one – explained both (Fisher, 1959). Nevertheless, the highly robust statistical studies won the day, as they should have (Stolley, 1991). A lesson of this case is that (generalized and unreasonable) scientific skepticism is certainly not a virtue when consequences are high. This lesson applies equally to skepticism about the role of fossil fuels in human-caused climate change today (Oreskes and Conway, 2011).

[17] See http://med.stanford.edu/phs/about.html.   [18] Dunn and Hayes (1999, p. S7).
[19] Valles (2018, p. 181).   [20] Simon (2019); cf. Schabath et al. (2016).
[21] Cerdeña et al. (2020).

posttraumatic arthritis, a population health perspective can make a large difference, with relatively little financial investment.

In fact, our best modern scientific knowledge tells us that most diseases are complex – they are the outcome of many genomic *and* environmental factors, which often interact. Notable exceptions, such as Huntington's disease, can be overblown in importance, leading to an overly exaggerated genomic paradigm for disease.[22] Huntington's disease and lung cancer differ because a single gene makes a difference to the expression of the former, while the frequency and duration of smoking highly predicts the development of the latter (even given genomic variation in susceptibility), and nonsmokers can also develop lung cancer for a variety of environmental reasons, including exposure to radon. But, in some deep sense, no disease – or phenotypic trait in general – is merely genetic. It seems reasonable to say that the disease phenotype associated with Huntington's can be caused only by the Huntington's dominant allele. But even so, an individual who will invariably develop Huntington's because she carries the allele must have the appropriate environmental factors, such as nourishment and social support, to even survive and live long enough to express the allele. Even for a strongly genetic disease, then, having the allele is *necessary* for disease expression, but it isn't *sufficient*. (And, as we have seen, in environmental or in complex diseases, carrying almost any single, particular allele would most likely not be necessary for expressing the disease as opposed to not expressing it.)

There is much else to say about meanings of *genetic* versus *environmental* disease, including with respect to statistical methodologies such as ANOVA, the increasingly genomic bent of cancer research, and ethnically correlated diseases. Examples of "genetic" diseases prevalent in certain populations include sickle-cell anemia and Tay–Sachs disease. A person homozygous for the sickle-cell anemia allele may experience pain and early death; meanwhile, heterozygotes (also known as *carriers*) are better able to fight the malaria parasite – thus, natural selection has favored the sickle-cell allele in areas such as western and central Africa, the Middle East, and India.[23] Tay–Sachs,

---

[22] This can be the case even when admitting a polygenic origin of disease, e.g., Gibson (2012) and Cerrone et al. (2019).

[23] See Hartl and Clark (1989, pp. 169–172). Due to the history of slavery in the United States, sickle-cell anemia prevails in the African American population (some 10% are heterozygote; the trait exists but is much less common in, for example, American Hispanic and Middle Eastern populations; on sickle-cell in the Middle East, see Burton, 2019). The disease is named after an

a neurodegenerative disease, is caused by a recessive allele that is not uncommon among Ashkenazi Jews.[24] In many of these cases, the prevalence of some single-locus genetic diseases that are strongly phenotypically recognizable within particular populations, such as Tay–Sachs, may be due in part to socially regulated patterns of marriage and reproduction – an interaction of culture and biology. Again, no gene creates a trait without environmental support, and no environment creates a trait without genetic support.

Intelligence, Female Orgasm, and Future Discovery

Human evolutionary genomics can illuminate a variety of other interesting and important features of the human condition, not just about biomedicine and disease, but also about more nebulous and fascinating features. Intelligence and the female orgasm stand as two separate, illustrative examples. When we focus on these topics using the tools of human evolutionary genomics, we learn about the challenges of the field and the competing complexities of its investigative objects.

In fact, looking more closely at the interplay between genetic influence and environment in the development and evolution of both intelligence and female orgasm shows that a "total" explanation of these respective complex and ambiguous phenotypes can be attempted, but may never be forthcoming. As we saw with Bajau freediving adaptations in Chapter 7, human evolutionary genomics explores the mechanistic action of individuated genes on different phenotypic traits. While the necessity of, and plasticity to, the developmental environment is granted, it is the causal arrow from genotype to phenotype that is of interest to our

allele that leads to red blood cells having an unusual "sickle" shape, much more so in homozygous individuals than in heterozygous individuals. The allele is a single-nucleotide mutation (on chromosome 11) that causes an abnormal hemoglobin structure since one amino acid, glutamic acid, is replaced with another – valine in the beta globin subunit of hemoglobin. Another family of hemoglobin disorders, *thalassemia*, also confers a selective advantage against malaria and is more frequent, compared to sickle-cell anemia, in the Mediterranean, India, and Southeast Asia. Thalassemia involves altered alpha or beta globin units of hemoglobin. (The genes for alpha globin are found on chromosome 16.) Individuals can suffer from both sickle-cell anemia and thalassemia. See https://ghr.nlm.nih.gov/condition/beta-thalassemia.

[24] See, e.g., Ostrer and Skorecki (2013). On Tay–Sachs in other populations, see, e.g., Kelly et al. (1975). No clear heterozygote advantage has been found for Tay–Sachs, although it might be associated with resistance to tuberculosis. Withrock et al. (2015) review a panoply of so-called genetic diseases and whether they confer resistance to a variety of infectious diseases, including cholera and viral infections.

field, and this I shall explore for both intelligence and female orgasm. However, unlike the case of the Bajau freedivers, and insofar as adaptation is the fit between organism and environment, and natural selection is something like a force (potentially mediated by culture) acting to produce adaptations, it is unclear how to measure or establish either intelligence or female orgasm as adaptations, or whether they even *are* adaptations. Thus, the two cases we now turn to illustrate that, in addition to adaptationism, other paradigms and deep explanatory factors must be taken into account, including mechanism and historicism, which we first met in Chapter 7.

Intelligence

When it comes to intelligence, three broad academic discourses and public debates exist about: (1) the genetics and mechanistic inheritance of cognitive capacities and skills; (2) the conceptualization and operationalization of intelligence, and whether IQ is an accurate measure of intelligence; and (3) the reality of race as a subdivision of *Homo sapiens*, and whether IQ (or intelligence, or both) differ systematically across races, and, if so, why. Interestingly, the genetics, measurement, and potential racial distribution of intelligence are not logically related topics. Their association in technical work and lay disputes since the 1960s is, bluntly put, not a necessary one, has changed over time, and was itself a historical accident. The third, in particular, can be (and is) addressed independently of the first. Moreover, there are many cognitive capacities in addition to intelligence, thereby disentangling the first and the second. Intelligence is itself varied, thereby problematizing the second (Figure 8.4). Finally, the three topics belong to different research domains, each with its own assumptions, measures, and practices: twin studies and heritability estimation; psychometrics and pedagogy; and anthropology and human taxonomy, respectively. But influential interlocutors such as Richard Lewontin, Stephen J. Gould, Arthur Jensen, and Charles Murray tied these topics together, and so it has become customary in the U.S. context to treat them as a set.[25]

Investigating intelligence can involve determining the heritability of cognitive traits such as intelligence. Of course, heritability differs from

---

[25] It did not have to be this way: One could interpret Lewontin (1970) as playing into the hands of Jensen (1969), for while Lewontin vehemently critiqued Jensen, he implicitly accepted Jensen's entanglement of the three themes.

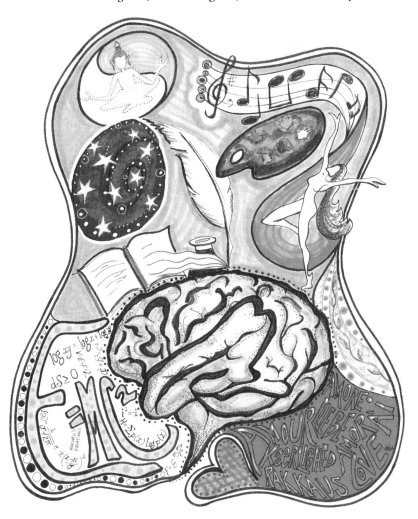

Figure 8.4. Varieties of Intelligence
Multiple types of intelligence exist, and the classifications are distinct and nonoverlapping. For instance, Gardner (1983) distinguishes various types, including linguistic, musical, bodily-kinesthetic, logical-mathematical, and interpersonal, whereas Sternberg (1985) defends a "triarchic theory," roughly: analytical, creative, and practical (that is, "street smarts"). Illustrated by Larisa DePalma. © 2021 Rasmus Grønfeldt Winther. (A black and white version of this figure will appear in some formats. For the color version, please refer to the plate section.)

inheritance: The former is a technical concept and measure (h^2, or *narrow sense heritability*) and refers to the relative amount of the total phenotypic variance that, in a phenotypic decomposition ANOVA, can be correlated to (more controversially, is explained by) the so-called *additive genetic variance*, as opposed to dominance variance, epistatic variance, environmental variance, genotype-by-environment interaction variance, or genotype–environment covariance. To oversimplify, and to zoom in on the first, heritability is the fraction of the total phenotypic variance of many individuals in a given population that can be theoretically explained by simply (and on average) varying the alleles of a given gene (or, metaphorically speaking, adjusting single, averaged genetic levers, one at a time). In measuring each gene's "additive effect" – also known as the "average effect of allele substitution" – we statistically abstract away from the levers of environmental factors, or from the interactive epistatic effects of alleles at other genes, which would be a kind of cross-gene interactive lever (a meta-lever?), or even, and importantly, from the potential interactive effects *within* a gene (across alleles) – i.e., dominance deviation.

In intelligence studies, IQ advocates simultaneously defend the coherence of heritability and believe that it is a large percentage of the total phenotypic variance decomposition of IQ.[26] In contrast, IQ skeptics deny both and sometimes also point out that there are multiple types of intelligence (Figure 8.4). The intensity of the debate, which continues with vigor today, speaks, in my view, to the equivocal and early-phase nature of the evidence.

Nonetheless, IQ advocates such as Jensen and Murray typically proffer an adaptive or adaptationist explanation. Such advocates believe that the

---

[26] See, e.g., Panizzon et al. (2014); for further details on heritability and the statistical decomposition of the total phenotypic variance into components such as additive genetic variance, dominance variance, epistatic variance, and so forth (see list above), see chapter 4 in Levins and Lewontin (1985), Wade (1992), chapters 8 and 10 in Falconer and Mackay (1996), Visscher et al. (2008), Winther et al. (2013), Lynch (2017), and Downes and Matthews (2020). As we saw in Chapters 2 and 3, this is Fisher's influential ANOVA phenotypic decomposition, first presented in Fisher (1918). Such *variance partitioning* methodology was later articulated by Cockerham in a different, but related context, as we saw in Chapter 5. For explicitly anticausal and critical interpretations of heritability, and of the heritability of IQ in particular, refer to, e.g., Lewontin (1970), Layzer (1972), Block (1995), and Sarkar (1998). For alternative "complex systems" frameworks in which cause cannot be uniquely decomposed, and in which *nature* and *nurture* interact and the distinction itself is critically examined, consult, e.g., Levins and Lewontin (1985), Oyama (2000a, 2000b), Wimsatt (2007), Winther (2008), and Walsh (2012, 2015). On "missing heritability," see Matthews and Turkheimer (2021).

presence of a trait in a species or a population is explained by natural selection, which Chapter 7 explored. Although advocates can point to a handful of studies for support, one simple problem undermines the adaptationist thesis: Plainly put, we do not know how to measure "intelligence," or even how to delimit the intelligence purportedly aiding survival and reproduction, either in the past or today.[27]

Some reasonable work has been done on at least moderately sized genetic correlations – not to say genetic causation – of intelligence.[28] For instance, Sniekers et al. used GWAS to identify dozens of genes statistically associated with variation in intelligence, individuated as a trait according to early twentieth-century English psychologist Charles Spearman's general intelligence factor, also known as the "*g* factor" or "Spearman's *g*."[29] Researchers found that many of these genes "are predominately expressed in brain tissue," with genes involved, for instance, in synapse formation and in regulating neuronal differentiation.[30] This is plausible evidence for some genetic basis for intelligence and for the potential inheritance of intelligence.

Even so, extant evidence for genomic effects on intelligence and IQ is far from incontrovertible, in part because of methodological issues concerning GWAS, on which measuring the heritability of genes associated with intelligence depends. Such issues include lack of clarity regarding the population ancestry and potential relatedness of experimental subjects, and too-small sample sizes, though some of these challenges are being overcome. There is also the matter of the "winner's curse": Among the many millions of nucleotide sites probed, the nucleotide

---

[27] In a provocative piece, Lewontin (1998) argued that we may never be able to answer any number of questions about how cognition evolved, including identifying plausible and empirically grounded adaptive scenarios or adaptationist narratives, and finding evidence of natural selection.

[28] See, e.g, Polderman et al. (2015), Sniekers et al. (2017), Plomin (2018), Plomin and von Stumm (2018), and Savage et al. (2018).

[29] See Sniekers et al. (2017). Bouchard (2009) provides some useful statistical background for the genetics of intelligence, conceptualized as Spearman's *g*. He builds causal models using Wright's path analysis (discussed in Chapter 10) for genetic and environmental factors, under different conditions – monozygotic twins reared together versus apart (e.g., Figure 1 in Bouchard, 2009, p. 529). One hope of IQ advocates is that these kinds of genetic studies and causal modeling strategies will shed light on the inheritance and development of intelligence, including across different human groups, although possibly not on the selection and history of intelligence in the deep past of the *Homo* genus, and beyond. As an IQ skeptic, Gould (1996 [1981]) critically discusses Spearman's *g* at some length (pp. 286–302) and provides reasons for thinking of it as a "reification" (pp. 295–299). Incidentally, that Gould took the "reification fallacy" seriously can be confirmed by consulting that index entry on p. 444.

[30] Sniekers et al. (2017, pp. 1107, 1110).

variants found that make a large enough difference to a given phenotypic trait often have overestimated effects, and are typically found because of luck (just as the winner of a timed race with many competitors is likely to be better than the average competitor but is also likely to have had a good day – her time that day exceeding her own average).[31]

More generally, there is some reason to be skeptical about the possibilities of intelligence research. For one, consider the work of behavioral geneticist Eric Turkheimer. He accepts the assumptions and limits of linear and structural statistical causal modeling, and points out that there are multifarious causal interactions both inside and outside a developing human. Each causal factor, and each interaction among potentially many causal factors, is only somewhat explanatory – and the aggregate, if measurable, would be inordinately complex. Turkheimer defends what he calls the "three laws of behavior genetics": "[1]. All human behavioral traits are heritable; [2]. The effect of being raised in the same family is smaller than the effect of genes; [and] [3]. A substantial portion of the variation in complex human behavioral traits is not accounted for by the effects of genes or families."[32] In excavating the third law in detail, Turkheimer describes a "gloomy prospect" for which neither environment nor genes – whether individually or in conjunction – illuminate very much at all in terms of explaining humans as cognitive and biological agents. Turkheimer effectively argues, here and elsewhere, that "the additive effect of genes may constitute what is predictable about human development, but what is predictable about human development is also what is least interesting about it."[33]

Against the backdrop of the tension between gene and environment, the ongoing and fraught study of intelligence is a test case for human evolutionary genomics attesting to the scientific and political complexity associated with a central trait (or set of traits) with a long history of study, use, and abuse. Similar to the early field of genetics itself, which relied on race (alternatively, human populations) as an analytical and interpretive (but also predetermined) object, the hereditary and comparative analysis

---

[31] Palmer and Pe'er (2017). Additionally, many variants found with GWAS may not have much biological relevance, and a new theoretical "omnigenic" framework may be required (Boyle et al., 2017). In the fullness of time, these matters could perhaps be solved (e.g., Eisenstadt, 2017). But we shall have to see. We are not there yet.

[32] Turkheimer (2000, p. 160). For a review, cautiously optimistic about the power and relevance of genetic explanations in behavioral genetics, see Plomin et al. (2016).

[33] Turkheimer (2000, p. 164). For a philosophical analysis of the concept of *interaction*, see Tabery (2014) and Longino (2021).

of intelligence has been marked by a riven past. In fact, the very history of statistics is tied not only to the history of genetics but also to the history of psychology and psychometrics – in particular to the history of intelligence studies – through figures such as Francis Galton, Alfred Binet (1857–1911), Charles Spearman (1863–1945), and Louis Leon Thurstone (1887–1955). Now, future genomic investigation may reveal that intelligence is genetically situated, but only very weakly, and is much more likely the result of environmental influences. Or genomic investigation may force researchers to redefine intelligence completely. Neither of these is a loss, of course, and we should and must be open to what empirical study might show. But I interpret the potentially ambiguous position of intelligence as an object of genomic study, and the likely impossibility of providing an *evolutionary* explanation of it, adaptationist or otherwise, in further light of Turkheimer's gloomy prospect, as underscoring a central thesis of *Our Genes*: Genomic study can tell us a lot, and yet its explanatory power is limited and highly contextual.

Female Orgasm

Human evolutionary genomics is a rich integrative field, and depending on the phenotypic trait(s) or gene(s) under study, the success of explanation will likely always vary. Therefore, let's digress to mythology for a moment. Tiresias's fateful encounter with Jupiter and Juno is recounted in Ovid's *Metamorphosis*, an important source of Roman mythology and history, written around the time of Jesus's birth. In Ovid's telling, Tiresias had been transformed from a man to a woman after striking two large snakes mating in a forest; seven years later, the same action transformed her back into a man. One evening, Jupiter and Juno were drunkenly arguing about whether women or men experienced the most sexual pleasure – Jupiter claimed women did, Juno asserted the opposite. Because he had lived as both, they called in Tiresias, who agreed with Jupiter. Irate, Juno struck Tiresias blind; Jupiter subsequently gave Tiresias oracular divinatory powers.[34] A key Greek source even has Tiresias uttering a ratio of 10:1 for how much more women enjoy sex compared to men – an immense order of magnitude difference.[35]

---

[34] Ovid, *Metamorphoses*, Bk III:316–338. See https://ovid.lib.virginia.edu/trans/Metamorph3 .htm#476975711.

[35] Pseudo-Apollodorus, 3.6.7. See www.theoi.com/Text/Apollodorus3.html.

Today, we know, with reasonable certainty, that there is truth to this knowledge of the ancients, even if 10:1 is excessive. As one source, according to Angier, the glans clitoris has approximately twice the number of nerve endings as the glans penis, although the external clitoris is smaller.[36] Additionally, while it is a different measurement, Shih et al. respected variation and measured the number of receptors per predetermined unit area in stained glans tissues of four female and four male cadavers. They found males had a much lower number of receptors per unit area than females, both as a spread and on average: 1, 2, 2, and 3 (males) and 1, 6, 7, and 14 (females).[37] Thus, whether the measure is the total number of receptors, or the number per unit area, women, indeed, are potentially equipped to enjoy sex more than men.

However, when it comes to the most overt expression of sexual pleasure – orgasm – the female version remains something of a scientific mystery. Upon investigation, we might find ourselves asking why there is a female orgasm at all. Male orgasm is tightly linked physiologically to ejaculation and hence has a clear evolutionary function, insofar as ejaculation is necessary for conception and reproductive success. It also occurs reliably with sexual intercourse.[38] In contrast, female orgasm has not yet been satisfactorily explained evolutionarily or genetically, even though orgasmic physiology is understood relatively well.

We have spent some time evaluating human evolutionary genomic explanations in spleen size as an adaptation in Bajau populations, and in intelligence. In the case of the Bajau from Chapter 7, we saw different explanatory components, some driven by paradigms such as mechanism or historicism, integrate into an overarching kind of total adaptationist explanation. In the case of intelligence, itself a complex and contentious trait, we found only partial and incomplete evidence for a genetic underpinning. We also saw that any kind of evolutionary explanation for intelligence, while of inordinate, generalized interest, adaptationist or otherwise, remains speculative and with unclear standards of evidence.

---

[36] Angier (1999) writes, "it is … a private joke, a divine secret, a Pandora's box packed not with sorrow but with laughter" (p. 64).

[37] See Table 1 in Shih et al. (2013, p. 1786); the unit area is approximately $1.9 \, mm^2$.

[38] Regarding the prevalence of male orgasm relative to female orgasm, based on results on 1931 U.S. adults who responded to the 2009 National Survey of Sexual Health and Behavior, Chalabi (2015) reports a much higher orgasm frequency in the last sexual encounter for men than for women (91% to 64%). There are male dysfunctions of various sorts, and some elements of male orgasmic and ejaculatory physiology remain to be discovered (cf. Alwaal et al., 2015). However, this is not the focus here as the evolutionary reason for male orgasm is clear.

Reaching for an evolutionary explanation of female orgasm provides a useful contrast to these two cases. It helps show the range of explanatory options available in the ever-growing field of human evolutionary genomics. Ultimately, there are good reasons to believe in a nonadaptationist, *developmental* (fitting comfortably within a mechanism paradigm) and *historical* explanation for the evolution of the trait of female orgasm. This appears to be so even if the genetics of female orgasm are hardly worked out.[39] Against the backdrop of the robust adaptationist evolutionary explanation for freediving traits in the Bajau, and overarching skepticism about evolutionary explanations in the case of intelligence, we have for female orgasm a highly plausible general developmental-historical evolutionary explanation. Yet, in terms of studying the underlying genetics, the case of the female orgasm is more like that of intelligence: We simply do not yet know very much about the genes underlying various features of (female) orgasm, or about potential gene–environment interactions.

Let's start simply: As with any other trait considered an object of evolutionary forces, and which possibly has some genetic basis, the female orgasm requires a working definition. Phenomenological trait individuation could involve asking individuals about the timing, frequency, and nature of orgasm, potentially in different sexual acts and contexts. Perhaps more positivist and replicable measures include ascertaining physiological states such as muscle movements or hormonal profiles, or averaging out individual variability by postulating qualitative models of different phases of sexual response.

I follow especially philosopher of science Elisabeth Lloyd in committing to "the relatively reductionistic biological descriptions of female orgasm," defining female orgasm as "involv[ing] a sensory-motor reflex including clonic contractions (spasms) of the pelvic and genital muscle groups." Lloyd's definition permits comparing human orgasms to those of nonhuman primates.[40] It also provides a much more operationalizable and robust measurement of orgasm than any trait definition bringing in emotional, experiential, or phenomenological features. Another reductionist set of criteria individuates female orgasm by hormonal – or, to a lesser extent, neurotransmitter – profiles. Certain hormones in the blood increase dramatically as a function of orgasm in both females and males,

---

[39] See, e.g., Dawood et al. (2005), Dunn et al. (2005), Burri and Spector (2008, 2011), Burri et al. (2009), and Jannini et al. (2015).

[40] Lloyd (2005a, pp. 21–23).

including prolactin and oxytocin, and certain neurotransmitters also increase dramatically with orgasm, such as dopamine (which can itself increase also with certain foods, musical experience, and drugs).[41] Finally, in their classic study, Masters and Johnson developed their four-phase model of the human sexual response: "(1) the excitement phase; (2) the plateau phase; (3) the orgasmic phase; and (4) the resolution phase."[42] Using this model, they diagram a single, typical male sexual response and three unique female sexual responses (Figure 8.1, chapter opener). Fortunately, all these reductionist definitions are highly correlated, precisely because (female) orgasm involves the muscle contractions favored by Lloyd, hormonal and neurotransmitter spikes, and the orgasmic phase singled out by Masters and Johnson.[43] Even so, such a reductionist strategy for defining and individuating the trait of female orgasm does not and cannot – by design – fully describe the female orgasm's multidimensional physiological, psychological, and emotional richness.

Now that we have individuated the trait of female orgasm reductionistically, let's turn to how and why it likely is not an adaptation. The action of natural selection, while strong and frequent for some human traits, is neither omnipresent nor omnipotent. As we have seen in earlier chapters, other evolutionary forces exist and act. Moreover, organisms are complex, with selection on one trait, in one sex, perhaps correlating with changes in another trait, or with that trait in the other sex, despite a distinct, though related, anatomy. There are indeed good empirical and theoretical reason to *not* think of female orgasm as an adaptation.

In fact, research fails to find *any* correlation between female orgasm and evolutionary fitness. Women easily and regularly get pregnant without orgasm, and sexual intercourse may be promoted by the wish for

---

[41] On prolactin, see Figure 2 in Krüger et al. (2005, p. 132) and Leeners et al. (2013); on prolactin and oxytocin, see Pavličev and Wagner (2016, pp. 4–7); and on various hormones and neurotransmitters, see Clarke (2019).

[42] Masters and Johnson (1966, p. 4).

[43] Other classifications of female orgasm exist, including clitoral versus vaginal (first articulated by Sigmund Freud), or orgasm by intercourse, partnered sexual contact other than intercourse, or masturbation (Dawood et al., 2005, p. 32). Harking back to the discussion in Chapter 4, these are "splitting" moves – seeing many traits where others only see one. There are good methodological reasons not to make these kinds of distinctions, both because the above three kinds of measures – muscle, hormonal/neurotransmitter, and phase – are correlated, and because, as I shall show, we can more effectively theorize evolutionarily and medically by not multiplying types of female orgasm beyond necessity. One website declaims: "Evidence suggests that the physiological and hormonal changes as well as the physical sensations experienced during orgasm are similar amongst the different 'types' of orgasm" (www.myvmc.com/lifestyles/female-orgasm/). Admittedly, this is controversial territory.

orgasm but hardly depends on it.[44] For instance, a well-designed twin study of over 8000 Finnish female nonidentical and identical twins based on self-reporting survey questionnaires found no genetic correlation between frequency of orgasm and number of offspring.[45] As the latter is a proxy for reproductive success, this immense study found no evidence for a link between orgasm and evolutionary fitness. Or consider Lloyd's convincing review of approximately 20 adaptationist theories for the evolution of female orgasm, which finds all of them lacking.[46]

One general reason, among several, for the difficulty with adaptationist hypotheses is the discordance or "gap" between intercourse and female orgasm. Two large twin studies published in 2005 find broadly similar patterns (Table 8.1): Many more women "always" orgasm by masturbating or through nonintercourse stimulation with partners (e.g., oral sex) than with sexual intercourse. Furthermore, for frequency of orgasm with sexual intercourse, there is a fairly flat distribution curve, with a significant number of women rarely or never orgasming. This is Lloyd's general finding, in analyzing 32 studies carried out throughout the twentieth century, presented in her Table 1 (Table 8.1, right-most column).[47] Female orgasm, she argues, is an evolutionarily unnecessary bonus.

Intercourse is simply not a very efficient way for women to orgasm. Other methods lead to much greater orgasmic success, even with the same partner (e.g., "always" row in Table 8.1). One website makes the point particularly well:

If the female orgasm was an evolutionary trait which assisted reproduction, one would expect orgasmic response to be highest during sexual acts which enabled reproduction (i.e. penetrative, vaginal sexual intercourse). On the contrary, most women emphasise the importance of clitoral stimulation for achieving orgasm, and the vast majority (>98%) employ at least some form of clitoral stimulation during masturbation. This evidence, combined with evidence from studies of female orgasm in primates, suggests that orgasm does not

---

[44] "No one is arguing that the clitoris – in its role of producing sexual excitement in the female, thereby promoting her to engage in sexual activity – does not play an important role in female fitness" (Lloyd, 2005a, p. 159); "the evolutionary account of female sexual excitement is distinct from an account that concerns orgasm, since sexual excitement often occurs in the absence of orgasm" (Lloyd, 2005a, p. 39).

[45] Zietsch and Santtila (2013).

[46] Eighteen of the studies are summarized in Table 2 in Lloyd (2005a, p. 104). See her chapter 3 on pair-bond accounts and chapter 7 on sperm competition and the uterine upsuck hypothesis.

[47] Lloyd (2005a, pp. 28–34).

Table 8.1. *Reported frequency of females experiencing orgasm per sexual act (heterosexual)*

	Dawood et al. (2005) $(N = 3080)$[a]			Dunn et al. (2005) $(N = 4037)$[b]		Lloyd (2005a)[c]	
	Sexual intercourse	With partner, other than intercourse	Masturbation	Sexual intercourse	Masturbation		Sexual intercourse
Always	5%	10%	27%	Always 14%	34%	Always	25%
Almost always	18%	17%	11%	>75% 23%	25%		
Usually (60–80%)	13%	11%	5%	51–75% 11%	8%	More than half the time	55%
Often (40–60%)	12%	10%	4%	About 50% 13%	7%		
Fairly often (20–40%)	14%	14%	5%	25–49% 8%	5%	Sometimes	23%
Rarely (<20%)	21%	20%	7%	<25% 16%	7%	Rarely or never	33%
Never	14%	14%	11%	Unsure/ never 16%	14%		
Do not do	4%	5%	31%			Never orgasm at all	5–10%

Dawood et al. (2005) surveyed twins in Australia, and Dunn et al. (2005) surveyed twins in the United Kingdom. The methodology is standard questionnaire self-reporting. Frequencies are per sexual act.[48]

[a] Rounded to the nearest integer; N is the total number of women responding to the questionnaire.
[b] N is the total number of women responding to the questionnaire.
[c] Metareview of 32 studies. Approximate composite means[49] – figures do not add up to 100% because different studies deploy different scales (e.g., "always" overlaps with "more than half the time" across studies).
Sources: Table 1 in Dawood et al. (2005, p. 29); Table 2 in Dunn et al. (2005, p. 261); and Table 1 in Lloyd (2005a, pp. 28–34).[50]

[48] See, e.g., Lloyd (2005a, pp. 39–43). In Dawood et al. (2005), respondents answered the following questions: "1) When you have sexual intercourse (i.e., during penetration of the penis), how frequently do you have an orgasm?; 2) How often do you have an orgasm with your sex partner, in ways other than sexual intercourse (e.g., during oral sex)?; and 3) When you masturbate, how frequently do you have an orgasm?" (p. 28). See Lloyd (2005b) for commentary on Dunn et al. (2005).

[49] Provided by Lloyd (2005a p. 36).

[50] See Symons (1979) for a similar summary of questionnaire data on Western peoples (pp. 83–84), highlighting the clear empirical evidence of the importance of clitoral stimulation to orgasm, and the relevance of ethnographic data (pp. 84–86, 91–92) – "cross-cultural variability in the occurrence of female orgasm is astonishing" (p. 84). Symons (1979, p. 86) cites Davenport (1977), who from the anthropological literature had abstracted out that "again and again, there are reports that coitus is primarily completed in terms of the man's passions and pleasures, with scant attention paid to the woman's response" (p. 149).

in fact occur in "normal" reproductive sex in which male orgasm and ejaculation are often considered the end point and of primary importance. Rather they can only occur when the sexual act is focused on providing the female partner sexual pleasure.[51]

Although future work may possibly elucidate some adaptive reason or selection scenario for female orgasm, currently female orgasm does not seem to meet two criteria for a total adaptationist explanation that we met in Chapter 7: adaptive scenario and evidence of selection. As I see it, this is precisely what makes the evolution of the female orgasm such an exciting – and even paradigmatic, under different, nonadaptationist paradigms – case for human evolutionary genomics. As we shall see, a robust nonadaptationist explanation, one that integrates developmental mechanisms and phylogenetic history, seems plausible for the evolution of female orgasm.

Given the *intercourse–female orgasm gap* and the *noncorrelation of female orgasm and fitness* found across the many studies that we reviewed above, and archived by anecdotal evidence, what, then, *is* the evolutionary explanation for female orgasm, if any? A developmental–historical explanation is by far the best candidate. As background for the developmental–historical explanation, the importance of one organ, the clitoris, should be considered. A wealth of evidence points to the clitoris as orgasm's epicenter. Masturbating until orgasm involves massaging the labia and stimulating the clitoris. Orgasm-oriented partner assistance during vaginal-penetrative sex almost invariably involves clitoral glans stimulation, whether manual, oral, or with toys. Or, as one respondent to sex researcher Shere Hite put it: "Sex in the best of all possible worlds? My clitoris would be in my vagina, for Christ's sake, so I could come when I fuck!"[52] As illustrated in Figure 8.5, the clitoris is a much larger organ than is immediately apparent. The clitoral glans with its thousands of nerve endings is only the proverbial tip of the iceberg. Potentially the controversial "G-spot" is precisely where the clitoral vestibular bulbs touch the vaginal canal, receiving pressure when engorged under sexual

---

[51] See www.myvmc.com/lifestyles/female-orgasm/.

[52] Hite (1976, p. 133), cited in Symons (1979, p. 88). As another example of the importance of the clitoris, and of basic biological parameters, for attainment of female orgasm, Wallen and Lloyd (2011) analyzed two classic historical empirical studies to find that there indeed is a strong correlation between, on the one hand, "likelihood" of orgasm during intercourse and, on the other hand, the *shorter* the distance is between the woman's clitoris and her urethral meatus (this distance is technically known as *CUMD*).

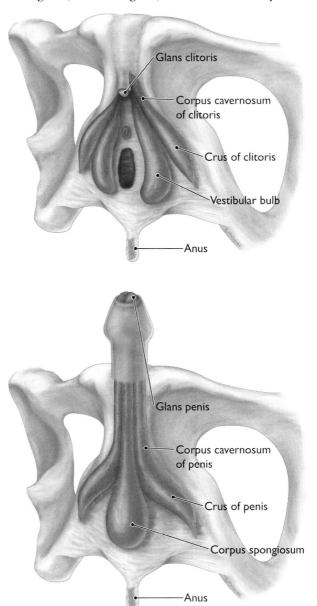

Figure 8.5. Adult human genitalia

Homologous structures of adult human genitalia in (top) females and (bottom) males include the glans, corpus cavernosum, and crus. The clitoris is much larger and more complex than externally apparent. Illustrated by Katelyn McDonald (CMI) with support from Rasmus Grønfeldt Winther, 2020. (A black and white version of this figure will appear in some formats. For the color version, please refer to the plate section.)

stimulation.[53] Essentially, the clitoris is the anatomical mechanism of female orgasm.

Given the importance of the clitoris to sexual stimulation, this organ is a useful lead for the likely evolutionary explanation of female orgasm. The clitoris and the penis are the "same" organ – they are homologous in structure (Figure 8.5), and are homologous in development (Figure 8.6), as they develop from the same early embryonic tissue.[54]

Specifically, both the clitoris and the penis develop from the ambisexual *genital tubercle*. Interestingly, this process is a function not only of genes but also of androgens: Genotypic male embryos can develop a clitoris "due to defects and or absence of the androgen receptor," and genotypic female embryos can develop penile morphology when, for example, they have autosomal recessive alleles for congenital adrenal hyperplasia, and thereby produce excess androgens in utero.[55] Only after eight or nine weeks of human embryonic development are typical female and male developmental pathways engaged and morphologically noticeable. Furthermore, the two organs are physiologically or functionally homologous vis-à-vis orgasm. Although the clitoris has more nerve endings per unit area – and probably more absolutely – the glans of the clitoris and penis have "indistinguishable" nerve endings.[56]

The anatomical and functional homology of clitoris and penis are central to the persuasive *byproduct account* of female orgasm. Donald Symons has presented ample evidence, across Western and non-Western cultures, and across primate and other mammalian species, for the relative rarity and nonadaptiveness of orgasm in mammalian females, and in human females in particular.[57] He considers orgasm "a potential all female mammals possess," realized more by humans than other mammals because of our greater interest in, and techniques of, foreplay (and, as it were, *afterplay*).[58] Symons's book baptized a strong hypothesis for the evolution of the female orgasm: "the potential for female orgasm can be understood as a byproduct of selection for male orgasm."[59] Stephen Jay

---

[53] Foldès and Buisson (2009). In a study of full clitoral anatomy, O'Connell and colleagues (2005) assert that "findings in cadavers and on MRI did not reveal any additional structure separate from the bulbs, glans or corpora of the clitoris, urethra and vagina that could be regarded as the G spot" (p. 1194). More informally, as O'Connell told the BBC, the vaginal wall *is* the clitoral bulbs (Leasca, 2017).

[54] See also Wallen and Lloyd (2008) and Pavličev and Wagner (2016).

[55] Baskin et al. (2018, p. 75).

[56] "It seems that genital corpuscular receptors are variations of a single theme: a terminal collection of sensory nerve fibers twisted in a skein-like fashion resulting in an increase in surface area for the perception of pain, temperature, vibration, and touch" (Shih et al., 2013, p. 1788).

[57] Chapter 3 in Symons (1979).   [58] Symons (1979, p. 89).   [59] Symons (1979, p. 94).

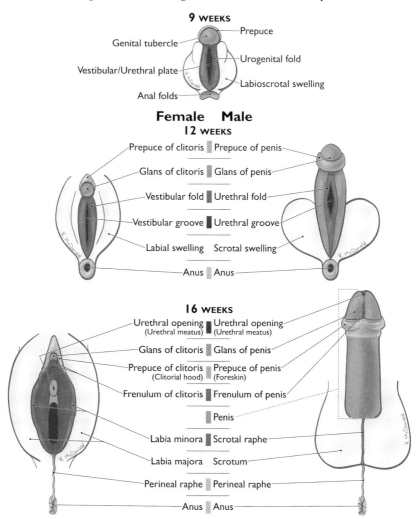

Figure 8.6. Development of human genitalia
Homologous structures at three developmental stages of human genitalia. The first represented stage is the ambisexual stage at roughly 9 weeks. At approximately 16 weeks the genitalia of the female and male embryo are reasonably differentiated. Illustrated by Katelyn McDonald (CMI) with support from Rasmus Grønfeldt Winther, 2020. (A black and white version of this figure will appear in some formats. For the color version, please refer to the plate section.)

Gould declaimed similarly in "Male Nipples and Clitoral Ripples": "The reason for a clitoral site of orgasm is simple – and exactly comparable with the nonpuzzle of male nipples. The clitoris is the homologue of the penis – it is the same organ, endowed with the same anatomical organization and capacity of response."[60] Developmental identity and similarity ground the multiple genital homologies of females and males, as shown visually in Figures 8.5 and 8.6. Under the byproduct account, because female orgasm is associated with the clitoris, and because male orgasm is associated with penile stimulation, and male orgasm is under direct selection, the evolution of the female orgasm can and should be explained as a consequence of homological genital structures across females and males, structures which share developmental pathways and mechanisms, as well as a phylogenetic history.

Lloyd articulates and defends this theory. Following Symons, she evaluates various kinds of evidence pertinent to the byproduct hypothesis of the evolution of female orgasm, including evidence from across cultures, and from other species, as well as evidence of variability in orgasmic responses across individuals. Significant differences in the frequency of female orgasm across cultures – as well as the enormous variability in frequency across individuals but within cultures – is taken by Lloyd as strong evidence for both cultural and biological influences on the trait. Regarding biology: "there has not been selection on [female] orgasm with intercourse."[61] The latter is meant in two ways: First, there has not been direct selection, in which, e.g., higher orgasmic rate or capacity leads to higher fitness; second, there has only been *indirect* selection – because of developmental and historical genital homologies, female orgasm is the evolutionary byproduct of direct selection on male orgasm. Male orgasm is strongly physiologically connected to ejaculation and sperm delivery.[62] Female orgasm is a byproduct, or in Lloyd's words, a "fantastic bonus."[63]

I take the byproduct account to be one among several possible theories or hypotheses in the nonadaptationist family of robust evolutionary explanations, both of female orgasm in particular, and of complex traits in general. A related, developmental–historical hypothesis of the evolution of female orgasm can be found in Pavličev and Wagner's *vestigial trait* account, which takes seriously Symons's suggestion of the orgasmic potential of *all* female mammals. As can be seen in Figures 8.7 and 8.8,

[60] Gould (1993, p. 86).   [61] Lloyd (2005a, p. 133).
[62] Lodé (2020) provides a useful taxonomic and phylogenetic distribution of intromittent organs in vertebrates. His discussion of orgasm is controversial. As we've seen, evolutionary discussions of these matters, including the above, reasonably assume a strong connection between male orgasm and fitness. This assumption does admittedly seem worthy of measurement, if not critique, despite methodological challenges such as how to measure variability in an association – orgasm–ejaculation – that is tightly and generally linked.
[63] Lloyd, personal communication, May 2020.

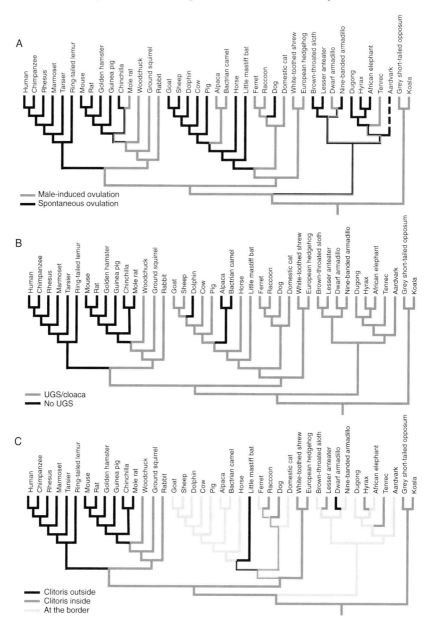

Figure 8.7. Phylogenetic distribution of mammalian female reproduction Across mammals, (A) modes of ovulation, (B) occurrence of the urogenital sinus (UGS; in basal species: cloaca), and (C) clitoral position in relation to the orifice of the UGS *or* vagina (inside, at the border, or outside). From Pavliček and Wagner, 2016, p. 5. © 2016 Wiley Periodicals, Inc. Reprinted with permission. Adapted by Mats Wedin and Rasmus Grønfeldt Winther.

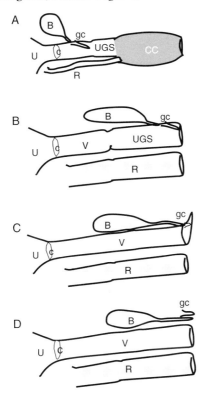

Figure 8.8. Female genital comparative anatomy
The comparative anatomical and phylogenetic trends of mammalian female genital anatomy: (A) a single anal and urogenital opening (i.e., cloaca), with an internal clitoris (e.g., koala); (B) anatomical distinction between rectal opening/anus and the (shortened) UGS, with the clitoris at the UGS orifice (e.g., pig, cow, cat); (C) clitoris at vaginal orifice (e.g., alpaca, dolphin); and (D) increasing phylogenetic "migration" of vaginal and urethral orifices, with clitoris outside (e.g., primates and rodents). Two combinations not shown here are (E) UGS with external clitoris (little mastiff bat and dwarf armadillo, only, in Figure 8.7) and (F) no UGS with internal clitoris (nonexistent in Figure 8.7). U = uterus; c = cervix; UGS = urogenital sinus; CC = cloaca; R = rectum; B = bladder; gc = glans clitoris, V = vagina. From Pavličev and Wagner, 2016, p. 8. © 2016 Wiley Periodicals, Inc. Reprinted with permission. Adapted by Mats Wedin and Rasmus Grønfeldt Winther.

Pavličev and Wagner collate phylogenetic, anatomical, and physiological evidence to defend their basic argument: *human female orgasm is a remnant of an ancestral neuroendocrine mechanism connecting a trigger (copulation) to a result (ovulation)*. This is a big-picture, historical thesis. It compares female genital anatomy, as well as the variety of types of ovulation causes, in different species of mammals.

To clarify and deepen their argument, and thereby provide more evidence and plausibility for a hypothesis that powerfully contrasts with the typically regnant adaptationist hypotheses, a little more detail is in order. First, ovulation can be triggered by environmental factors such as seasonally warmer temperatures; by the act of copulation, also known as *male-induced ovulation*; or through a well-defined physiologically based hormonal cycle, as happens in primates including humans, also known as *spontaneous ovulation*.[64] Seasonal shifts can affect both the male-induced and spontaneous ovulation causal types, and Pavličev and Wagner focus on comparing male-induced and spontaneous ovulation (Figure 8.7, A).[65]

To provide evidence for their vestigial trait hypothesis, Pavličev and Wagner needed to collate information about female reproductive comparative anatomy – in particular the relative location of the clitoris, the reproductive canal, and genital orifice (Figure 8.8). On the mammalian phylogenetic tree, they mapped out whether the clitoris was in one of three places: inside the canal; at the border of the canal and orifice; or outside and far, as it were, from the canal and orifice, as it is in humans (Figure 8.8, A–D; Figure 8.7, B and C). Pavličev and Wagner saw a generalized trend toward "increasing compartmentalization" in female urogenital anatomy across mammals.[66] This may be true, but if one takes a primate-centric perspective, it appears especially true.

At a minimum, it is important to observe the fairly neat phylogenetic matching of ovulation types with two features of mammalian female urogenital anatomy: separation of the reproductive canal from urinary or defecatory canals and orifices (Figure 8.7, B), and clitoral location (Figure 8.7, C). Such phylogenetic matching – not to say mapping – shows that lineages and species with the excitable glans clitoris *inside* the cloaca or urogenital tube tend *also* to exhibit male-induced ovulation (Figure 8.7, A and C).[67] There, the glans clitoris sits as a kind of detector of male penetration, waiting to trigger ovulation through the neuroendocrine

---

[64] The distinction is hardly absolute. Even in spontaneous ovulators like humans, there may be copulatory influence on, e.g., implantation of fertilized eggs. Pavličev and Wagner (2016) cite a study of relative success of implantation of in vitro fertilized patients in the group asked to engage in intercourse (with male ejaculation) briefly before the fertilized egg transfer, and the group "instructed to abstain from sex" before the procedure (p. 6, citing Tremellen et al., 2000).

[65] Pavličev and Wagner (2016, pp. 3–4).   [66] Pavličev and Wagner (2016, p. 8).

[67] In contrast, those lineages and species with the clitoris at the border tend to possess spontaneous ovulation. To be precise, of the 12 "clitoris inside" species, 9 (75%) are "male-induced ovulation," while of the 15 "at the border" species, 11 (73.3%) are "spontaneous ovulation."

reflex. In this role, female orgasm is putatively adaptive, although it has lost that function in humans and other primates.[68]

To complete their argument for their vestigial trait account, Pavliček and Wagner worked to ground their historical hypothesis in mechanistic evidence. While more research remains to be done, Pavliček and Wagner take seriously the controversial hypothesis of female orgasm and associated hormonal profiles in nonprimate species being somewhat similar to human female orgasm profiles, citing evidence from Kinsey that female ferrets, cats, and rabbits (all of which are male-induced ovulators; Figure 8.7, A) can climax. Females of other mammalian species may reach orgasm, or perhaps just feel, as it were, intense pleasure through copulation, with the clitoris inside the reproductive canal. Robust evidence is also presented that the hormonal profile of orgasm in human females, including spikes in prolactin and oxytocin, closely resembles the hormonal profile of copulation-induced ovulation.[69] In later work, Pavliček, Wagner, and other collaborators adduced experimental evidence in favor of a consequence or prediction of their hypothesis: Reducing orgasm intensity or frequency in copulation-induced ovulators such as rabbits also reduces ovulation.[70] Thus, significant evidence has been accumulating for the vestigial trait account. Again, this shows that adaptationist explanations are not the only evolutionary explanations in town, and that in their explicative capacity they always need to be supplemented with more context anyway (as in the Bajau freediving case).

Ultimately, the vestigial trait account is an intriguing, broad-strokes hypothesis for the origin and function of female orgasm in mammals. There may be multiple selective reasons for the emergence of the triggering function of orgasm for ovulation in ancestral mammals – for

---

[68] While accepting the noncorrelation of female orgasm and fitness, Pavliček and Wagner (2016) remain open to the possibility that female orgasm in humans took on new functions or adaptations, "such as [pair] bonding or [muscle] contractions, that are specific to the primate lineage" (p. 9) – that is, new selection regimes kicked in in our lineage. This stands in contrast to the byproduct account's explicit and thorough anti-adaptationist stance. Even so, the explanatory work of the vestigial trait account is done by conceptual and theoretical resources within mechanism and historicism paradigms.

[69] To a much lesser extent, it also resembles the menstrual hormonal cycle, where, for example, luteinizing hormone is key (Pavliček and Wagner, 2016, pp. 4–7; on luteinizing hormone, see Ray, 2018). On ferrets, etc., see Pavliček and Wagner (2016, p. 2).

[70] Pavliček et al. (2019) performed appropriate controls, to help establish that the experimental protocol of giving orgasm-suppressing Prozac to rabbits actually reduced the ovulation rate via the central nervous system, which is a conduit for the orgasm trigger, rather than by direct action on the ovaries.

example, the availability of males or the requirement of appropriate environmental conditions. At least some of the future discussion for developmental–historical hypotheses, I believe, must address how to operationalize and measure "sameness" of profiles of both hormones and neurotransmitters, for orgasm and for ovulation, across species with different ovulation cause types, and across females and males. Future work could also assess mechanistic and functional reasons and primacy of whether ovulation type (e.g., spontaneous) evolved prior to, simultaneously with, or subsequently to the anatomical externalization of the clitoris. The case of female orgasm illustrates that comparability and temporality of complex systems – namely, organisms and in particular the human organism – are essential features of empirical study and philosophical reflection in human evolutionary genomics.

So far, the evidence for evaluating female orgasm as an adaptation stems primarily from survey questionnaires, comparative studies across species, and developmental, anatomical, and physiological studies. But to establish a genetic basis for female orgasm requires establishing a clear candidate gene – or set thereof – using molecular and statistical methods. There is no doubt that genes are relevant here. Journalist and scientist Olivia Judson's critical review of Lloyd's work points out that the causes of variation in orgasmic experience in women could be due to variation in any variety of factors, including phenomenological or cultural ones.[71] While there is truth to this, studies of many sorts indicate the clear relevance of biological (hence, partly at least, also genetic) factors, showing a significant relationship between higher orgasmic experience and shorter distance of the clitoris from the urethra;[72] a clear phylogenetic signal of migration of the clitoris out of the reproductive canal in primates and rodents;[73] and significantly higher orgasmic rates in women through masturbation than sexual intercourse (see Table 8.1).

The case of the female orgasm helps clarify that some traits can have a genetic basis and yet exist independently of natural selection.[74] Identical (monozygotic) twins are much more genetically similar than fraternal (dizygotic) twins, holding all (rather than half, on average – just like regular, full siblings) segregating genetic variation in common.

---

[71] Judson (2005, p. 917).   [72] Wallen and Lloyd (2011).   [73] Pavlićev and Wagner (2016).

[74] Lest we fall into adaptive just-so storytelling per Gould and Lewontin (1979), we must acknowledge that there can be evidence *against* adaptation, not just lack of evidence that a given trait is an adaptation. Making adaptation not only the null hypothesis, but also the imperialist, ubiquitous explanation is an extreme form of adaptationism.

Moreover, we can divide the environment of twins raised in the same family into shared environment and unique (or nonshared) environment.[75] If the environmental effects on sets of identical twins equal those for fraternal twins (i.e., the "equal environments" assumption is met),[76] then there is no genotype–environment covariance, and if all non-additive components of genetic variance, and genotype-by-environment interaction, are also negligible, then the additive genetic variance fully accounts for differences in the degree of resemblance between the two classes of twins. Only when these problematic assumptions hold may a simple comparison of differences in the correlations within classes of twins be used to estimate the narrow sense heritability. As for female orgasm, Dawood and colleagues' twin study found commensurable amounts of additive genetic variance compared to variance in shared environment for orgasm during masturbation, orgasm during intercourse, and orgasm during sexual contact other than intercourse.[77] They thereby provided a *full* explanatory model that fits the data and suggests both genetic and environmental reasons for the distribution curve of female orgasm.

Most generally, research into the genetic and environmental influence on the female orgasm shows how human evolutionary genomics can continue to ask provocative questions and explore meaningful, genetically based answers about origins and identity. However, work on the female orgasm also shows the dangers – of simplification, of reification – inherent to the field. For instance, due to its focus on deep history, cross-species comparison, and narrow biological function and definition, the evolutionary study of female orgasm can and does abstract away from emotional and cultural aspects. Moreover, as this exploration helps illustrate, adaptationism is not, and should not be, the gold standard methodological heuristic and ontological assumption in evolutionary integrative explanations.

The Distinction Between Gene and Environment, and the Adaptationist Paradigm

The distinction between gene and environment both illuminates and blinds. While it can shed light on different kinds of disease etiology and suggest biomedical strategies, it can also overly biologize and narrow

---

[75] See Plomin and Daniels (1987), Turkheimer and Waldron (2000), and Polderman et al. (2015).
[76] But see Lewontin (1995 [1982], pp. 97–103) and Falconer and Mackay (1996, pp. 171–174).
[77] Figure 1 in Dawood et al. (2005, p. 30).

Figure 8.9. Infinitree of Life
Improvisation on the recursive, profoundly interconnected, and dense tree or network of life, even as we proclaim the absolute centrality of *Homo sapiens*. *Infinitree of Life* by Alan Kennedy. © 2016 Aberrantt.com.

ambiguous traits. Furthermore, natural selection, and its associated adaptationist paradigm, is hardly the only explanation in the evolutionary toolbox. We witness the limits of adaptationism by the unreliability – and indeed potential unmeasurability – of adaptation in studies of intelligence and IQ, and by the likelihood of nonadaptationist developmental-historical explanations (i.e., the byproduct and the vestigial trait hypotheses) in studies of the evolution of female orgasm. Future philosophical work would include finessing and developing new rubrics, as it were, of total explanation for mechanism and historicism paradigms in human evolutionary genomics, beyond the six-component rubric for adaptationist total explanation considered in Chapter 7.

Moreover, Chapter 7 discussed the interpenetration – and the emergent possibility of neither–nor or synthesis – of distinctions such as gene and environment, nature and nurture, and even adaptationism and mechanism, or adaptationism and historicism. Ambiguous and historically entrenched, these distinctions persist to reframe our discussions. They are lenses through which we genomically analyze other species, connecting or alienating us from other members of the "Infinitree of Life" (Figure 8.9). In the complex and recursive set of biological phenomena, distinctions organize our thinking and practices of study, sometimes in a way that seems zero-sum or mutually exclusionary. Integration platforms are possible, however, not only in studies simultaneously drawing from adaptationist, mechanistic, and historical paradigms (such as the Bajau freedivers of Chapter 7), but also in a recent effort toward "a community-maintained standard library of population genetic models."[78]

Even if conclusive answers are scarce, we gain a better view on evolutionary theory and its predictive power by exploring the diversity of findings in the study of human evolution and human genomics. By shining various *spotlights* (distinctions, paradigms) – both serially and in combination – we may create a great *floodlight* that illuminates our many similarities, as well as our few meaningful differences, bringing us closer to the bigger answers we seek.

---

[78] Adrion et al. (2020).

Figure 9.1. Race
Representation of the fractured, cubistic, and fraught relations among
individuals, and populations, of *Homo sapiens*. Illustrated by Larisa DePalma.
© 2021 Rasmus Grønfeldt Winther.

9 · *Is Race Real?*

As we have learned throughout these pages, defining, investigating, and interpreting identity is complex. Many of our questions about *Who am I?* refer us to our local, geographic population(s) of origins, but just as many refer us to each other, as a whole, and to the human genome we each carry. This chapter zooms out from case study analysis to consider another form of identity that genomics helps us define and investigate – race (Figure 9.1, chapter opener). A book that approaches human evolutionary genomics through a philosophical lens would be incomplete without considering the particularly contentious analyses and interpretations associated with the genomic study of race. Conceived of as a social indicator of varying importance, race is linked to social structures and, thus, politics. Racism is the process of frequent and systemic judgments and prejudices based on perceived racial differences leading to differential access to social goods, including dignity, trust, and opportunity. We can work to ignore race, insist race does not matter, and strive for the post-racial future some of us may long to inhabit, but we do not live there yet, and a variety of futures are possible.

Given the capacity of human evolutionary genomics to identify and potentially negotiate the relationships between individuals and groups, and between different groups, the field occupies a powerful position in the forum of race debates. However, while genomics stands to uncover meaningful genetic indicators by which individuals can be identified as part of or apart from particular groups, it also accentuates race as a complicated concept, and only a limited set of the concept's multivalent meanings is suitable for genomic study. For example, researchers can use heterozygosity-based measures of genetic variation such as F_{ST} to find out how different certain populations of a large geographic region are from each other. Yet these findings (and others like them) can only tell us about biologically and, in particular, genomically grounded notions of identity. Such findings tell us nothing about race as a social indicator of individual or group identity.[1] What genomics tells us about race is therefore fundamentally ambiguous: We can use its tools and its associated disciplines to separate *Homo sapiens* into large clusters or populations of populations, but such aggregates gain meaning only through the interpretive values associated with them.

As presaged in Chapters 1 and 2, it is difficult to fully dissociate genomics from its historical development involving the intense dedication of nineteenth- and early twentieth-century students of heredity to eugenics. We cannot ignore the social, political, and economic abuse of marginalized populations deemed "unfit" through biased interpretations based on motivated pursuits. However, modern genomics, including some of the equations, metrics, and models mentioned in this book, may be a suitable aid in understanding race in an appropriately narrow sense. For example, heterozygosity and F_{ST} can be used for biogenomic racial taxonomic efforts, and genetic differentiation metrics can be deployed for inferring human history, as we saw in Chapter 4. Moreover, as we investigated in Chapter 7, integrating explanatory paradigms for certain populations adapted to extreme environments, such as the Bajau, can perhaps inspire the creation of effective biomedical interventions.

This chapter maps out realism, antirealism, and conventionalism at three levels of race: biogenomic cluster/race, biological race, and social race. It does so by exploring A.W.F. Edwards's 2003 provocative and somewhat influential response to Lewontin's classic 1972 paper "The Apportionment of Human Diversity," and then by turning to contemporary discussions. The chapter builds a nuanced conception of race, one that stands a chance of being meaningfully informed by genomics but that can also tell us something about each other and ourselves. The chapter ends by interrogating the relevance of these scientific discussions for political positions and possible futures.

Philosophers of Race Speak

But let's start with how some influential philosophers of race speak about, well, race. Scientific and technical details matter to understanding race and the social and academic discourse around race, and this is particularly the case when we also consider insights from philosophy. In fact, bringing genomics to bear on questions about race can conciliate the following complex, even contradictory claims:

---

[1] TallBear (2013) resonates with my analysis.

"The lack of fixed traits for each so-called race means that race cannot be inherited as is popularly thought. Rather, the specific physical characteristics variably associated with races in cultural contexts are inherited through family descent as is the rest of human biology. Race, therefore, supervenes on human genealogy or family inheritance."[2]

"There are no racial genes responsible for the complex morphologies and cultural patterns we associate with different races."[3]

[The] "logical core" of "the ordinary *conception* of race" involves identifying "a group of human beings" "(1) ... distinguished from other groups ... by visible physical features of the relevant kind," "(2) ... whose members are linked by a common ancestry," and "(3) ... who originate from a distinctive geographic location."[4]

The genomic lens gives us one, and only one, particular way of conceptualizing and thinking about race: We must supplement this perspective with other lenses.

Before we look more closely at the genomic lens, let's consider if identifying a "racial gene" could even work. Would it be a gene – or, more accurately, an allele – present in all individuals in a particular human population (or cluster of populations) but not in any other? Would its presence thus define a genetically unified race? Or would there be a lower bound of frequency (say, 90% or 60%), under which we could no longer conceptually delimit a race or, if that allele was present in a given individual, no longer clearly identify that individual as a member of that race, or both? And even if racial genes, as it were, do not exist, what about the correlational structure across loci, per empirical pattern #5 of Chapter 6? Could such correlational structure not produce clusters, not to say races? Already, we begin to see ambiguities requiring context and interpretation: Whereas splitters would argue for delimitation and definition, lumpers would likely set the lower bound to a very high value, implicitly arguing that no population is an island unto itself.

What would be the purpose of defining a racial gene? If, for example, the purpose of the racial gene depends on population identification, then would the races need to be defined relative to one another? For example, must the racial allele's frequency then also be 0% (or perhaps 10% or some other low number) in *other* races (or groups, or ethnicities, or populations)? To further complicate such an investigation, how different

---

[2] Zack (1999, p. 84).   [3] Haslanger (2000, p. 43).   [4] Hardimon (2003, pp. 451–452).

across the genome must populations be to count as distinct? Would population differences in allele frequencies of one gene with multiple alleles be sufficient; or should races be identified by 2, 5, or 15 distinct genes, together with their respective, associated alleles, and starkly distinct allele frequencies in different races?

Already, we must take a step back and ask to what we imagine the concept of *race* even refers? Is it the standard conception of U.S.-based racial categories? Or do we mean other racial or ethnic categories as conceptualized elsewhere, in any variety of languages and cultures (India, the Caribbean, or South Africa, to take just a few examples)? After all, if we're going to define and investigate race, in pursuit of better understanding and interpreting the possibility of a racial gene, how many distinct (social) classificatory systems of slicing up *Homo sapiens* into races or ethnicities should we countenance?[5]

In point of fact, human population groups as defined by human evolutionary genomicists need not match up with any of our conventional categories. Researchers sometimes use different categories altogether, such as aboriginal populations or theoretical clusters. Terminology is important because, as we learned in Chapter 5, conventional categories often and perhaps irrevocably theoretically *infect* – to use a provocative concept from the philosophy of science – the data collected and managed in variance partitioning and clustering analysis of the global human genome. It falls upon philosophers who are interested in both the genomic and biological aspects of race and the metaphysics and ontology of race to factor out different kinds of racial categorization – that is, to perform an assumption archaeology on the very concept, in the spheres of discourse of both human evolutionary genomics and the social sciences.

It seems especially important, when critically thinking about race, to emphasize that we still have much to learn, and much conceptual analysis remains ahead of us. CRISPR biotechnological applications, or results about genetic mechanisms, as illuminated with statistical analysis from genome-wide association studies, are likely to surprise many – researchers and the broad public alike, as well as philosophers and other critical thinkers.[6] For instance, light might be shed on disease etiology, or on the evolutionary mystery of the female orgasm, as we explored in Chapter 8.

---

[5] These are the kinds of questions that motivated me to co-edit Lorusso and Winther (2022).

[6] The acronym CRISPR is more widely familiar than what it stands for: clustered regularly interspaced short palindromic repeats – specialized strings of DNA serving as an adaptive immune system in bacteria and other prokaryotes. CRISPR technology makes easy gene editing possible.

Moreover, discovery about potential, plausible genetic underpinnings of at least some cognitive features may upset a variety of interlocutors from across the political spectrum. There is a trend among researchers in human genomics and philosophers studying biological aspects of race to underscore similarity at the expense of acknowledging the reality of some genomic differences. As a critical thinker, I resist this trend: I believe *we must do what we can to bolster the legal frameworks and moral justifications of human equality and justice regardless of the existence of systemic genetic differences not just among but also within large populations.*

The Reality and Reification of Race

Sharp ontological lines are continually drawn in the debates surrounding genomics and race. As I started to reference in Chapter 4, some claim that our finest genomic data and methodology indicate that human races are biologically *real* entities (e.g., Robin Andreasen, Charles Murray, Neven Sesardić, and Quayshawn Spencer). Others defend the well-established *antirealist* perspective that has been developed following Lewontin (1972) (e.g., Joshua Glasgow, Adam Hochman, and Naomi Zack). In this view, races are something like social fictions that do not exist biologically. This chapter articulates, in more philosophical and historical detail, a *constructivist conventionalism* about race in the context of genomic research. That is, realism or antirealism about biogenomic race are optional and open – contingent on various measures and models. Indeed, a given researcher, research program, or critical thinker can reasonably take a certain racialized taxonomy of *Homo sapiens* to be real or not. There is significant flexibility in the construction of biogenomically racialized classification. I argue that our best genomics will always underdetermine whether we should believe in the existence (or not) of genomically real human races.[7] Given particular choices and norms about how to interpret the genomic data and about which mathematical measures and methodologies to use, either realism or antirealism can be justified. As we have seen throughout *Our Genes*, this is in part because we are somewhere along the Galápagos-Writ-Large to Planet Unity spectrum, albeit closer to the latter. Therefore, in our search for identity and for belonging – to our current or ancestral populations, or both – each of us, lay reader included, must

---

[7] Weiss and Fullerton (2005) resonates with my argument.

be critical thinkers *ourselves* and learn genomic facts and methods to decide how real the populations to which we ascribe membership are. Differently put, using genomic knowledge to answer *Who am I?* always already requires our participation and (distributed) construction.

This chapter distinguishes three kinds of racial realism:

- *Biogenomic cluster/racial realism*[8] claims that population structure exists in *Homo sapiens*, assessed with genomic measures such as F_{ST}.
- *Biological racial realism*[9] affirms that a stable mapping exists between the groups characterized genomically and the social groups identified as races. That the groups are biological populations – interpreted as having certain averages and ranges of phenotypic features, or understood in an essentialist manner, or both – explains why these particular social groups, rather than others, are so identified, and also helps stabilize social races in terms of corresponding, inherent genetic characteristics. Furthermore, for some biological racial realists – let us call them *the hereditarians* – the existence of biological populations (and of the genetically grounded properties of their constituent individuals) explains and justifies at least some social inequalities.[10]
- *Social racial realism* defends the existence of distinct human populations in our ordinary discourse and social interactions. Such groups are often identified and stabilized by surface factors, as it were, such as skin color or facial features. Moreover, while this is not strictly necessary for social racial realism, group membership is often correlated with access (or not) to goods, services, and wealth.

---

[8] The forward slash here indicates that this kind of realism is not necessarily about a race concept. As made evident in the workshop on "Genomics and Philosophy of Race," some influential and socially responsible evolutionary geneticists have no desire to become involved in debates over race (Winther et al., 2015). However, genomic work of this sort is often *taken to be* about race, and it is not clear that the semantic slippage between "cluster" and "race" is avoidable (consult Reardon, 2005; Feldman, 2010; Morning, 2011; Happe, 2013; Donovan, 2014; and Donovan et al., 2019).

[9] I call this *biological racial realism* because the concept, and term, has a history in these debates. Some of the confusion in the current literature stems from failing to distinguish biogenomic cluster/race from what has often been called *biological race*. Acknowledging the existence of genetic population structure need not in any way imply a hereditarian commitment to the reality of genetically based properties and differences constituting (or explaining) either the existence of socially identified races, or, especially, the "racial" characteristics about which debate revolves, such as IQ or health disparities.

[10] See, e.g., Jensen (1969), Herrnstein and Murray (1995), Rushton (1995), Lynn and Vanhanen (2002), and Murray (2020).

Whether it is legitimate to divide the human species into smaller, genetically distinct populations depends on the purposes, measures, and models at play, many of which were investigated in detail in Chapters 3–5. I thus argue that *constructivist conventionalism* rather than realism is the proper stance toward investigating and understanding *biogenomic cluster/race*.

Biological races exist when a correlational – or, more accurately, a stable causal – mapping can be drawn between genetic group differences *and* the variety of distinct social groups based on those. Biological racial realism demands a one-to-one mapping between genomic groups (sometimes defined in an essentialist manner) and social groups. For instance, in discussions about intelligence and race, which we started exploring in Chapter 8, it is commonplace to follow conventions in the United States of a three- or five-race taxonomy and hierarchize accordingly.[11] However, I reject the existence of a strong correspondence or mapping between biogenomic and social classifications. That is, I am an *antirealist* about biological race.

Trying to determine (and argue about) the existence (or not) of biogenomic cluster/race would be a wholly intellectualist and detached endeavor if there were no political, ideological, and morally relevant consequences.[12] If putative group membership only determined relatively insignificant characteristics, such as toenail width, normative concerns would hardly be as important. The sticking point is about the reality (or not) of *biological* race. The entire issue of biogenomic groups, clusters, populations, or races, if you will, would not be politically, socially, or morally challenging if nothing social or moral depended on it.[13]

Commitments to realism, antirealism, constructivist conventionalism, or pernicious reification about either biogenomic cluster/race or biological race are broadly independent of questions regarding the reality of races understood as populations with socially ascribed meanings – entities that are real because those populations are socially identified, entrenched, and maintained. Such *social races* exist when there are psychologically and

---

[11] African American, East Asian, Euro American; sometimes with Indigenous American and Pacific Islander added – with, for example, Latin American or Middle Eastern somehow, somewhat falling outside of the taxonomy, though these are also standard social race groupings in U.S. public discourse.

[12] Discussed in various ways in, e.g., Lewontin (1970), Longino (2002, 2013), Hacking (2005), and Kitcher (2007).

[13] Helen Longino provided constructive feedback here.

communally perceived stable kinds of racialized people, often leading to profound and pervasive discrimination and oppression.[14] While a *post-racial* future might be possible and perhaps desirable, for now, *realism* about social races is the best description of the practices, expectations, and norms of many contemporary societies (though possibly not all). Social racial realism can imply a prescriptive point – which I make explicit – of wishing to resist current racist power structures in order to attain a post-racial society, which remains an elusive social vision to characterize. Indeed, a post-racial society is really only accessible through rigorous, transparent, and ultimately limited investigation and interpretation of both the biological and the philosophical aspects of race.

In short, I defend constructivist conventionalism about biogenomic cluster/race, antirealism about biological race, and realism about social race.[15]

In line with broad themes from *Our Genes*, the constructivist conventionalist nature of biogenomic cluster/race provides the imaginative and discursive space, I believe, to explore questions about our own origins and identities in an open-minded and critical manner that is responsive and attentive to our collective futures. Thus, perhaps ironically, the future of social race might be seen to depend on the intrinsically ambiguous and changing nature of biogenomic cluster/race.

Lewontin's Distribution and "Lewontin's Fallacy": When the Single Shannon Just Isn't Enough

While the reality of race relies on explicit definitions and context to determine an applicable meaning, it also requires statistical methodologies for exploring the global human genome (e.g., variance partitioning

---

[14] See Mills (1997, 1998), Haslanger (2000, 2008), Fanon (2008 [1952], 2004 [1961]), Hacking (2005), Coulthard (2014), Taylor (2014), Alcoff (2015), Ásta (2018), and Kendi (2019).

[15] More generally, each of these concepts of race and racial realism corresponds to a set of debates and discourses on race. Biogenomic cluster/race concerns, for instance, which evolutionary genetic measures and models should be used to identify human populations, clusters, and groups, whether there is an appropriate level of human population structure that makes sense to privilege and emphasize, and whether a subdivision or continuum approach better describes human genomic population structure. Biological race discourse addresses the genomic correlates of social racial groupings and social inequalities. Finally, discourse about social race is about the influence that social identification and treatment have on inequality – issues of affirmative action, "color-blindness," and "identity politics" are, for instance, important in this third set of debates. Future projects could investigate these discourses rather than the narrower commitments of realism, antirealism, and conventionalism about each of the three concepts.

and clustering analysis), as well as a normative stance regarding the relevance and import of Galápagos-Writ-Large and Planet Unity patterns of human genomic variation. The ongoing debate about Edwards's critique of standard interpretations of Lewontin's distribution helps make clear the political ramifications of realist and antirealist conceptions of biogenomic cluster/race. Such themes are now here foregrounded because the methodological differences between Lewontin and Edwards may not be as significant as some believe: They seem to be asking different, but complementary questions, as we saw in Chapter 5 and with empirical patterns #4 and #5 in Chapter 6. Recall Lewontin's distribution. Most genetic variation is within groups rather than among groups: Approximately 86% of genetic variation is among individuals within populations, 7% is found across populations within continental regions, and 7% is found across continental regions. Edwards concurs and adds that in spite of relatively small group differences at individual loci, pooling information across loci allows for efficient clustering and classification. Lewontin subsequently affirmed that using information from many loci to cluster – e.g., Rosenberg et al. (2002) – is fundamentally a sound procedure: "The continental clustering in these large sets of data derives mainly from small differences in allele frequencies at large numbers of markers, not from diagnostic genotypes. This clustering reflects the history of human migrations."[16] However, this was never the question about "the average *amount* of genetic diversification between and within geographical groups."[17] Their broad methodological agreement, as well as their focus on different questions, suggests that the most important disagreements between Lewontin and Edwards lie in normative domains, beyond data and statistical methods.

Lewontin contrasted the vast importance tied to social ascriptions of racial identity with the meager amount of genetic difference actually found among broad races as typically defined. His finding that only a small fraction of total human genetic variation could be assigned to or accounted by what we usually think of as races has been taken, by many, to undermine our ordinary understanding of racial differences as biological. Racial categories are thus of "virtually no genetic or taxonomic significance."[18] The normativity is explicit in both Lewontin's 1972 article and 1974 book:

---

[16] Feldman and Lewontin (2008, p. 92).   [17] Feldman and Lewontin (2008, p. 90).
[18] Lewontin (1972, p. 397).

Human racial classification is of no social value and is positively destructive of social and human relations. Since such racial classification is now seen to be of virtually no genetic or taxonomic significance either, no justification can be offered for its continuance.[19]

The taxonomic division of the human species into races places a completely disproportionate emphasis on a very small fraction of the total of human diversity. That scientists as well as nonscientists nevertheless continue to emphasize these genetically minor differences and find new "scientific" justifications for doing so is an indication of the power of socioeconomically based ideology over the supposed objectivity of knowledge.[20]

That is, genetics cannot and should not be made responsible for the creation, maintenance, and importance of our current socially important racial categories. The contingency and weakness of the mapping between genomic differences and social races undermine the continued use of genetics in explaining and justifying social races.[21] Politics and science intertwine. Biological racial realism fails, under Lewontin's analysis, because there is no one-to-one mapping from biogenomic clusters/races to social races.

Lewontin's sustained ire on this topic has not been aimed so much at biologists interested in human genetic variation[22] as at the hereditarians.[23] Hereditarians argue that many important current social and political inequalities, both within and among nations, ethnicities, and other human groupings, are due in large part to hereditary differences in the (average) "native" abilities between races as usually conceived (e.g., hereditarians often appeal to alleged, if not real, genetic variance for intelligence). They endorse a Galápagos-Writ-Large picture of biological race. As social scientists, they are less concerned with the details of biogenomic cluster/race.[24]

Lewontin's critiques of the hereditarians were trenchant. His Marxism and antiracism should also not be forgotten in this context. Lewontin interpreted the market penetration of biological racial realism in

---

[19] Lewontin (1972, p. 397).   [20] Lewontin (1974, p. 156).

[21] See Lewontin (1970, 1972, 1974, 1993, 1995 [1982]), Feldman and Lewontin (1975, 2008), and Lewontin et al. (1984); see also Kaplan (2000).

[22] E.g., Lewontin (1978) is a brief response to Mitton (1977).

[23] Consult Jensen (1969), Herrnstein and Murray (1995), Lynn and Vanhanen (2002), Wade (2014), and Murray (2020). See also the controversy surrounding Jason Richwine's association with a Heritage Foundation report against immigration, itself partly based on his dissertation's claims regarding the lower IQs of "Hispanic" immigrants (Parker and Preston, 2013).

[24] But see Murray (2020).

intellectual and political life as evidence of "the power of socioeconomically based ideology"[25] and wished to resist it, in part because of its presumed scant genetic justification.[26] According to sociologist of science Jenny Reardon, "Lewontin himself conceived his article in the midst of his anti-racist political work with the Black Panthers in Chicago," even if, in correspondence with Reardon, Lewontin also "explained that he probably would have written the piece anyway, given that its 'technical side' ... would have interested him even if he had not been doing any political work."[27] Finally, his paper and politics had broad influence, as we also saw in Chapter 6.[28]

Edwards takes antirealists to task for letting their politics influence their science. He submits that premising – or at least mapping – moral equality on genetic similarity runs the risk of backfiring, should genetic dissimilarity among broad human populations turn out to exist. He pithily argues:

It is a dangerous mistake to premise the moral equality of human beings on biological similarity because dissimilarity, once revealed, then becomes an argument for moral inequality.[29]

To avoid this consequence, Edwards strongly distinguishes discovering biogenomic cluster/race from any social or political uses to which one could put such knowledge: "A proper analysis of human data reveals a substantial amount of information about genetic differences. What use, if any, one makes of it is quite another matter."[30] Edwards's analysis seems motivated by a high-level normative concern to not allow moral or political positions to impact science. Given this, and following interviews with Edwards,[31] I interpret Edwards as defending a biogenomic cluster/racial realism while remaining ontologically noncommittal about – and perhaps uninterested in – biological race or social race.

---

[25] Lewontin (1974, p. 156).

[26] Similarly, Coop et al. (2014) is a letter signed by more than 100 geneticists rejecting biological race and denouncing Wade's (2014) "guesswork" and "conjectures."

[27] Reardon (2005, p. 184).

[28] Cf. Mukhopadhyay and Moses (1997, p. 519) and Reardon (2005, p. 35). Two statements on race and racism by the American Association of Physical Anthropologists (AAPA) resonated with and were influenced by Lewontin's results (AAPA, 1996; Fuentes et al., 2019). For an earlier, analogous discourse, see UNESCO (1952), a statement that R.A. Fisher felt he could not endorse (Edwards, 2021).

[29] Edwards (2003, p. 801), as reprinted in Winther (2018a, p. 253).

[30] Edwards (2003, p. 801), as reprinted in Winther (2018a, p. 253).

[31] Transcribed in Winther (2018a).

Rereading this debate with the distinction between biogenomic cluster/race and biological race in mind, and focusing on ethical and normative concerns, helps clarify many outstanding issues. Lewontin did not deny that there is population structure in humans or even that biogenomic clusters/races can be found[32] but, rather, tried to block the usual inference from genetic differences among populations and groups to justifications for the existence and importance of social races. Edwards seems motivated by moral and political principles as well (including about how science should be practiced) but worried that Lewontin adopted the wrong strategy in not keeping his science and his political principles about race clearly separate.

An ongoing brainteaser emerging out of Edwards's baptism of "Lewontin's fallacy" is what we could call the *small percentage racial conundrum*. Are the genetic differences among races indeed small? Even if they are, in the small differences might there lie alleles and genes playing at least *some role* in the development of cognitive or behavioral characters, or in the susceptibility to disease? Genes like *Duffy*, *EDAR*, and *FOXP2* seem relevant in this context, as we saw in Chapter 7. Moreover, further work is required to critically discuss and consider the large range of the variance components of Lewontin's distribution (including the among races apportionment for mitochondrial DNA and Y-chromosomal haplotypes), as well as the meaning, limits, and assumptions of F_{ST}.[33] All of this is very much to the chagrin of the *liberal consensus* heralding humans as basically the same, and much to the joy of hereditarians who wish to locate relevant potential genetic differences among human groups. Finally, reasonable differences of interpretation can exist about, for example, the genetics of intelligence, as we saw in Chapter 8.

Moreover, this debate is vexed by *burden of proof arguments*: Both sides seem to argue that the burden of proof lies with the other side. The ideological left (the liberal consensus) might utter, "The hereditarians must show that there is a gene of major effect to cognitive development, or at least that certain genes associated with cognitive differences in each population act causally in the same way, in different populations," while the ideological right (many hereditarians) might declaim, "The liberal consensus must show that in fact *none* of the genetic differences between

---

[32] Cf. Feldman and Lewontin (2008), which seems to support a realism about biogenomic cluster/race.

[33] Lewontin's distribution is empirical pattern #4 in Chapter 6; regarding F_{ST} in this context, see note 22 in Chapter 4, and the "F_{ST} Reification" section in Chapter 5.

populations is associated with any cognitive differences that are stable across human populations."

The Song Remains the Same

Questions of race are potentially relevant because our currently dominant cultures have colonialist and racist ongoing politics and historical trajectories, but they are also relevant to our conception of ourselves. The answers to the questions with which I began this book – *Who am I? Who are you? Who are we?* – depend in large part on whether we understand ourselves to live on Planet Unity or Galápagos-Writ-Large. The biogenomic cluster/racial realist is concerned with whether we, as researchers in human evolutionary genomics, ought to recognize human populations as legitimate biological entities and, thereby, endorse a kind of Galápagos-Writ-Large. Using the same standards biologists employ for other species and research domains (e.g., conservation biology) to identify populations worthy of biological attention, can we pick out legitimate human populations?

In the contemporary literature, Sesardić writes as if this were the primary question of interest for biological racial realism, though I believe that more properly this question pertains to biogenomic cluster/racial realism.[34] The difference between the two is evident in the fact that discussions repeatedly turn solely to the level of genetic and genomic evidence, as we saw above. Spencer, however, should also be interpreted as a biogenomic cluster/racial realist, because he attempts to deploy the practices and results of genomic clustering to justify the identification and naming of racial groups, at least in the United States, and because he distances his position from social concerns.[35] In stark contrast, Hochman is a biogenomic antirealist: Following Templeton, he denies the reality of human races by noting that the population structure found in humans would hardly force the identification of distinct populations in nonhuman species.[36] By critiquing biogenomic cluster/race, Hochman implicitly endorses a Planet Unity scenario. Fujimura et al. peer onto human genomic variation through a sociological window and are also effectively biogenomic cluster/racial antirealists.[37]

[34] Sesardić (2005, 2010, 2013).
[35] Spencer (2013, 2014, 2015, 2019); see also Risch et al. (2002).
[36] Hochman (2013, 2016); see Templeton (1999).
[37] Fujimura et al. (2014) (Lewontin is a coauthor).

An interesting "middle position" can be gleaned from the more technical literature on heterozygosity measures and F_{ST}. Recall from Chapters 4 and 5 a discussion of the argument that F_{ST} validly represents human population structure only under certain assumptions (e.g., same expected genetic diversity across populations), which are not, in fact, met in our species.[38] Long and collaborators interpret human F_{ST} values as "moderately great," but by using generalized hierarchical models they also show that a racial classificatory system in which African, East Asian, and European populations are grouped together respectively does not account for human population genetic structure: "Race fails!"[39] Other researchers argue that surprisingly low estimates of F_{ST} among high-diversity African populations say less about actual biological phenomena of the population structure of natural, extant populations and more about "the intrinsic mathematical dependence" of F_{ST} on levels of genetic diversity and allele frequencies across different (theoretical) populations.[40] Thus, the putatively low F_{ST} values of humans may already be describing a more Galápagos–Writ-Large picture than we typically accept.[41] In light of these subtle, technical arguments, our Planet Unity and Galápagos–Writ-Large thought experiment can indeed serve as a starting point for comprehending human genomic variation, and for guiding future research, philosophical and biological alike. Even so, we need not necessarily see these two extremes as dual, mutually exclusive poles, since interpenetration and synthesis, per Chapter 7, remain interpretative options. Future discussion of the realism, antirealism, and constructivist conventionalism of biogenomic cluster/race must take these mathematical arguments seriously.

This entire debate can be, and seems to be, practiced mostly independently of concerns about the robust, socially ascribed populations usually thought of as races in ordinary discourse. That is, a realist about biogenomic cluster/race merely affirms that there are populations of humans that are biologically legitimate populations, in the (weak) sense that, using genetic data and measures, biologists would pick them out as worthy of interest in nonhuman species. Here, two interpretive approaches questioning such realism emerge. The first is the

[38] Long and Kittles (2003), Long (2009), and Long et al. (2009).
[39] Long (2009, pp. 801–802); see also Long et al. (2009, Figures 2 and 4, pp. 28 and 30).
[40] Jakobsson et al. (2013, p. 515) and Edge and Rosenberg (2014).
[41] This is not to mention new empirical data on introgression into the modern human genome of DNA from archaic humans including, among others, Denisovans, especially in contemporary Oceanian populations (Bergström et al., 2020), among other relevant evidence.

constructivist conventionalism approach, in which race as an object of study is impacted by the measures and models chosen to investigate it and by the interpretation to which investigation is applied. Under this analysis, the reality (or not) of biogenomic clusters/races depends on the variety of "explanatory interests" deployed,[42] and the measures and models used, as we saw especially in Chapters 4 and 5. According to this analysis, some biologists (i.e., splitters) would recognize the kinds of subdivisions found within the human species as interesting and worthy of attention were they similarly found in a nonhuman species; other biologists (i.e., lumpers) would not recognize these population subdivisions, whether in humans or nonhumans. Second, the *pernicious reification* of race approach, identified in Chapter 1 and deployed in Chapter 5, highlighting the ever-present danger wherein theoretical expectations and constructs are inappropriately and perniciously conflated with the complex world, also emerges.[43] In this analysis, "what is cultural or social is represented as natural or biological, and what is dynamic, relative, and continuous is represented as static, absolute, and discrete"[44]; or, alternatively, we engage in "conflating and confusing" mathematical models with the world.[45] (The pernicious reification approach can be applied to both biogenomic clusters/races and biological races.)

But all of this skirts the issue of whether populations picked out thusly deserve to be called races. Because *race* has an established use when referring to human populations, we should refrain from using the concept where that application is not the one intended. Indeed, as should be clear by now, slippage between biogenomic usages and social usages has serious consequences. The *biological* racial realist argument here is that, while the match is not perfect, populations socially identified as races overlap significantly with some populations that can be identified using modern genomic clustering techniques, and hence at least some biogenomic clusters/races are also social races. For instance, Shiao et al. make this argument from a sociology of science perspective.[46]

---

[42] Ludwig (2015).   [43] See also Gannett (2004) and Nielsen (2022); critique in Spencer (2013).
[44] Gannett (2004, p. 340).   [45] Winther (2020a, p. 94).
[46] Shiao et al. (2012). Another, very different, biological racial realist argument notes that insofar as our social categories themselves create and reinforce biological and biomedical differences, those social categories will be populations of real biological interest, although here the causal arrow points from the social to the biological. See Gravlee (2009), Kaplan (2010), and Fuentes et al. (2019). Finally, see Pigliucci and Kaplan (2003) for a defense, based on "ecotypes," of how "biologically meaningful human races might exist," although "these will not correspond to folk racial categories" (p. 1161).

Some biological racial antirealists argue that many populations picked out by biological approaches will not resemble social races. That is, many biogenomic clusters/races will be much smaller or narrower populations than races as usually conceived, and some will simply be different. And any associations between biogenomic clusters/races and social races will thus be unstable since social racial classifications change over time.[47] Other biological (and, simultaneously, biogenomic) antirealists about race stress the frequently clinal nature of genetic variation, where allele frequencies tend to increase (or decrease) in a somewhat smooth manner across neighboring populations, within regions of continents. Clines are themselves an informative signal in the global human genome, representing Out of Africa migration, with founder effect sampling, random genetic drift, and localized adaptation, as well as historical migration among geographically close populations.[48] Such antirealists generally deny that the clusters or races found are sufficiently independent of the statistical methods used to develop or discover them to properly count as biologically meaningful in a robust way.[49]

Finally, because socially manufactured racial ascriptions have enormous impacts on the lives of the people so ascribed but do not map neatly onto biogenomic cluster/race, many social racial realists and biological racial antirealists disregard and simply ignore the debates about biogenomic cluster/race. Essentially all socially important goods and services are distributed unequally with respect to populations picked out on the basis of racial ascriptions as embodied in our everyday discourse, in the United States, Europe, and globally. Given this fact, to ask whether these populations are also biogenomically legitimate populations, under some understanding of how genomic practices properly pick out populations, is interpreted as pointless. That is, identifying biogenomic clusters/races sheds no light on the importance of socially ascribed racial categories in our lives.

In the end, the hope that genomics and biology would permit us to determine, once and for all, whether races are biological entities (or not) is dashed. This does not represent a failure on the part of our best biology, but a failure to take seriously that the questions posed to our best genomic and biological practices are, in this case, inextricably mired

---

[47] See, e.g., Weiss and Fullerton (2005).

[48] Regarding clines, see Hartl and Clark (1989, pp. 281, 317–319) and Nielsen (2022).

[49] See, e.g., Livingstone (1962) and Serre and Pääbo (2004); but see response in, respectively, Dobzhansky (1962) and Rosenberg et al. (2005); cf. Fujimura et al. (2014).

in assumptions, practices, and values from other disciplines and from society at large. The answer we give to the question *Are "races" biologically real?* does tell us something about ourselves – and about each other – but the thing it tells us reveals more about our own beliefs (about what the question means, about what work the answer is supposed to do, and about which assumptions of all kinds we make) than it does about the biological nature of populations. Even our biogenomic cluster/racial classifications may turn out to be (often ugly) depictions of our socially entrenched biases and expectations, rather than pristine reflections of objective biological reality.

Political Consequences and Possible Futures

One of the consequences of philosophical analysis into race as an object of genomic study is the recognition that the racial category can, and perhaps even must, be redefined and refined to account for both realist and antirealist approaches. Charles Mills provides a useful taxonomy of some of the positions one might hold regarding the existence of races; these range from the denial of their existence (e.g., Glasgow, 2009) to the belief in biologically real deep racial "essences."[50] Mills's "objective constructivist" position seems the most plausible: (social) races are real, and their reality emerges from, and is continually reinforced by, our (contingent, but no less powerful for that) social practices. Ásta provides an analogous and highly resonant "conferralist" framework for analyzing the real, as it were, construction of race, and other social categories such as sex and gender. For Ásta, a social category such as race is defined by a social property, which "is a social status consisting in constraints on and enablements to the individual's behavior in a context."[51] Finally, few scholars have been more influential in showing the reality and consequences of social race (e.g., structural racism), in the context of European colonialism, than the French philosopher and psychiatrist Frantz Fanon.[52]

---

[50] Mills (1998). [51] Ásta (2018, p. 2).

[52] Fanon (2008 [1952], 2004 [1961]). The literature on the importance and inertia of race as socially ascribed categories (not dependent for their force on biological correlates) is vast. In terms of the more philosophical works, in addition to Mills, Ásta, and Fanon, I motion toward, e.g., Omi and Winant (1986), Appiah (1989), Appiah and Gutmann (1996), and Taylor (2022). The contingency of the categories that are identified as races and the ways in which a person's race, as a socially meaningful classification, can (and does) depend on their culture and spatiotemporal location (including global context, see e.g., López Beltrán, 2011; Lorusso and Winther, 2022), further complicates this picture (see, e.g., Hudson, 1996; Harawa and Ford, 2009; and Ousley et al., 2009).

Crucially, the social construction of race must be understood as historical and contingent.[53] Racial categories could always be otherwise. In *When Maps Become the World*, I recommended a practice of *imagining "What if…?"*:

In posing this most capacious question, philosophy opens up a space for memories, feelings, hopes, and imagination. When we ask "What if…?," we swap one set of assumptions for another and follow the world-making consequences of each, whether in the future or in potential existence more generally. Perhaps this is a kind of future-oriented pluralistic ontologizing … What if social relations were structured with institutions, values, and behaviors dramatically different from those in place here, today?[54]

One "What if…?" possibility is that of post-racial societal futures. In a post-racial future, perceived membership in a race no longer correlates with differential access to social goods such as dignity, trust, and opportunity. Such a future requires ripping up what Mills refers to as the *racial contract*. According to Mills, the abstract and seemingly neutral social contract political theory tradition of Thomas Hobbes, John Locke, Jean-Jacques Rousseau, Immanuel Kant, and John Rawls has effectively been a philosophical analysis "not [of] a contract between everybody ('we the people'), but between just the people who count, the people who really are people ('we the white people'). So it is a Racial Contract." Plausibly, Mills argues that "White supremacy is the unnamed political system that has made the modern world what it is today." A post-racial future, then, will require struggles for recognition and for reclaiming power, land, and a decent quality of life on the part of oppressed and exploited peoples. Allies must be found among those with power, wealth, and influence. It will also require taking Indigenous peoples seriously, and diminishing harm to them, by protecting their land rights, and by curtailing the brutal visible hand of extractive colonialism and neoliberal imperialism prevalent in our increasingly globalized world.[55]

---

[53] Hacking (1999) provides a refreshing discussion of realism and constructivism. See also Haslanger (2003) and the last chapter, "Map Thinking Science and Philosophy," of Winther (2020a).

[54] Winther (2020a, p. 253), references omitted.

[55] Mills (1997, pp. 1, 3). Critical analysis of whiteness, racism, and colonialism, as well as suggestions for the antiracist struggles for recognition and power required under our post-racial "What if…?" reflections (and actions) can be found in, e.g., Galeano (1973 [1971]) Fanon (2004 [1961], 2008 [1952]), Coulthard (2014), Alcoff (2015), and Kendi (2019); but see McWhorter (2021). Happe (2013) and TallBear (2013) show how genomics could undermine such struggles.

Perhaps post-racial societies are both possible and desirable. But they would require you and me, and us, to do the hard, critical work of identifying, not only with our respective groups, but also with *all of us* and perhaps with *all of life* – that is, with all species and all ecosystems. In so doing, we must also work not to perniciously reify and naturalize the current social expectations and prejudices of our ties. The result may not be societies in which people no longer "see" race; perhaps differences will be recognized only to be celebrated, as many already view the world's diversity in culture, music, or food. Whether a post-racial society should then reward people based on individual qualities such as work ethic or charm – that is, whether it should be an individualistic meritocracy – is a separate matter requiring further "What if. . .?" discussion.[56] In considering such a society, we must bear in mind the possibility that genomics, which provides a limited group-based understanding of the individual, may have to be reevaluated, reinvestigated, and pushed forward to accommodate a post-racial social paradigm.[57]

With respect to the existence of population structure in *Homo sapiens*, and the best way(s) to determine and classify that structure, I have sought, in this chapter, to defend a constructivist conventionalism. Proceeding from the intersection of genomics and philosophy, we can and should ask a number of important questions using data, measures, and models. Answers to clearly stated questions, once well specified, are not arbitrary. For instance, there is a nonambiguous, correct answer to a clustering analysis given data and a sampling scheme, a particular distance metric, clustering methodology (with explicit assumptions), concrete populations, and cluster number, as we saw in Chapters 5 and 6. However, nothing makes one question and answer set, rather

---

[56] A starting point for such a discussion might be Sandel (2020).

[57] Another perspective on a possible future of social race is provided by philosopher Chike Jeffers. In considering philosophical analyses of social race, Jeffers distinguishes "between a focus on politics or culture" (Jeffers, 2013, p. 403). He accepts the social relevance of – and prevalence of work on – the former, but wishes to carve out a space for the latter (Jeffers 2013, 2015, 2019). In a conference talk engaging with the earlier, second edition of Taylor (2022), Jeffers (2015) proclaims: "If we take culture to be of limited importance and social hierarchy to be essential to race, then we will seek to end race and bring about a post-racial society. If, on the other hand, we view cultural difference as of at least equal importance and as potentially independent of social hierarchy, then we will have reason to strive not for the end of race but for the goal of racial equality – that is, a future world in which racial difference no longer implies social hierarchy but in which it may nevertheless persist as a form of cultural diversity" (p. 9); cf. McGary (2012). Ian Peebles and Alexander Tolbert provided constructive feedback here.

than another, the obviously correct one.[58] This is neither silly nor simple relativism – rather it is honest and informed attention to tailoring appropriate data, methods, and theory to particular, clear research questions, assumptions, and purposes.

Conventionalism about biogenomic cluster/race does not imply that conventionalism about related positions is required, however. First, just as there is no straightforward link between a defensible realism about biogenomic clusters/races and a defensible biological racial realism, so conventionalism about the former does not demand conventionalism about the latter. The positions are logically independent. By maintaining their independence, I clarify these as two distinct levels of discourse and research practices that should be kept apart. In addition, the biological and sociological evidence reviewed in this chapter and throughout *Our Genes* strongly suggests that antirealism, not conventionalism, is the appropriate ontological stance toward biological races. At a global scale, contemporary racialist social configurations of unequal access to social goods are often based on a nebulous notion that somehow biological race is real. But there is no good mapping from ambiguous and changing multidimensional genomic patterns of population structure *to* the socialized (and pluralized, in different societies on a global scale) classifications of race.

Nor does conventionalism about biogenomic cluster/race imply conventionalism about social race. One can – and should – be a realist about social race as an objective (albeit socially constructed and historically contingent) fact about the social world, as Mills, Ásta, and Fanon, among others, have made clear, while maintaining that discoveries surrounding the population structure of humans are largely irrelevant to the major issues engendered by the social categories of race. Again, such an argument about the current state of affairs need not commit one to any position about possible future states of affairs.

Ultimately, genomics thus tells an *ambiguous* story about how to carve up the world into human populations (or clusters or groups thereof). This is one reason why genomics and politics do not mix well – there is no unequivocal genomic answer to whether humans are the same or different in their genetic potential, a frustrating ambiguous state of affairs for political advocates of equality (sameness) and hierarchy (difference)

---

[58] The biogenomic constructivist conventionalism here defended has affinities with the "deep conventionalism" position with respect to biological categories more generally, as articulated by Barker and Velasco (2013).

alike. Genomics and biology more broadly also seem to be telling us a *changing* story about biological race. While contemporary evidence favors the antirealist argument about biological race, the small percentage racial conundrum remains, and we might be forced to adopt a partial realism downstream. That is, we may end up discovering genes, or gene combinations, varying dramatically in frequency across conventionally defined races (or smaller groups) that really do have biological meaning for intelligence or other politically charged phenotypes. Finally, we have come a long way toward the legal and moral recognition of the fundamental equality of all human beings, and, in my view, even of all species. None of this depends on what we learn in biology, because moral and legal status does not derive from statistical properties of populations. Even our best genomic science should not dictate how we should treat one another. We need to bolster moral arguments and legal frameworks to guarantee equality and protection to all, independent of how the genomic story unfolds. At any rate, this is my take on the "What if…?" thought experiment of imagining a post-racial future.

Is Race Real?

So, is race real? *Socially*, yes – but *biologically*, no. And *biogenomic cluster/race is a matter of convention*. There are no simple answers to our complex question. Again, we must take multiple analytical perspectives. While a careful study of genomics helps negotiate and inform the relationship between individuals and populations, its stories are also fundamentally ambiguous. We can choose to see this as a limitation, if not failure – and perhaps it is – but I also like to see it as a blank but textured page upon which we could write a post-racial future, with a clear vision of equality and sustainability.

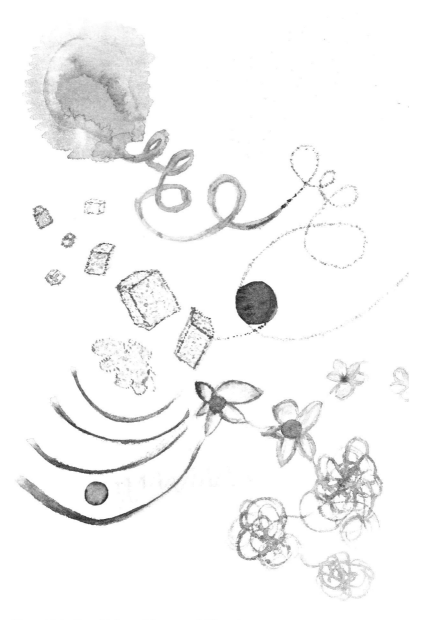

Figure 10.1. Soup Universe (Unstructured Change)
Rendition of the chancy and disorderly soup universe, as described in the text.
Processes and objects link together, but in unpredictable and unexplainable manners.
(Co-written with Raffn.) Illustrated by Marie Raffn. © 2021 Rasmus
Grønfeldt Winther.

10 · The Conscious Universe: Genes in Complex Systems

This chapter synopsizes our discussions by taking methodological step back to statistics, and to the irreducible entanglement of explanatory paradigms. We will never fully resolve the philosophical themes central to *Our Genes*, but we can certainly reflect on them. So far, we have investigated themes related to causation and explanation; the multivalent, ambiguous role of theoretical assumptions; the importance of data and empiricism; and the power of philosophy – for example, conceptual analysis, tracking ethics and power, critique, and imagining "What if. . .?" – in scientific investigation.

As we have discovered, in the twenty-first century, statistics and probability theory are essential tools for reading and interpreting often messy empirical data. Despite general efforts and the strong desire to produce data under controlled experimental conditions, we often must make do with noncontrolled and nonrandomized findings. In human evolutionary genomics, statistics provides the means for using such data to delineate *causal analysis* within complex systems, systems informed by the gene–environment nexus pertinent to all biological individuals. To understand explanations in human evolutionary genomics, then, we must grasp statistics as a tool that helps us get at causes and helps us understand how causes lead to effects.

Consequently, this chapter revisits some of the statistical practices and assumptions introduced and illustrated in earlier chapters in order to contextualize statistics as part of a larger philosophical and multidisciplinary predictive and explanatory approach to human evolutionary genomics. I seek to communicate the particular utility served by statistics, not just in our field but in our world. We begin, therefore, by investigating causal logic in a three-universe thought experiment. By considering causality across a law-governed clockwork universe, a disordered soup universe, and a balanced conscious universe, we start to understand the variety of ways in which causation can exist and operate. We also equip ourselves with information we need to investigate our own universe's uniqueness.

The three-universe thought experiment helps us identify the world required by our scientific methodologies. We ultimately ask: *What must our world be like in order to permit our best sciences to discover the world's structure?* In our universe, which is closest to the *conscious universe*, statistics makes sense of causal regularities. In fact, and as we shall see, in our conscious universe, researchers use statistical methodologies related to the *general linear model* (here abbreviated as "GLM"), such as *regression analysis* and *path analysis*, to synthesize determinism and indeterminism.[1] Researchers can reliably simplify complex systems and do so to make possible – and license – causal inferences. However, we also find that statistics is still limited, and a long methodological journey awaits those who wish to understand interaction and feedback in complex systems. For instance, in order to understand the way genes and environment interact, we must develop appropriate statistical methods and concepts to describe and analyze the subtle causal order of nature. This chapter starts building a statistical compass for overarching explanation and causal analysis in human evolutionary genomics (and other sciences). To that end, I start by sketching a preliminary map to guide our exploration.

Clockwork, Soup, and Consciousness

Consider first two extreme universes: a *clockwork* universe and a *soup* universe. In the severe clockwork universe, there are independent and inexorable laws, causal regularities, and (fundamental) kinds of objects and properties; moreover, there are relatively few such laws and causes, objects and properties. This view was once powerful across Western scientific traditions (e.g., eighteenth-century English Deism) and thinkers (e.g., the Leipzigian intellectual and coinventor of calculus G.W. Leibniz [1646–1716] and the Enlightenment French mathematician and philosopher Laplace). In this tradition, and among these thinkers, the universe was like an immense clock, consisting of gears, pendulums, weights, and so forth. Although perhaps initially wound up by a (Christian) God, it was then left to function according to its own well-designed and clear devices, ticking away according to simple materialist and mechanistic laws and causal regularities.

---

[1] On the GLM and regression analysis, see the "Statistics in a Conscious Universe" section below. On path analysis, consult also Wright (1920, 1921, 1922, 1931b, 1934) and Pearl and Mackenzie (2018).

We can imagine a clockwork universe by considering a universe that consists of two identical, smooth, and solid spheres in empty space. A single law operates: gravity. The spheres may be white and warm, or black and cold, or shiny and wet. We do not know, and it does not matter.[2] The only causal regularity in this universe is uniform motion around a common center of gravity, equidistant between the centers of mass of the two spheres. In its essentials, a clockwork universe is taken to be deterministic; orderly and regular (across space and time); (potentially) long lasting; simple; and (potentially) totally explainable in that all causes can be postulated and identified, and their effects fully mapped out.

Now, let's contrast this with an utterly disordered soup universe. In this universe, there are no reliable laws or regularities, nor stable types of objects or properties (whether fundamental or emergent). Such wild chaos and utter chance may be more likely objectively, as it were, than clockwork. According to arguments from the "anthropic principle," universes can have any combination of fundamental constants and laws of nature. However, only a relatively narrow range and combination of such constants and laws support the kind of long-lived and predictable universes in which intelligent life and humans could even emerge, let alone talk, study, or plan.[3] In other words, many possible universes of the multiverse are short-lived and unstable. These can be thought of as "soupy," or otherwise chaotic and irregular.

To imagine a soup universe, consider a world filled with what looks like mist. We can neither measure nor assess this mist for its physical properties or chemical composition. But in it, we observe a swirling mass of sand that behaves erratically and unpredictably. It can become a perfectly shaped sphere of solid granite; at other times, it can irregularly morph into thousands of fist-sized cubes of crystals that emit sounds. Over time, some crystals transform into identifiable and measurable flowers, hairballs, gold, and other familiar objects and materials (Figure 10.1, chapter opener). Unfamiliar shape-shifting may occur elsewhere in this universe, while other parts never change. All there "is" is unstructured change. Whether following scientific arguments about multiverses and the anthropic principle, or using our imagination

---

[2] It might be irrelevant for one of three reasons: first, because there simply are no optical or thermal laws or properties – this is a bare universe indeed. Second, because any such "laws" or "properties" are clockwork-like anyway. Finally, and somewhat paradoxically, because there is no consciousness present to measure it, or our consciousness only cares about clearly postulated laws, e.g., gravity, or both.

[3] See, e.g., Bostrom (2002) and Lightman (2013).

to build universes, a soup universe is stochastic and chancy; disorderly; likely short-lived and unstable; so utterly complex that any instruments and measurement protocols attempting to measure it are useless; and unexplainable because regular and robust causes do not seem to exist.

Neither a clockwork nor a soup universe seems capable of provoking enough interest to motivate our efforts. The universes are simple and lack hierarchy because nothing reliably exists on which to build or organize: Potential building blocks lack complexity in clockwork, stability in soup. Further, all the laws, causes, and kinds in the clockwork universe can be written down on a small napkin. It is boringly regular and simple. Meanwhile, no predictive or stable laws, causes, or kinds exist in the soup universe. It is chaotic and basically unknowable.

But what if there were a universe in the middle "ground" or middle "distance" of clockwork and soup?[4] Let's consider another universe with numerous, varied, and robust laws as well as stable and complex hierarchical objects. Such a universe is a synthesis or extension of clock and chaos. Essential to it is the existence of intelligent beings, with abilities to intentionally inquire into the nature of the universe in general, including asking about the very role they play in it. They are also able to grow tools for analyzing and theorizing about that universe, tools which in the case of human intelligence include the mathematics (and statistics!) and experiments of science, and the creative reflections of philosophy. This is what I call the *conscious* universe. In it, both laws and chance operate: Causes in complex systems can be delineated using the simplifying and abstracting methods of statistics and experimentation; structure all the way up to consciousness exists.

Chance, Law, and the Limits of Total Explanation

My thought experiment of distinguishing clockwork, soup, and conscious universes helps us consider *our* universe. The three analogies do not purport to illuminate actual, basic properties of the universe, as described by physicists.[5] (See Figure 10.2 for one such artistic interpretation of the entire history of our actual universe.) Rather, they help

---

[4] See Smith (1996) for an intriguing discussion of the importance of observers maintaining a "middle distance" from what they are observing, in order to articulate "the registered world, including all of ontology" (pp. 293ff.).

[5] On analogies, see Hesse (1966, 1974), Bartha (2010), and Winther (2020a). For actual, interesting properties of or ideas about our universe, the reader may wish to learn about "Boltzmann brain," "quantum information," or "spacetime."

illustrate how we understand and reflect on causal analysis and statistical practice in our universe. We can begin to think more clearly about our hierarchical universe by following the theories of chance and probabilities that intensified in the late nineteenth century. This journey helps us perceive our universe as filled with imperfectly acting laws and causes that can be understood using statistically motivated experimental analysis. It was no accident that the questions driving the development of human evolutionary genomics and statistics included questions about causation.

The hierarchical potentiality of our universe and its characteristic consciousness were classically described in the nineteenth century. Laplace, for example, offered a statement of laws and determinism in his influential *A Philosophical Essay on Probabilities*:

All events, even those which on account of their insignificance do not seem to follow the great laws of nature, are a result of it just as necessarily as the revolutions of the sun.[6]

With this metaphysical determinism, Laplace argued against "final causes" (teleology) and "hazard" (chance), considering both to be "imaginary causes [that] have gradually receded with the widening bounds of knowledge and disappear entirely before sound philosophy, which sees in them only the expression of our ignorance of the true causes."[7] Indeed, he also alluded to a superior intelligence, of which human intelligence was only the weakest reflection, and which was later baptized *Laplace's demon*:

an intelligence which could comprehend all the forces by which nature is animated and the respective situation of the beings who compose it – an intelligence sufficiently vast to submit these data to analysis – it would embrace in the same formula the movements of the greatest bodies of the universe and those of the lightest atom; for it, nothing would be uncertain and the future, as the past, would be present to its eyes.[8]

In some ways, Laplace seems to say, laws comfort.

Even so, chance also pervades our universe, as was *also* recognized in the nineteenth century. Savant and philosopher Charles Sanders Peirce eloquently articulated the power of chance thus:

[6] Laplace (1902 [1814], p. 3).   [7] Laplace (1902 [1814], p. 3).   [8] Laplace (1902 [1814], p. 4).

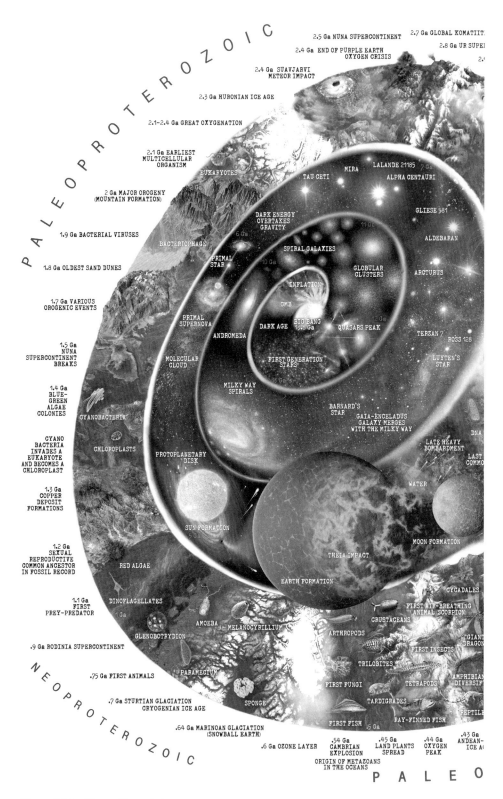

Figure 10.2. Nature Timespiral

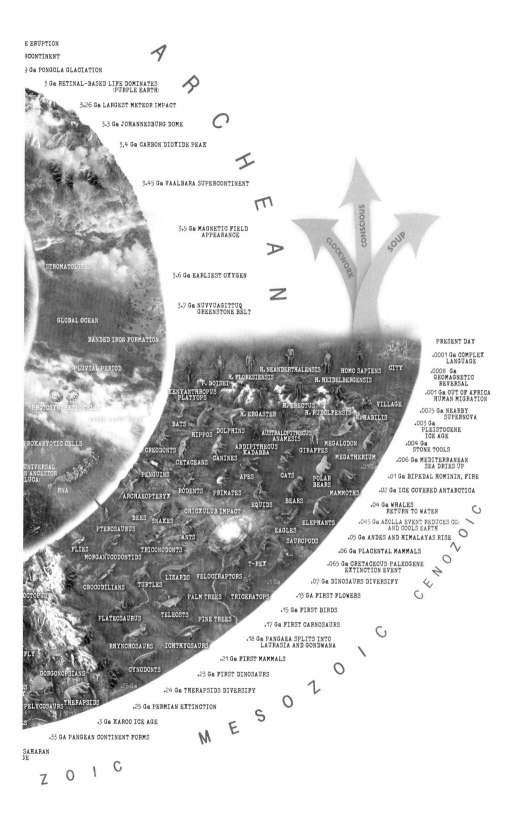

For a long time, I myself strove to make chance that diversity in the universe which laws leave room for, instead of a violation of law, or lawlessness.[9]

In a kind of synthesis strategy regarding the nature, history, and future of our universe, Peirce opposed determinism ("anancasm") to "absolute chance," which was a metaphysical thesis of small, ubiquitous, and irreducible chanciness ("tychasm"). Peirce accepted aspects of both determinism and chance as descriptive and prescriptive of our changing universe. But he hypothesized, defended, and came to believe in a synthetic, third option: dynamic laws of nature evolve, creatively growing to fit with one another and with the universe's chanciness ("evolutionary love" or "agapasm").[10] According to Hacking, Peirce:

opened his eyes, and chance poured in . . . His working days of experimental routine, and his voyages of the mind, took place in a new kind of world that his century had been manufacturing: a world made of probabilities.[11]

Complementing Peirce's metaphysical argument about the centrality of chance in our universe stands fellow pragmatist John Dewey's epistemological critique of necessity and determinism from the same 1893 journal volume:

Caption for Figure 10.2. (cont.) "The history of nature from the Big Bang to the present day shown graphically in a spiral with notable events annotated. Every billion years (Ga) is represented by 90 degrees of rotation of the spiral. The last 500 million years are represented in a 90-degree stretch for more detail on our recent history. Some of the events depicted are the emergence of cosmic structures . . . [;] the emergence of the solar system, the Earth and the Moon[;] important geological events . . . [;] [the] emergence and evolution of living beings . . . [; and] the evolution of hominid species and important events in human evolution." From: www.pablocarlosbudassi.com/2021/02/nature-timespiral.html. Nature Timespiral by Pablo Carlos Budassi and Rasmus Grønfeldt Winther (2021). Reprinted with permission. (A black and white version of this figure will appear in some formats. For the color version, please refer to the plate section.)

---

[9] Peirce (1893b, p. 544).

[10] Peirce (1893a, 1893b), Hacking (1990), and Hausman (1993). Hausman (1993) states: "Agapasm is evolution that includes chance and necessity and something else: it is the synthesis of chance and necessity, which is not reducible to either or to both simply added together" (p. 174). Incidentally, recall synthesis from the metaphilosophy of distinctions of Chapter 7.

[11] Hacking (1990, p. 201).

When we say something or other *must* be so and so, the "must" does not indicate anything in the nature of the fact itself, but a trait in our *judgment* of that fact.[12]

Dewey diagnosed a variety of fallacies associated with determinists imposing inappropriate parts and causes.[13] He cared deeply about the role of knowledge-producing agents in our quest to use scientific and statistical tools to understand, act on, and, ultimately, change our world. Dewey's action-oriented, intervention-focused *epistemological antideterminism* should be distinguished from *epistemological indeterminism*, which is the claim that we simply do not – and perhaps cannot – know whether statistical laws and regularities are fundamentally deterministic or absolutely chancy. Chance was riding a crest of influence at the end of the nineteenth century that flooded into the twentieth century. It is precisely on this wave that statistics rose.

In the conscious universe, law and chance interweave, Laplace and Peirce meet. Here is Galton, a by-now familiar key early mover of statistics and the genetic paradigm, in *Natural Inheritance*:

I know of scarcely anything so apt to impress the imagination as the wonderful form of cosmic order expressed by the "Law of Frequency of Error." The law would have been personified by the Greeks and deified, if they had known of it. It reigns with serenity and in complete self-effacement amidst the wildest confusion. The huger the mob, and the greater the apparent anarchy, the more perfect is its sway. It is the supreme law of Unreason. Whenever a large sample of chaotic elements are taken in hand and marshalled in the order of their magnitude, an unsuspected and most beautiful form of regularity proves to have been latent all along.[14]

Galton was here poetically describing what is more generally known as the "bell curve" or the normal distribution. Error becomes law.[15] In many areas of knowledge, especially in the biological and behavioral sciences, reliable statistical correlations became potential causal regularities. In the conscious universe, whether statistical laws are fundamentally

---

[12] Dewey (1893, p. 363).

[13] I consider Dewey's analysis and critique of what I call *pernicious reification* in Winther (2014).

[14] Galton (1889, p. 66).

[15] More precisely, the "law of frequency of errors" maps the size of observation errors to their frequency, when many meticulous measurements of a given quantity are made. Small errors are common, large errors (relatively) rare, and the map or distribution is normal or Gaussian, a bell curve (e.g., chapter 1 in Fisher, 1925).

stochastic at some basement level (i.e., metaphysical indeterminism) or whether laws are statistical because of our ignorance (i.e., epistemological indeterminism) is less important than the fact that statistical laws now have an "autonomy" in that they can *"be used not only for the prediction of phenomena but also for their explanation."*[16] Chancy laws can have tremendous explanatory power. In the imaginary of the conscious universe, as constructed toward the end of the nineteenth century, and as represented and applied in our best statistical methodologies today, determinism and chance no longer conflict.[17]

Let's return to our thought experiment with this history in mind. Because our universe is structured into molecules, multicellular organisms, societies, and planets, among other objects and processes, and because our universe is a conscious universe, scientists can make *predictions*, offer *effective explanations* furthering understanding, and recommend *interventions* (e.g., biomedical, economic, pedagogical, and ecological). For instance, because of the fuzzy hierarchy and statistical laws our universe exhibits, even evolutionary theorists can make novel, surprising, and correct predictions of bacterial or endosymbiotic evolution.[18] In fact, *conscious entities in our universe seek and produce explanations.*[19] Ultimately, in our universe, intervention is produced by goal-oriented action,[20] responds to social learning and organization,[21] and is driven by community and institutional needs.[22] Thus, in contrast to clockwork and soup universes, a conscious universe is characterized by structure and incompletely acting laws and causes; emergent and fuzzy, yet stable, hierarchies; a (potential) long life; consciousness; complexity; and the possibility of interesting – yet incomplete and multiple – causal explanation.

More simply put, in the conscious universe, questions about the mutual coherence and applicability of laws of nature and causal regularities, and about the uniqueness and classifications of kinds of objects and properties, can be asked, in part because such questions can be conceptualized and, on occasion, even answered. Precisely because there is (at

---

[16] Hacking (1990, p. 182). On epistemological indeterminism, consult Hacking (1983b), Suppes (1984), Dupré (1993), and Glennan (1997).

[17] Of course, the relationship remains complex. In a letter to Max Born dated September 7, 1944, Albert Einstein wrote, "We have become Antipodean in our scientific expectations. You believe in the God who plays dice, and I in complete law and order in a world which objectively exists, and which I, in a wildly speculative way, am trying to capture" (cited in Born et al., 2005 [1971], p. 146). On "probabilistic causation," see Hitchcock (2012).

[18] See Margulis (1975), Williams (1982), and Winther (2009b).   [19] Woodward (2011).

[20] Hacking (1983a) and Walsh (2013, 2015).   [21] Dewey (1929).   [22] Longino (2002, 2013).

least human) intelligence in our universe, humans have started to provide a total causal explanation of its parts, levels, and dynamics.

Although I desire a more Laplacian chance guided by consciousness, meaning a universe in which formal tools can be applied to understand the natural, psychological, and social architecture of that universe, I also recognize our interpretative limitations. Some philosophical interlocutors critique the possibility of universal laws, unique classifications, and total explanations.[23] Other philosophers and scientists argue that the universe is a simulation or a complex mathematical structure, or both, in which total explanation is not only possible but *necessary*.[24] Then, of course – and perhaps most interestingly – if we extend Ian Hacking's *looping effect*,[25] and assume consciousness itself affects explanation, then total explanations, of even purely material processes, that bracket explainers and explainers' needs, histories, and assumptions, would not be forthcoming.[26] Put differently, pragmatic factors will always enter explanations; our desires and dreams will alter the explanation itself.

For this and many other reasons, controversies about the possibility of total explanations in a conscious universe remain strong. This matters to human evolutionary genomics because such controversies help us think about how we can trace multiple, potentially interacting, causal pathways in the complex system constituting the individual, and its ongoing development, as we started doing in Chapter 8. They also assist us in understanding why integrating a variety of explanatory paradigms matters, as we did in Chapter 7. Humans are always already subject to many genetic and environmental influences, and we have intentionality and can engage in feedback – both as individuals and as collectivities – in order to influence and modulate genetic and environmental causes. In a conscious universe with emergence, hierarchy, and agency, such as we commonly – but not always – take ours to be, explanation is interesting, and intervention important.

Statistics in a Conscious Universe

The conscious universe is the philosophical – not to say logical – consequence of the growth and implementation of statistics and probability

---

[23] Cartwright (1983, 1999), Dupré (1993), Hacking (2007b), and Wimsatt (2007).
[24] Wolfram (2002), Bostrom (2003), and Tegmark (2014); but see Frenkel (2014) and Hacking (2014).
[25] Hacking (1995, 2007a).   [26] See, e.g., Longino (2002) and Winther (2021a).

theory, in a universe where such engagement is possible and powerful. Indeed, in this conscious universe, formal studies of chance make probability, according to Ian Hacking in his 1990 book *The Taming of Chance*, "*the* philosophical success story of the first half of the twentieth century." Probability is:

A quadruple success: metaphysical, epistemological, logical and ethical.

Metaphysics is the science of the ultimate states of the universe. There, the probabilities of quantum mechanics have displaced universal Cartesian causation.

Epistemology is the theory of knowledge and belief. Nowadays we use evidence, analyse data, design experiments and assess credibility in terms of probabilities.

Logic is the theory of inference and argument. For this purpose we use the deductive and often tautological unravelling of axioms provided by pure mathematics, but also, and for most practical affairs, we now employ – sometimes precisely, sometimes informally – the logic of statistical inference.

Ethics is in part the study of what to do. Probability cannot dictate values, but it now lies at the basis of all reasonable choice made by officials. No public decision, no risk analysis, no environmental impact, no military strategy can be conducted without decision theory couched in terms of probabilities. By covering opinion with a veneer of objectivity, we replace judgement by computation.[27]

Hacking's list could be further generalized by describing probability as a "*scientific* success story." While formal studies of chance are central to the biological and behavioral sciences, statistics is particularly crucial to human evolutionary genomics because statistical assumptions and protocols permit the abstraction and decomposition of a complex system.

Evidence abounds. Early evolutionary geneticists and statisticians such as Galton, Pearson, Fisher, and Wright were driven by questions of ancestry and futurity. They measured and modeled separate and combined genetic and environmental causal factors, using statistics (and its simplifying assumptions) as their fundamental tool. Indeed, modern statistics grew, to a significant extent, out of measuring and modeling genetic and environmental factors hypothesized to produce phenotypic traits. We saw evidence for this in the first five chapters of *Our Genes*.

Statistical tools permit reliable measurement of – and experimental and theoretical access to – complex, yet relatively well-behaved, phenomena

---

[27] Hacking (1990, p. 4).

and processes. Statistical thinking also helps with inference, theory confirmation, and decision-making, as we saw with phylogenetic inference in Chapter 4, with detecting the signature of natural selection in Chapter 7, and with causal, genetic analysis of disease etiology (and the whole slew of ethical questions associated with clinical etiological analysis) in Chapter 8. While the explosion of statistical thinking in modern society might very well have gone beyond R.A. Fisher's wildest dreams, it is not, I dare say, utterly surprising given the nature of the conscious universe as one hybridizing law and chance, in light of (experimentally and mathematically oriented) intelligence. Fisher's ghost is likely smiling.[28]

In our conscious universe then, statistics plays a central role, as described in the prefatory words of *Statistics: A Very Short Introduction*:

Statistical ideas and methods underlie just about every aspect of modern life. Sometimes the role of statistics is obvious, but often the statistical ideas and tools are hidden in the background ... Statistics suffers from an unfortunate but fundamental misconception which misleads people about its essential nature. This mistaken belief is that it requires extensive tedious arithmetic manipulation, and that, as a consequence, it is a dry and dusty discipline, devoid of imagination, creativity, or excitement ... [But] *statistics is the most exciting of disciplines*.[29]

The preceding chapters offer relevant testimony. Diagnosing causation in a complex system, such as identifying the genetic basis of spleen adaptations in Bajau freedivers, or finding the allele variants associated with lactase persistence, requires abstracting out individual factors and gauging their relative single effects. It also requires understanding the way factors combine and interact with other factors, including those found within other, diverse explanatory paradigms (e.g., adaptationism, mechanism, and historicism). The ideal and most straightforward way to investigate causation, and produce causal explanations, is to deploy randomized experiments together with statistical analysis.

In many cases, however, randomized experiments are impractical (as in the case of comparing the macroeconomies of different countries), unethical (as in breeding experiments in humans), or both. Although under such nonideal conditions, uncontrolled "experiments" or observational data, and complex statistical procedures, can lead to conflicting data, meaningful interpretations (if not dialogue and progress) remain

---

[28] For a statistically informed perspective on Fisher's long reach, see Efron (1998).
[29] Hand (2008, preface and p. 1).

possible, as illustrated in prior pages.[30] This is because fields like human evolutionary genomics that depend on diagnosing causation in a complex system still accumulate knowledge by abstracting out potentially causal factors from large data sets of individuals with different combinations of those factors, and investigating their hypothesized effects. The GLM stands as a case in point: Much of the statistics behind this model of causal analysis was born out of assumptions and methods developed in the late nineteenth and early twentieth centuries and associated with the overarching genetic paradigm and human evolutionary genetics.[31]

Today, GLM statistical procedures, especially regression analysis and ANOVA, and the more complex methods following from them, such as path analysis or structural equation modeling, are met by most students in their statistics courses. We have already visited these in the pages of *Our Genes*. However, the GLM and its associated procedures are not only required tools in the study of human evolutionary genomics. They also offer a paradigmatic illustration of the usefulness – and the necessity – of modern statistics in the conscious universe. Therefore, an extended example of the GLM is warranted.

The GLM, in its simplest form, starts with a mathematical model relating an independent variable to a dependent variable. For instance, we could imagine hanging up a spring and attaching serially heavier weights, measuring the extension of the spring associated with each weight. Our model for the experiment might propose that the extension we measure is a combination of a law-based, deterministic function of the weight we applied and a random "error" term that stochastically – but in a known and statistically explicit way – generates deviations from the deterministic prediction. If we represent spring length by Y, applied weight by X_1, and random error by e_1, we could write

$$Y = B_0 + B_1 X_1 + e_1. \qquad \text{[Eq. 10.1]}$$

---

[30] See also Blalock (1964) and Lieberson (1987).

[31] Trail-blazing works on these methods include Fisher (1918, 1921, 1922, 1925, 1926, 1935), Wright (1921, 1931b, 1934, 1968), Neyman (1923 [1990]), Haavelmo (1943), Simon (1953), Blalock (1964), Duncan (1966), Holland (1986), Cartwright (1989, 2007), Spirtes et al. (1993), Pearl (2000), and Pearl and MacKenzie (2018). For introductions to GLM statistical methodology (and some of its history and assumptions), see Herr (1980), Wahlsten (1990), Wade (1992), Sokal and Rohlf (1995), Hocking (1996), Shipley (2000), Muller and Fetterman (2002), Kline (2016), Stigler (2016), Harville (2018), Poldrack (2018), Edge (2019), and Winther (2020a). For a critique of causal interpretations of statistical analyses, consult Levins and Lewontin (1985) and Freedman (1991, 1994, 2009).

This is a linear regression model for the spring. Essentially, we are saying that the outcome or observation is a function of a *deterministic model plus error*. Scientists are often interested in terms like B_1. This, a *regression coefficient,* captures the functional relationship between the dependent (response) and the independent (explanatory) variables – in this case, the amount by which the length of the spring changes when we change the weight applied to the spring. If we know that the equation of a line is $y = mx + b$, we can grasp that Equation 10.1 essentially specifies a family of regression lines, where the y-intercept (i.e., b) of a given line is B_0 and the slope (i.e., m) is B_1. In this particular case, applied weight is the x-axis and spring length is the y-axis. (Note also that the spring has baseline length B_0 when no weight is applied to it.)

But, because we must also account for chance, or the "plus error" referenced above, a chancy random variable *error term*, e_1 (also known as, though not equivalent to, a *residual*) is built into this GLM regression model. The error term effectively states that there will be a spread or "cloud" of predicted (or actual) measurements around the true (or estimated) regression line. To simplify, the smaller the error term, the tighter the cloud – i.e., the closer each and every measurement of spring length for an applied weight is to the regression line – and the more a given regression line explains all the (variance in the) measurements. In our case of weights on a spring, provided we set up our experiment appropriately, we are likely to have a very small error term indeed, and a highly explanatory regression line (i.e., a line with what statisticians call a *coefficient of determination* near its maximum of 1). To use a concept from Chapter 7, the linear, deterministic component of Equation 10.1 *interpenetrates* with the error, chancy part. (Laplace meets Peirce.) Each requires the other in our search for the best regression line to explain as much of the data as possible. Indeed, the interdependence points to an irony undergirding statistics, to which we return below.

We gain a clearer view of determinism, chance, and explanation in the GLM by exploring the following standard, theoretical assumptions the GLM makes, also in relation to our spring experiment:[32]

1. *Linear relation*. Explanatory and dependent variables are related linearly. Examples include the mass and weight of an object, or voltage

---

[32] For explicit statements of such assumptions, articulated in different ways, see, e.g., Muller and Fetterman (2002, pp. 10–14), Freedman (2009, pp. 98–99), Kline (2016, pp. 33–35), and Winther (2020a, pp. 201–202).

and current across a resistor, or the applied force and length of a spring (Hooke's law, $F = -kx$, where k is the spring constant and serves as our regression coefficient B_1 in the case of a spring). Linear correlations are common in biology (e.g., the additive effects of different genes, intercellular chemical gradients produced differentially by cells during early development), although *threshold phenomena* (e.g., digital nerve signals or "action potentials" are sent only once a certain charge has accumulated) are also important.

2. *Normal distribution of errors or residuals with constant variance.* The deviations around the regression line (i.e., *errors* or *residuals*) have means of 0 and constant variances at each of the levels or values of traits or properties being measured, for each independent variable (e.g., across many different extensions of our spring; constant variance is known as *homoscedasticity*). While not among the barest assumptions of the GLM, it simplifies mathematics and improves estimation to also assume that the error term e_1 for a single independent variable follows a Gaussian or normal distribution with mean 0 and variance σ^2 (written: $N(0, \sigma^2)$), and to assume that, for multiple linear regressions (i.e., linear regressions with more than one independent variable), the mean and variance of the various error terms of the different variables are the same (i.e., the errors are *independent and identically distributed*), and indeed also Gaussian.

3. *Cross-variable explanatory impotence.* In cases of multiple linear regression, the behavior of any explanatory variable cannot be predicted from any of the others. A gigantic probabilistic machine, the statistical causal model needs multiple independent variables. Each variable carries explanatory or causal information. (For instance, think of multiple genes for height or for any number of other complex traits. Here our spring example is less helpful – as a physical law, more in line with a clockwork universe, it calls out one explanatory variable, applied weight, practically totally explanatory by itself.)

4. *Ignored outliers.* When any data point seems exceedingly different from the sample, ignore it or minimize its importance. In the case of our spring, if we attach too heavy a weight we will approach or exceed the elastic limit of the material, and will start deviating from Hooke's law.

That aforementioned irony – if not paradox – of statistics is clear here: *That which we wish to know, i.e., population parameters, is unobservable and immeasurable, while that which we can measure, i.e., sample statistics, is merely an estimate of that which we wish to know.* Illustrated in the distinction between two interpretations of Equation 10.1, this irony is also inherent to the first and second GLM assumptions.[33] Although we must set aside the detailed mathematics here (see sources in note 31), there is nonetheless a first, theoretical and ideal interpretation where B_1 is a population parameter describing the true regression line's slope and e_1 is an unobservable random variable error term producing the theoretical, predicted deviations around the true regression line. Yet a question lurks: "If one assumes that the theoretical model I have written is a true description of the spring, what is the best possible guess or estimate for the values of B_1 and e_1?" That is, we can take one spring and manipulate it by adding different weights to it, exposing it to different temperatures, and so forth. Our measurements, our data, can thus be used, according to certain well-known statistical methods such as the *sum of squared residuals*, to estimate B_1 and e_1, now considered as sample statistics. Informally, we are drawing the best straight line through a field of points that, to the extent possible, "balances" the vertical distance (i.e., the residual) from each and every point to the line (i.e., the best line minimizes the sum of squared residuals). This is the second, data analysis interpretation of Equation 10.1 – actual measurements provide an estimated regression line, with associated residuals. In other words, data are used to estimate the (unobservable) true regression line and error term. One of the more elegant consequences of statistics is the development of robust methods of estimation. Such methods are possible in part because the key distinction between statistics and parameters is rigorously held: The two are neither conflated nor confused.

While the role played by statistical analyses under the GLM becomes clear in the spring experiment, it becomes even clearer when considered against the backdrop of the three universes. In a clockwork universe, the error terms are likely small, or even nonexistent, although the deterministic, linear part of the GLM remains useful for understanding and measuring the applicability of laws. In a soup universe, error terms will fluctuate widely and wildly over time and magnitude of the measured variable and cannot be investigated. Consequently, the GLM yields no meaningful information. In our conscious universe, however, in which both clockwork and soup elements operate, the GLM imposes elements of clockwork linear law onto soup-like chance to yield deep insights and

---

[33] Recall the distinction between sample statistic and population parameter from Chapter 4, as well as the danger of *statistic–parameter reification* from Chapter 5.

far-reaching meanings. The deterministic part of the GLM is simple, orderly, and separable into distinct influences of different independent variables (captured by B_1, B_2, ... B_n).[34] Even the corresponding error terms for each variable (e_1, e_2, ... e_n), the disordered and chancy part of the regression, are rigidified.[35] How so? When a model with single or multiple explanatory factors is fit to real data using general estimation methods such as *ordinary least squares*, inferences of the true regression coefficients and error terms are typically based on the standard GLM assumptions outlined above. Very much in line with Galton's veneration of the "Law of Frequency of Error," errors are modeled as random in the GLM, but the randomness is structured and domesticated for purposes of estimating the true regression line. Chanciness here is neither metaphysical nor epistemological indeterminism. Indeed, *methodological* determinism might be a better description and concept of how the conscious universe makes chance more clockwork-like by taming error terms.[36]

In evolutionary genetics, Fisher's own development of ANOVA offers a key case in the field's efforts to domesticate error. Fisher, perhaps acknowledging environmental influences as a kind of well-domesticated error, worked to simplify the complex gene–environment individual–population system by focusing primarily on the influence of genetic factors, with their corresponding allele variation. (And in his earliest papers on the subject, environmental factors were themselves considered a kind of error, adding noise to the genetic signal on the phenotype.[37]) Subsequent to Fisher, ANOVA, itself deeply connected to linear regression, was expanded as a general analytical system for decomposing sources of variation – read: for identifying explanatory factors – in any

---

[34] Thinking in terms of the three universes, and trying to explicitly avoid statistic–parameter reification, has helped me clarify and tighten my list of GLM assumptions compared to my earlier attempt (Winther, 2020a, pp. 201–202), where I also multiplied assumptions beyond what is necessary.

[35] Typically, we employ models with various explanatory variables and corresponding residuals or error terms. Such multiple regression models might look like this: $Y = B_0 + (B_1X_1 + B_2X_2 + ... + B_nX_n) + (e_1 + e_2 + ... + e_n)$ [Eq. 10.1n]. (Recall from the second GLM assumption above that the error terms are assumed to be Gaussian, and independent and identically distributed.) The regression line would thus be drawn through multidimensional space.

[36] Experimental protocol in human evolutionary genomics and elsewhere helps this conscious forcing of clockwork on soup. Experiments involving procedures such as randomizing, replicating, blocking, and blinding force GLM error term assumptions to hold, thereby making estimation possible and reliable. Carefully designed experiments domesticate disorder also by helping us avoid *confounding* (namely, spurious or obscure) external factors being co-associated with the factors or variables studied.

[37] See, e.g., Fisher (1918, pp. 400, 432–433).

variety of complex systems, including the economy, institutions, and ecosystems. In human evolutionary genomics, ANOVA is the central precursor to the statistical methodology grounding GWAS.[38]

While the force behind intelligent and deliberate statistical analysis is consciousness, consciousness is also a source of error. Unsurprisingly, the scientist's statistical training and analytical goals are satisfied and further entrenched when disorder is domesticated, explanation and understanding produced, and informed intervention suggested. However, given our notorious bias as observers, and our systematic unreliability as statistical thinkers, scientists *must* attempt, via proper experimental protocol (e.g., blinding and controlling), to eliminate the partly pernicious footprints of our own consciousness.[39] This is why we use, to take just one example, double-blind procedures in medical randomized controlled trials to help eliminate behavior adaptations between medical professionals and patients, and to thereby help analyze the causes of diseases such as cancer, as examined in Chapter 8.[40]

Clearly, we cannot escape the political and ethical implications of statistical analyses. We have seen how this might play out in the preceding chapters. For instance, the way global genomic data are collected and owned and shared is ripe for legal and ethical analysis.[41] Or, given the worsening social context of the Bajau freedivers in Chapter 7 (and of many Indigenous peoples), questions about who gets to use statistically inferred causal claims – and for what – must be answered. Tracking ethics and power is imperative in human evolutionary genomics.

Finally, recall the old chestnut "correlation is not causation." The ability of statistical analysis to provide causal information of our conscious universe is in question. Many researchers in econometrics, behavioral psychology, human evolutionary genomics, and other fields maintain that causation can be inferred from data as long as proper statistical safeguards are in place, but others are more cautious. In fact, the justification of causal inference from even the most rigorous and sophisticated data is highly controversial. For instance, statistician David A. Freedman declaims:

---

[38] Consult Visscher and Goddard (2019). Of course, contemporary statistics keeps growing, and simplification via linearity is less important today for other statistical methods, such as bootstrapping or Bayesian multilevel modeling, although simplification, generalization, and abstraction remain (consult Gelman and Vehtari, 2021).

[39] See, e.g., Kahneman (2002, 2011).

[40] See, e.g., Solomon et al. (2009); but see Deaton and Cartwright (2017).

[41] Race, Ethnicity, and Genetics Working Group (2005), Vitti et al. (2012), TallBear (2013), and Kulynych and Greely (2017).

To derive a regression model, we need an elaborate theory that specifies the variables in the system, their causal interconnections, the functional form of the relationships, and the statistical properties of the error terms – independence, exogeneity, etc.[42]

In other words, claiming causation requires *already* assuming a causal model.[43] Is this kind of theoretical bootstrapping to be permitted? Are the causal stories inferred by scientists and used by the powers that be objectively true or are they more like a beautiful fairy tale?

Philosophies of Causation

Indeed, philosophical analysis is essential to the overarching dialogue about causation, complexity, and the conscious universe explored in this chapter. Philosophical debates – both today's and yesteryear's – can help us nuance notions of causation. Does cause exist in the world as such, independently of us, or is it at least partially an outcome of mental structures that invariably find pattern and meaning, as they do with Rorschach tests? To Immanuel Kant, causation was something like a built-in expectation that we, as rational human inquirers, could not help but see and perceive in the world. Just like we are capable of observing only a narrow range of the light spectrum, our other limitations and proclivities mean we experience the world in finite ways – e.g., as a causal world. This does not reflect the potentially infinite variety of experiences: There are many ways to experience the world, including the fundamentally different ways that other creatures and species experience it. Other philosophers, however, especially today within so-called analytical metaphysics, understand causation as an intrinsic feature of the material world. They worry about what exactly the causal relation is, in reality, and what kinds of objects, events, or processes it exists between.[44] Finally, a tradition skeptical of causal analysis as such, and even of the very concept of *cause,* also exists, and includes philosophers and scientists such as David Hume, Bertrand Russell, and R.A. Fisher. Is causation a mental expectation or a property of the world, or is it perhaps a mere mirage? The debates rage.

The power and ubiquity of consciousness is not limited to how it operates on and with the fundamental category of causation. Consider

---

[42] Freedman (1991, pp. 292–293).   [43] See Holland (1986) and chapter 2 in Cartwright (1989).
[44] Schaffer (2016).

that we perceive time serially. Consciousness, in most of its activities, is inherently chronological – it remembers the past and anticipates the future from the point of view of the present, it forms sequences of arguments that proceed in order of imputation, and it imagines causes leading to effects and not the reverse. Indeed, given the importance of time to most physical law, and most causation, does the temporal predilection of consciousness shape the kinds of laws, causes, and processes that we take to exist, and model through our statistics? Turning to complexity, how could we even decide what kind of system is complex (or simple), of what parts it consists, and what is inside or outside, without the action of consciousness as judge and jury of how to individuate and decompose systems? My own rational bet is to accept that the mental, cognitive, and theoretical abilities of our remarkable ape species exert significant creative power in our conscious universe.

Philosophical work here deserves to happen, given the intertwined technical, conceptual, and ethical nature of the matters at stake. Analogous to quantum mechanics and relativity theory, human evolutionary genomics gives us one window into causation, complexity, statistics, and choice and value in our conscious universe, a window that was opened wide in the beginning of the twentieth century, and that we can open yet wider today.

Explanatory Paradigms

I don't wish to readjudicate questions posed in preceding chapters in the court of Clockwork, Soup, and Consciousness, but let us consider some of the potentially relevant interdisciplinary applications of integrating different kinds of explanatory paradigms into our understanding of the universe. To that end, let us also return to our example of the springs. We may find that we gain a more complete picture of the springs and their actions and interactions by asking *other* questions about these springs, even if we are interested only in the linear regression consistent with Hooke's law. We could, for instance, ask about the history of the springs, or about their composition and organization. We may find that the history reveals clues about the kinds of materials or design used, that influence the way the springs function. We may find that the composition and organization of the springs affect the regression coefficient, k. We may find, if we lack direct data about the materials the springs are made of and the way in which they are coiled, that we can gain some fallible knowledge by having information about the factory in Germany or Thailand that constructed the springs, or about the mines from which the springs' iron and chromium (or copper) ore originated. That is, other kinds of data, and methodologies for acquiring such data, may very well be pertinent to our statistical explanation of the behavior of springs. Moreover, such diverse and contextual data are central when we are mainly interested in explaining the etiology or chemical makeup of a particular spring.

In fact, knowing about the origins (historicism paradigm) and organization (mechanism paradigm) of a system helps specify and fine-tune the statistical model – regressions and path analysis models are not imposed from a vacuum. The variables and relationships that appear in the regression model must be reasonable in the context of the theoretical background the researcher brings to the problem under study. That is, statistical practice must be embedded into a larger theoretical framework, where it may not always fit comfortably.[45] As Plato and other philosophers have suggested, theory must cut nature at its joints, and statistical models should reflect and apply that carving of, for example, causal factors.

These questions, and the explanations providing partial answers, are also important in another very simple way. The universe is, from our perspective, complex. Therefore, we *must* approach it from multiple perspectives. If we wish to understand the full range of causes and outcomes in the temporal evolution of complex systems, and if we furthermore desire to intervene intelligently in reality in order to change it for the better, scientists will need to simultaneously investigate broad biological, psychological, and social aspects of reality. While statistics remains ubiquitous to measurement and data collection and analysis, even for mechanism and historicism paradigms, statistical models of causation, per se, give us just one tool by which we widen our window onto human evolutionary genomics and the complex reality it helps explain.

We must recall, of course, that we seek to understand the universe in part as an extension of our ongoing project to understand ourselves. In the case of the Bajau freedivers, lactase persistence, and human–virus coevolution, we saw that historical and mechanistic explanations were important for understanding the purposes, modes of functioning, and emergence of humans. However, we also saw that a third perspective is required to get to the heart of the theoretical organism of human evolutionary genomics and thus of *Our Genes*. This, as made clear in

---

[45] See, e.g., Meehl (1978).

Chapter 7 on natural selection, is the perspective of *adaptation*. Adaptation is the "fit" of early humans to their environments, and the associated genetic inheritance of selected phenotypes, potentially including behaviors and cognitive capacities, as well as humans' adaptation to our contemporary environment – a kind of fit heavily mediated by technology and culture as well as by our genome.

More simply put, the adaptationist paradigm refers to understanding what works: not just in terms of what part is adapted to what part, or what trait is adapted to which environmental feature, but also when natural selection might have had an effect on our genome, and in what measure natural selection acted in relationship to other factors, such as random genetic drift and developmental constraints. Understanding the mechanistic integration of organisms at different levels is necessary for explaining their complex functionality, but we also need to see how natural selection gives direction to this functionality. Self-organization alone is not sufficient to explain the evolution of complexity.[46] Natural selection shapes and directs the emergence and refinement of complexity over evolutionary time, as fitter organisms gain advantages in survival and reproduction, and as deleterious mutations are purged. Moreover, a historical perspective is *also* essential in all of this. Fortunately, human evolutionary genomics is particularly effective at tying together and modeling the outcomes of many evolutionary forces acting over many generations, and even over much longer periods of time. By not losing sight of other important features such as development, physiological mechanisms, and phylogenetic and demographic history (features which I believe can constrain, if not overpower, the action of natural selection), we could replace spotlight visions of reality with an overarching floodlight vision.[47]

An Invitation to the Reader

Amid our philosophical investigations of statistics and causation, of important political and ethical considerations, and of entangled explanatory paradigms, I invite the reader to engage in an even larger set of three-universe thought experiments. These experiments require us to ask even more difficult questions about perspective, and its variable applicability and utility in the three universes. In a clockwork universe, the

---

[46] On self-organization, consult Kauffman (1993), Goodwin (2001), Forgacs and Newman (2005).
[47] See, e.g., Winther (2011, 2020a).

question of perspective is neither interesting nor important: We can extract the full explanatory story from a single mathematical law perspective. In a soup universe, perspectives are not even possible. However, the conscious universe accommodates, and indeed requires, perspective. From the vantage point of a conscious universe, we can therefore work to "soften" – "soupify," if you will – questions about the clockwork universe to ensure that explanation is partial and perspectival, requiring multiple types of explanations and a plurality of explanatory paradigms. By extension, we can imagine scientific investigations into a soup universe to which some structure (e.g., hierarchy; more robust and stable objects and processes) has been added, or which is now inhabited by agency or consciousness, with its concomitant use of statistics and experimentation (Figure 10.3). This work might push us to ask even more difficult questions about perspective: When might ethical implications in such universes enter? Would it be when agency is free, or when there are agents who, while possibly not free, can suffer or experience pleasure and happiness? In short, how do we soup the clockwork universe, and how can we clockwork the soup universe? And what might, if anything, be unique and emergent about our conscious universe vis-à-vis a soupified clockwork or a clockworked soup universe?

I have no easy answers. However, it is particularly useful and crucial to understand, from a metacognitive and philosophical point of view, the inordinately complex relation among statistics, experiment, causation, explanatory paradigms, perspective, action and intervention, and the political and ethical context of such thought experiments – all in the context of the complex interrelation among technical and ethical questions in human evolutionary genomics.

The Grand Scheme of Life

Just as we have used three kinds of universes to help think through statistics and causation, we can zoom out from the phylogenetic picture of human evolution of Chapter 2 to the grand scheme of all life on Earth.

In the case of human evolution, we sketched the branching, deep ancestry of *Homo sapiens* at multiple fractal levels: our species (and immediate sister subspecies), hominins, and – most generally, and deepest historically – catarrhines (i.e., great apes, gibbons, and Old World monkeys). However, let's consider the highest scale possible for life on Earth (Figures 10.4 and 10.5). In a view first clearly articulated and championed by Charles Darwin, all life originated from some first

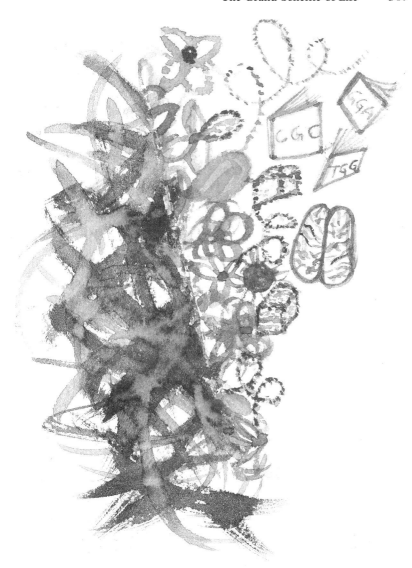

Figure 10.3. Close-Up of Soup Universe
Zooming in on the fine structure of a stochastic and unstable universe – where, remarkably, some clockwork organization and consciousness can also be found. Thus, apparent chaos gives rise to brains, books, and codons. Illustrated by Marie Raffn. © 2021 Rasmus Grønfeldt Winther.

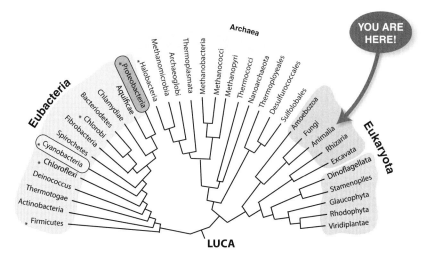

Figure 10.4. Branching life
"Schematic showing the broad phylogenetic relationships among the prokaryotes (Eubacteria and Archaea) and the eukaryotes (Eukaryota consisting of algae, plants, animals, and fungi) that descended from a last universal common ancestor (LUCA). The cyanobacteria and the proteobacteria are highlighted because, according to the endosymbiotic theory, cyanobacteria-like and proteobacteria-like organisms respectively evolved into modern day chloroplasts and mitochondria. Clades within the Eubacteria and Archaea that have photosynthetic representatives are denoted by an asterisk." From Niklas, 2016, p. 33. Adapted by Rasmus Grønfeldt Winther and Mats Wedin. Reprinted with permission. (A black and white version of this figure will appear in some formats. For the color version, please refer to the plate section.)

progenitor, today called LUCA (last universal common ancestor) – recently identified as potentially possessing some 350 proteins (and associated coding genes), necessary for life, and likely living in thermophilic vents at ocean bottoms some 4 billion years ago.[48] According to Figure 10.4, all of life descended from LUCA, branching endlessly. In fact, the general representational emphasis is one of branching and dichotomization, a powerful and suggestive way of thinking about speciation and lineage separation, as we explored in Chapters 2 and 5.

Yet this is too simple. Just as even human evolution is not just a branching ordeal (see, for example, Figures 2.11 and 2.12) but also involves introgression and merging of sister (sub)species (see Figure 2.10) also across pedigrees (see Figures 2.5–2.7), so the entire

---

[48] Weiss et al. (2016).

evolution of life is a story of merging and interweaving, and in a much more fundamental manner. For instance, the serial endosymbiotic theory (Figure 10.5) indicates that lineages have frequently crossed boundaries and borders, merging in complex series of endosymbiotic events within cells, and within cells within cells, and so forth. This goes much beyond the evolutionary history of cellular respiration (mitochondria) and photosynthesis (chloroplasts).[49] Thus, while human evolution may be more branching than merging (admixture, introgression), and while the entire history of life is likely at least as much merging (endosymbiosis, horizontal gene transfer) as branching, both scales of the evolutionary process are a mix of the two basic topologies.

Ultimately, although it may be a cliché, it is true: All of life is connected. This is not a trivial point. Whether there is a single last common ancestor or perhaps a network or community of primordial cells that then crisscrossed and networked up through the immense, branching tree of life, the historical connection of all life is undeniable. At each stage of this deep history, there is a broad ecological perspective in which life always requires other life for energy and metabolism, and for empathy and reproduction. The genetic paradigm we have investigated throughout *Our Genes* gives us a way to see life, in all its connectivity, across generations and across geological time, perhaps all the way up to Gaia.[50] Genetically and ecologically, *Homo sapiens* is connected to all creatures. Thinking through Figures 10.4 and 10.5 shines a light on the responsibility of stewardship we all have.

The Final Paradox

Ultimately a paradox lives within this book's pages. Our genes can tell us a lot about ourselves and our ancestry, and perhaps less about our future medical and cognitive conditions. But our genes also tell us, well, next to nothing, about the way we view the world, the societies we grow up in, and the languages we speak. Although evolutionary genomics and evolutionary psychology have something to say about why human societies have some fairly similar, shared gross structures, culture is flexible and can

---

[49] Lynn Margulis, known for her detailed account of endosymbiotic events during the history of life, was also a staunch defender of the five-kingdom system, even in light of Carl Woese's work on the three-domain system of life (e.g., Woese et al. 1990). Nevertheless, the five-kingdom system has generally been subsumed under the three-domain system.

[50] Lovelock (1979).

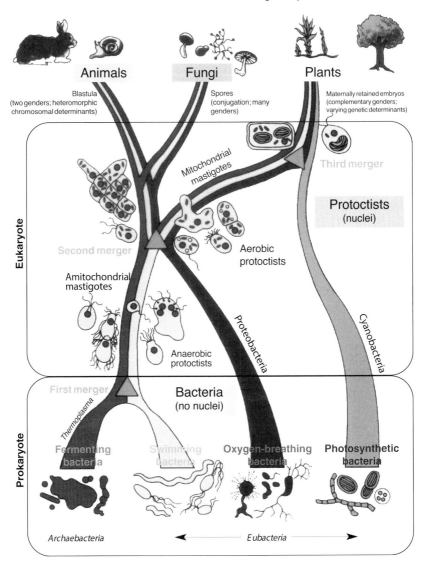

Figure 10.5. Symbiotic life

"The minimal four prokaryotes of the plant cell. Swimming eubacteria (1) merged with sulfidogenic archaebacteria (2) and formed archaeprotists (amitochondriate mastigotes). O_2-respiring eubacteria (3) in the second merger produced ancestors to eukaryotic heterotrophs. Some acquired cyanobacteria (4) as undigested food and became algae with the third merger. All eukaryotes evolved from symbiotic mergers, whereas prokaryotes did not. Past (at bottom) to present (at top) is represented by the Archean Eon [4–2.5 billion years ago (Ga)] dominated by prokaryotes, the Proterozoic Eon [2.5–0.54 Ga] of protoctists and the upper level Phanerozoic

stretch, and is explanatory in its own right, utterly independent of genetics. I have not hidden this paradox. In fact, I seek, with *Our Genes*, to emphasize it: Genes are highly informative, and not at all informative. They tell us something about who I am, who you are, who we are, in relation to each other, where we were, and where we are going. They also tell us hardly anything at all.

Despite even the scant – if rich – information that genes provide, we cannot chart a path toward equality and freedom using only scientific information. We must move forward according to the methods, paradigms, and philosophies of morality and law, which are premised on intrinsic equality, fairness, and empathy, not genes. By casting the questions that help situate and inform human evolutionary genomics against a moral and legal background, we stand to gain greater access to the field's lasting relevance and consequences, and to possible future work.

Caption for Figure 10.5. (*cont.*) Eon [0.54 Ga–present], marked by the abundance of plants, animals and fungi." From Margulis, 2004, p. 173. Illustration by Kathy Delisle. Reprinted with permission. (A black and white version of this figure will appear in some formats. For the color version, please refer to the plate section.)

References

1000 Genomes Project Consortium. 2015. A Global Reference for Human Genetic Variation. *Nature* 526(7571): 68–74.

AAPA (American Association of Physical Anthropologists). 1996. AAPA Statement on Biological Aspects of Race. *American Journal of Physical Anthropology* 101(4): 569–570.

Acciaioli, G., Brunt, H., and Clifton, J. 2017. Foreigners Everywhere, Nationals Nowhere: Exclusion, Irregularity, and Invisibility of Stateless Bajau Laut in Eastern Sabah, Malaysia. *Journal of Immigrant & Refugee Studies* 15(3): 232–249.

Adams, J.U. 2008. Human Evolutionary Tree. *Nature Education* 1(1): art. 145.

Adrion, J.R., Cole, C.B., Dukler, N., Galloway, J.G., Gladstein, A.L., et al. 2020. A Community-Maintained Standard Library of Population Genetic Models. *Elife* 9: e54967.

Agrawal, S., and Khan, F. 2005. Reconstructing Recent Human Phylogenies with Forensic STR Loci: A Statistical Approach. *BMC Genetics* 6: art. 47.

Alcoff, L.M. 2015. *The Future of Whiteness*. Cambridge: Polity.

Aldrich, J. 1997. R.A. Fisher and the Making of Maximum Likelihood 1912–1922. *Statistical Science* 12(3): 162–176.

Alexander, D.H., Novembre, J., and Lange, K. 2009. Fast Model-Based Estimation of Ancestry in Unrelated Individuals. *Genome Research* 19(9): 1655–1664.

Allendorf, F.W., Luikart, G.H., and Aitken, S.N. 2013. *Conservation and the Genetics of Populations*. 2nd ed. Chichester: Wiley-Blackwell.

Allentoft, M.E., Sikora, M., Sjögren, K.-G., Rasmussen, S, Rasmussen, M., et al. 2015. Population Genomics of Bronze Age Eurasia. *Nature* 522(7555): 167–172.

Alwaal, A., Breyer, B.N., and Lue, T.F. 2015. Normal Male Sexual Function: Emphasis on Orgasm and Ejaculation. *Fertility and Sterility* 104(5): 1051–1060.

Anderson, E.C., and Dunham, K.K. 2008. The Influence of Family Groups on Inferences Made with the Program Structure. *Molecular Ecology Resources* 8: 1219–1229.

Andreasen, R.O. 2000. Race: Biological Reality or Social Construct? *Philosophy of Science* 67(Proceedings, Part II): S653–S666.

Andreasen, R.O. 2004. The Cladistic Race Concept: A Defense. *Biology and Philosophy* 19: 425–442.

Andreasen, R.O. 2007. Biological Conceptions of Race. In *Philosophy of Biology*, ed. M. Matthen and C. Stephens. Amsterdam: Elsevier, pp. 455–481.

Angier, N. 1999. *Woman: An Intimate Geography*. New York: Houghton Mifflin.

Ankeny, R.A., and Leonelli, S. 2011. What's so Special about Model Organisms? *Studies in History and Philosophy of Science, Part A* 42(2): 313–323.

Anthony, D.W., and Ringe, D. 2015. The Indo-European Homeland from Linguistic and Archaeological Perspectives. *Annual Review of Linguistics* 1: 199–219.

Appiah, K.A. 1989. The Conservation of "Race." *Black American Literature Forum* 23(1): 37–60.

Appiah, K.A., and Gutmann, A. 1996. *Color Conscious: The Political Morality of Race*. Princeton, NJ: Princeton University Press.

Arias, L., Schröder, R., Hübner, A., Barreto, G., Stoneking, M., and Pakendorf, B. 2018. Cultural Innovations Influence Patterns of Genetic Diversity in Northwestern Amazonia. *Molecular Biology and Evolution* 35(11): 2719–2735.

Arunotai, N. 2017. "Hopeless at Sea, Landless on Shore": Contextualising the Sea Nomad's Dilemma in Thailand. In *Austrian Academy of Sciences Working Papers in Social Anthropology*, Volume 31, edited by Eva-Maria Knoll. Available: https://epub.oeaw.ac.at/0xc1aa5576%200x0036cd03.pdf Accessed April 2020.

Ásta. 2018. *Categories We Live By: The Construction of Sex, Gender, Race, and Other Social Categories*. New York: Oxford University Press.

Bae, C.J., Douka, K., and Petraglia, M.D. 2017. On the Origin of Modern Humans: Asian Perspectives. *Science* 358(6368): art. eaai9067.

Baković, D., Eterović, D., Saratlija-Novaković, Ž., Palada, I., Valic, Z., et al. 2005. Effect of Human Splenic Contraction on Variation in Circulating Blood Cell Counts. *Clinical and Experimental Pharmacology and Physiology* 32(11): 944–951.

Bamshad, M.J., Wooding, S., Watkins, W.S., Ostler, C.T., Batzer, M.A., and Jorde, L.B. 2003. Human Population Genetic Structure and Inference of Group Membership. *The American Journal of Human Genetics* 72(3): 578–589.

Baranova, T.I., Berlov, D.N., Glotov, O.S., Korf, E.A., Minigalin, A.D., et al. 2017. Genetic Determination of the Vascular Reactions in Humans in Response to the Diving Reflex. *American Journal of Physiology: Heart and Circulatory Physiology* 312(3): H622–H631.

Barbujani, G., and Colonna, V. 2010. Human Genome Diversity: Frequently Asked Questions. *Trends in Genetics* 26(7): 285–295.

Barbujani, G., Magagni, A., Minch, E., and Cavalli-Sforza, L.L. 1997. An Apportionment of Human DNA Diversity. *Proceedings of the National Academy of Sciences of the United States of America* 94(9): 4516–4519.

Barker, M.J., and Velasco, J.D. 2013. Deep Conventionalism about Evolutionary Groups. *Philosophy of Science* 80(5): 971–982.

Bartha, P.F.A. 2010. *By Parallel Reasoning: The Construction and Evaluation of Analogical Arguments*. New York: Oxford University Press.

Baskin, L., Shen, J., Sinclair, A., Cao, M., Liu, X., et al. 2018. Development of the Human Penis and Clitoris. *Differentiation* 103: 74–85.

Batzoglou, S., Pachter, L., Mesirov, J.P., Berger, B., and Lander, E.S. 2000. Human and Mouse Gene Structure: Comparative Analysis and Application to Exon Prediction. *Genome Research* 10(7): 950–958.

Beatty, J. 1987. Weighing the Risks: Stalemate in the Classical/Balance Controversy. *Journal of the History of Biology* 20(3): 289–319.

Beatty, J. 1997. Why Do Biologists Argue Like They Do? *Philosophy of Science* 64(Proceedings, Part II): S432–S443.

Beerli, P., and Felsenstein, J. 1999. Maximum-Likelihood Estimation of Migration Rates and Effective Population Numbers in Two Populations Using a Coalescent Approach. *Genetics* 152(2): 763–773.

Begun, D.R. 2016. *The Real Planet of the Apes: A New Story of Human Origins.* Princeton, NJ: Princeton University Press.

Bellwood, P. 2013. *First Migrants: Ancient Migration in Global Perspective.* Chichester: Wiley-Blackwell.

Bellwood, P. 2017. *First Islanders: Prehistory and Human Migration in Island Southeast Asia.* Hoboken, NJ: John Wiley.

Berger, M.F., and Mardis, E.R. 2018. The Emerging Clinical Relevance of Genomics in Cancer Medicine. *Nature Reviews Clinical Oncology* 15(6): 353–365.

Bergström, A., McCarthy, S.A., Hui, R., Almarri, M.A., Ayub, Q., et al. 2020. Insights into Human Genetic Variation and Population History from 929 Diverse Genomes. *Science* 367(6484): art. eaay5012.

Bergstrom, C.T., and Rosvall, M. 2011. The Transmission Sense of Information. *Biology and Philosophy* 26(2): 159–176.

Biddanda, A., Rice, D.P., and Novembre, J. 2020. A Variant-Centric Perspective on Geographic Patterns of Human Allele Frequency Variation. *eLife* 9: art. e60107.

Blalock, H.M. Jr. 1964. *Causal Inferences in Nonexperimental Research.* Chapel Hill, NC: University of North Carolina Press.

Block, N. 1995. How Heritability Misleads about Race. *Cognition* 56(2): 99–128.

Bodmer, W. 2018. Anthony Edwards's Seminal Contributions to Phylogenetics, Likelihood, and Understanding R.A. Fisher and the History of Genetics. In *Phylogenetic Inference, Selection Theory, and History of Science: Selected Papers of A.W.F. Edwards with Commentaries*, ed. R.G. Winther. Cambridge: Cambridge University Press, pp. 317–324.

Böhme, M., Spassov, N., Fuss, J., Tröscher, A., Deane, A.S., et al. 2019. A New Miocene Ape and Locomotion in the Ancestor of Great Apes and Humans. *Nature* 575(7783): 489–493.

Bolnick, D.A. 2008. Individual Ancestry Inference and the Reification of Race as a Biological Phenomenon. In *Revisiting Race in a Genomic Age*, ed. B.A. Koenig, S.S.-J. Lee, and S.S. Richardson. New Brunswick, NJ: Rutgers University Press, pp. 70–85.

Booker, T.R., Jackson, B.C., and Keightley, P.D. 2017. Detecting Positive Selection in the Genome. *BMC Biology* 15: art. 98.

Boose, K., White, F., Brand, C., Meinelt, A., and Snodgrass, J. 2018. Infant Handling in Bonobos (*Pan paniscus*): Exploring Functional Hypotheses and the Relationship to Oxytocin. *Physiology & Behavior* 193(Part A): 154–166.

Born, M. (ed.). 2005 [1971]. *The Born–Einstein Letters 1916–1955: Friendship, Politics and Physics in Uncertain Times*, transl. I. Born, commentary by M. Born, foreword by B. Russell, introduction by W. Heisenberg, new preface by D. Buchwald and K.S. Thorne. New York: Macmillan.

Bostrom, N. 2002. *Anthropic Bias: Observation Selection Effects in Science and Philosophy.* New York: Routledge.

Bostrom, N. 2003. Are You Living in a Computer Simulation? *Philosophical Quarterly* 53(211): 243–255.

Bouchard, T.J., Jr. 2009. Genetic Influence on Human Intelligence (Spearman's *g*): How Much? *Annals of Human Biology* 36(5): 527–544.

Bowker, G.C., and Star, S.L. 1999. *Sorting Things Out: Classification and Its Consequences*. Cambridge, MA: MIT Press.

Bowler, P.J. 1989. *The Mendelian Revolution: The Emergence of Hereditarian Concepts in Modern Science and Society*. London: Athlone Press.

Box, G.E.P. 1976. Science and Statistics. *Journal of the American Statistical Association* 71(356): 791–799.

Box, J. 1980. R.A. Fisher and the Design of Experiments, 1922–1926. *The American Statistician* 34(1): 1–7.

Boyle, E.A., Li, Y.I., and Pritchard, J.K. 2017. An Expanded View of Complex Traits: From Polygenic to Omnigenic. *Cell* 169(7): 1177–1186.

Bradshaw, C.J.A., Ehrlich, P.R., Beattie, A., Ceballos, G., Crist, E., et al. 2021. Underestimating the Challenges of Avoiding a Ghastly Future. *Frontiers in Conservation Science* 1: art. 615419.

Brandon, R.N. 2022. Five Advantages of the Phylogenetic Race Concept. In *Remapping Race in a Global Context*, ed. L. Lorusso and R.G. Winther. London: Routledge, pp. 227–243.

Bresnick, J. 2018. What Are Precision Medicine and Personalized Medicine? *Health IT Analytics*. Available: https://healthitanalytics.com/features/what-are-precision-medicine-and-personalized-medicine Accessed May 2020.

Bridget, A., and Moorjani, P. 2017. DNA Dating. *The Conversation*. Available: https://theconversation.com/dna-dating-how-molecular-clocks-are-refining-human-evolutions-timeline-65606 Accessed May 2020.

Brisbin, A. 2010. *Linkage Analysis for Categorical Traits and Ancestry Assignment in Admixed Individuals*. Doctoral Dissertation, Applied Mathematics, Cornell University, Ithaca, NY.

Brown, G.R., Gill, G.P., Kuntz, R.J., Langley, C.H., and Neale, D.B. 2004. Nucleotide Diversity and Linkage Disequilibrium in Loblolly Pine. *Proceedings of the National Academy of Sciences of the United States of America* 101(42): 15255–15260.

Brown, R.A., and Armelagos, G.J. 2001. Apportionment of Racial Diversity: A Review. *Evolutionary Anthropology* 10(1): 34–40.

Burchard, E.G., Ziv, E., Coyle, N., Gomez, S.L., Tang, H., et al. 2003. The Importance of Race and Ethnic Background in Biomedical Research and Clinical Practice. *New England Journal of Medicine* 348(12): 1170–1175.

Burri, A.V., and Spector, T.D. 2008. The Genetics of Female Sexual Behavior. *Sexuality, Reproduction and Menopause* 6(2): 22–27.

Burri, A.V., and Spector, T.D. 2011. An Epidemiological Survey of Post-Coital Psychological Symptoms in a UK Population Sample of Female Twins. *Twin Research and Human Genetics* 14(3): 240–248.

Burri, A.V., Cherkas, L.M., and Spector, T.D. 2009. The Genetics and Epidemiology of Female Sexual Dysfunction: A Review. *The Journal of Sexual Medicine* 6(3): 646–657.

Burton, E.K. 2019. Red Crescents: Race, Genetics, and Sickle Cell Disease in the Middle East. *Isis* 110(2): 250–269.

Cain, A.J., and Sheppard, P.M. 1950. Selection in the Polymorphic Land Snail *Cepæa nemoralis*. *Heredity* 4(3): 275–294.

Callaway, E. 2020. The Race for Coronavirus Vaccines: A Graphical Guide. *Nature* 580: 576–577.

Campbell, M.C., and Tishkoff, S.A. 2008. African Genetic Diversity: Implications for Human Demographic History, Modern Human Origins, and Complex Disease Mapping. *Annual Review of Genomics and Human Genetics* 9: 403–433.

Cann, R., Stoneking, M., and Wilson, A. 1987. Mitochondrial DNA and Human Evolution. *Nature* 325: 31–36.

Cartwright, N. 1983. *How the Laws of Physics Lie*. New York: Oxford University Press.

Cartwright, N. 1989. *Nature's Capacities and Their Measurement*. New York: Oxford University Press.

Cartwright, N. 1999. *The Dappled World: A Study of the Boundaries of Science*. Cambridge: Cambridge University Press.

Cartwright, N. 2007. *Hunting Causes and Using Them: Approaches in Philosophy and Economics*. New York: Cambridge University Press.

Cavalli-Sforza, L.L. 1966. Population Structure and Human Evolution. *Proceedings of the Royal Society of London. Series B, Biological Sciences* 164(995): 362–379.

Cavalli-Sforza, L.L. 2005. The Human Genome Diversity Project: Past, Present and Future. *Nature Reviews Genetics* 6(4): 333–340.

Cavalli-Sforza, L.L., and Bodmer, W.F. 1971. *The Genetics of Human Populations*. San Francisco, CA: Freeman.

Cavalli-Sforza, L.L., and Edwards, A.W.F. 1967. Phylogenetic Analysis: Models and Estimation Procedures. *Evolution* 21(3): 550–570. Reprinted as Paper 37 in Winther 2018a, pp. 82–103.

Cavalli-Sforza, L.L., and Feldman, M.W. 1981. *Cultural Transmission and Evolution: A Quantitative Approach*. Princeton, NJ: Princeton University Press.

Cavalli-Sforza, L.L., and Feldman, M.W. 2003. The Application of Molecular Genetic Approaches to the Study of Human Evolution. *Nature Genetics* 33: 266–275.

Cavalli-Sforza, L.L., Piazza, A., Menozzi, P., and Mountain, J. 1988. Reconstruction of Human Evolution: Bringing Together Genetic, Archaeological, and Linguistic Data. *Proceedings of the National Academy of Sciences of the United States of America* 85(16): 6002–6006.

Cavalli-Sforza, L.L., Menozzi, P., and Piazza, A. 1994. *The History and Geography of Human Genes*. Princeton, NJ: Princeton University Press.

Cerdeña, J.P., Plaisime, M.V., and Tsai, J. 2020. From Race-Based to Race-Conscious Medicine: How Anti-Racist Uprisings Call Us to Act. *The Lancet* 396(10257): 1125–1128.

Cerrone, M., Remme, C.A., Tadros, R., Bezzina, C.R., and Delmar, M. 2019. Beyond the One Gene–One Disease Paradigm: Complex Genetics and Pleiotropy in Inheritable Cardiac Disorders. *Circulation* 140(7): 595–610.

Chalabi, M. 2015. The Gender Orgasm Gap. *FiveThirtyEight*. Available: https://fivethirtyeight.com/features/the-gender-orgasm-gap/ Accessed June 2020.

Chang, J.T. 1999. Recent Common Ancestors of All Present-Day Individuals. *Advances in Applied Probability* 31(4): 1002–1026.

Charati, H., Peng, MS., Chen, W., Yang, X.-Y., Ori, R.J., et al. 2019. The Evolutionary Genetics of Lactase Persistence in Seven Ethnic Groups Across the Iranian Plateau. *Human Genomics* 13: art. 7.

Chen, F.C., and Li, W.H. 2001. Genomic Divergences Between Humans and Other Hominoids and the Effective Population Size of the Common Ancestor of Humans and Chimpanzees. *The American Journal of Human Genetics* 68(2): 444–456.

Chen, L., Wolf, A.B., Fu, W., Li, L., and Akey, J. 2020. Identifying and Interpreting Apparent Neanderthal Ancestry in African Individuals. *Cell* 180(4): 677–687.

Chorowicz, J. 2005. The East African Rift System. *Journal of African Earth Sciences* 43(1–3): 379–410.

Chow, K.U., Luxembourg, B., Seifried, E., and Bonig, H. 2016. Spleen Size Is Significantly Influenced by Body Height and Sex: Establishment of Normal Values for Spleen Size at US with a Cohort of 1200 Healthy Individuals. *Radiology* 279(1): 306–313.

Churchill, F.B. 2015. *August Weismann: Development, Heredity, and Evolution.* Cambridge, MA: Harvard University Press.

Clark, A.G., Nielsen, R., Signorovitch, J., Matise, T.C., Glanowski, S., et al. 2003. Linkage Disequilibrium and Inference of Ancestral Recombination in 538 Single-Nucleotide Polymorphism Clusters Across the Human Genome. *The American Journal of Human Genetics* 73(2): 285–300.

Clarke, L., Fairley, S., Zheng-Bradley, X., Streeter, I., Perry, E., et al. 2017. The International Genome Sample Resource (IGSR): A Worldwide Collection of Genome Variation Incorporating the 1000 Genomes Project Data. *Nucleic Acids Research* 45(D1): D854–D859.

Clarke, M. 2019. What's Going on with Hormones and Neurotransmitters during Sex. *Atlasbiomed Blog.* Available: https://atlasbiomed.com/blog/whats-going-on-with-hormones-and-neurotransmitters-during-sex/ Accessed June 2020.

Coates, D.J., Byrne, M., and Moritz, C. 2018. Genetic Diversity and Conservation Units: Dealing with the Species–Population Continuum in the Age of Genomics. *Frontiers in Ecology and Evolution* 6: art. 165.

Cockerham, C.C. 1969. Variance of Gene Frequencies. *Evolution* 23(1): 72–84.

Cockerham, C.C. 1973. Analyses of Gene Frequencies. *Genetics* 74(4): 679–700.

Collins, F.S., and Varmus, H. 2015. A New Initiative on Precision Medicine. *New England Journal of Medicine* 372(9): 793–795.

Coop, G., Eisen, M.B., Nielsen, R., Przeworski, M., and Rosenberg, N. (with many more signatories online). 2014. Letters: "A Troublesome Inheritance." *The New York Times.* Available: www.nytimes.com/2014/08/10/books/review/letters-a-troublesome-inheritance.html Accessed March 2015.

Cordell, H.J. 2002. Epistasis: What It Means, What It Doesn't Mean, and Statistical Methods to Detect It in Humans. *Human Molecular Genetics* 11(20): 2463–2468.

Cornfield, J., Haenszel, W., Hammond, E.C., Lilienfeld, A.M., Shimkin, M.B., and Wynder, E.L. 2009 [1959]. Smoking and Lung Cancer: Recent Evidence and a Discussion of Some Questions. *International Journal of Epidemiology* 38(5):

1175–1191. Originally published in 1959 in the *Journal of the National Cancer Institute* 22(1): 173–203.

Corum, J., and Zimmer, C. 2020. Bad News Wrapped in Protein: Inside the Coronavirus Genome. *The New York Times*. Available: www.nytimes.com/interactive/2020/04/03/science/coronavirus-genome-bad-news-wrapped-in-protein.html Accessed June 2021.

Cote, S.M. 2004. Origins of the African Hominoids: An Assessment of the Palaeobiogeographical Evidence. *Comptes Rendus Palevol* 3(4): 323–340.

Coulthard, G.S. 2014. *Red Skin, White Masks: Rejecting the Colonial Politics of Recognition*. Minneapolis, MN: University of Minnesota Press.

Craver, C.F. 2007. *Explaining the Brain: Mechanisms and the Mosaic Unity of Neuroscience*. New York: Oxford University Press.

Crombie, A.C. 1994. *Styles of Scientific Thinking in the European Tradition* (Vols. 1–3). London: Duckworth.

Crow J., and Kimura, M. 1970. *An Introduction to Population Genetics Theory*. New York: Harper and Row.

Danchin, É., Pocheville, A., and Huneman, P. 2019. Early in Life Effects and Heredity: Reconciling Neo-Darwinism with Neo-Lamarckism Under the Banner of the Inclusive Evolutionary Synthesis. *Philosophical Transactions of the Royal Society B: Biological Sciences* 374(1770): art. 20180113.

Darwin, C.R. 1964 [1859]. *On the Origin of Species by Means of Natural Selection, or the Preservation of Favoured Races in the Struggle for Life*. Cambridge, MA: Harvard University Press.

Darwin, C.R. 1874 [1871]. *The Descent of Man, and Selection in Relation to Sex*. 2nd ed. New York: D. Appleton & Company.

Davenport, W.H. 1977. Sex in Cross-Cultural Perspective. In *Human Sexuality in Four Perspectives*, ed. F.A. Beach. Baltimore, MD: Johns Hopkins University Press, pp. 115–163.

Dawood, K., Kirk, K.M., Bailey, J.M., Andrews, P.W., and Martin, N.G. 2005. Genetic and Environmental Influences on the Frequency of Orgasm in Women. *Twin Research and Human Genetics* 8(1): 27–33.

Dayalu, P., and Albin, R.L. 2015. Huntington's Disease: Pathogenesis and Treatment. *Neurologic Clinics* 33(1): 101–114.

de Waal, F. 2013. *The Bonobo and the Atheist: In Search of Humanism Among the Primates*. New York: W.W. Norton.

de Waal, F. 2019. *Mama's Last Hug: Animal Emotions and What They Tell Us about Ourselves*. New York: W.W. Norton.

Dear, P. 2019. *Revolutionizing the Sciences: European Knowledge in Transition, 1500–1700*. 3rd ed. Princeton, NJ: Princeton University Press.

Deaton, A., and Cartwright, N. 2017. Understanding and Misunderstanding Randomized Controlled Trials. National Bureau of Economic Research Available: www.nber.org/system/files/working_papers/w22595/w22595.pdf Accessed August 2021.

Denayer, T., Stöhr, T., and Van Roy, M. 2014. Animal Models in Translational Medicine: Validation and Prediction. *New Horizons in Translational Medicine* 2(1): 5–11.

Dewey, J. 1893. The Superstition of Necessity. *The Monist* 3(3): 362–379.

Dewey, J. 1929. *The Quest for Certainty*. In *The Later Works of John Dewey, 1925–1953, volume 4: 1929 (The Collected Works of John Dewey, 1882–1953)*, ed. J.A. Boydston. Carbondale, IL: University of Southern Illinois Press, pp. 1–250.

Dewey, J. 1985 [1931]. "Context and Thought." In *The Later Works, 1925–1953, volume 6: 1931–1932 (The Collected Works of John Dewey, 1882–1953)*, ed. J.A. Boydston. Carbondale, IL: University of Southern Illinois Press, pp. 3–21.

Dicks, J., and Savva, G. 2007. Comparative Genomics. In *Handbook of Statistical Genetics*, ed. D. J. Balding, M. Bishop, and C. Cannings. 3rd ed. Chichester: John Wiley & Sons, pp. 160–199.

Dietrich, M.R. 1994. The Origins of the Neutral Theory of Molecular Evolution. *Journal of the History of Biology* 27(1): 21–59.

Dobzhansky, T. 1955. A Review of Some Fundamental Concepts and Problems of Population Genetics. *Cold Spring Harbor Symposia on Quantitative Biology* 20: 1–15.

Dobzhansky, T. 1962. On the Non-existence of Human Races (Comment). *Current Anthropology* 3(3): 279–280.

Doll, R. 2002. Proof of Causality: Deduction from Epidemiological Observation. *Perspectives in Biology and Medicine* 45(4): 499–515.

Doll, R., and Hill, A.B. 1950. Smoking and Carcinoma of the Lung: Preliminary Report. *The British Medical Journal* 2(4682): 739–748.

Donovan, B.M. 2014. Playing with Fire? The Impact of the Hidden Curriculum in School Genetics on Essentialist Conceptions of Race. *Journal of Research in Science Teaching* 51(4): 462–496.

Donovan, B.M., Semmens, R., Keck, P., Brimhall, E., Busch, K.C., et al. 2019. Toward a More Humane Genetics Education: Learning about the Social and Quantitative Complexities of Human Genetic Variation Research Could Reduce Racial Bias in Adolescent and Adult Populations. *Science Education* 103(3): 529–560.

Doolittle, W.F. 1999. Phylogenetic Classification and the Universal Tree. *Science* 284(5423): 2124–2128.

Downes, S.M., and Matthews, L. 2020. Heritability. In *The Stanford Encyclopedia of Philosophy* (Spring 2020 Edition), ed. E.N. Zalta. Available: https://plato.stanford.edu/archives/spr2020/entries/heredity/ Accessed September 2020.

Drori, J. 2018. Conservation and Sustainable Use of Plant Genetic Resources. Available: https://files.nettsteder.regjeringen.no/wpuploads01/blogs.dir/221/files/2018/04/Seed-Vault-Summit-Conclusions.pdf Accessed September 2021.

Duncan, O.D. 1966. Path Analysis: Sociological Examples. *American Journal of Sociology* 72(1): 1–16.

Dunn, J.M., and Hayes, M.V. 1999. Toward a Lexicon of Population Health. *Canadian Journal of Public Health* 90(Suppl. 1): S7–S10.

Dunn, K.M., Cherkas, L.F., and Spector, T.D. 2005. Genetic Influences on Variation in Female Orgasmic Function: A Twin Study. *Biology Letters* 1(3): 260–263.

Dupré, J. 1993. *The Disorder of Things: Metaphysical Foundations of the Disunity of Science*. Cambridge, MA: Harvard University Press.

Dupré, J. 2004. Understanding Contemporary Genomics. *Perspectives on Science* 12(3): 320–338.

Earnshaw-Whyte, E. 2012. Increasingly Radical Claims about Heredity and Fitness. *Philosophy of Science* 79(3): 396–412.

Earnshaw-Whyte, E. 2018. *Modelling Evolution: A New Dynamic Account.* London: Routledge.

Edge, M.D. 2019. *Statistical Thinking from Scratch: A Primer for Scientists.* New York: Oxford University Press.

Edge, M.D., and Coop, G. 2018. How Lucky Was the Genetic Investigation in the Golden State Killer Case? Blog Post. Available: https://gcbias.org/2018/05/07/how-lucky-was-the-genetic-investigation-in-the-golden-state-killer-case/ Accessed June 2019.

Edge, M.D., and Coop, G. 2019. Reconstructing the History of Polygenic Scores Using Coalescent Trees. *Genetics* 211(1): 235–262.

Edge, M.D., and Rosenberg, N.A. 2014. Upper Bounds on F_{ST} in Terms of the Frequency of the Most Frequent Allele and Total Homozygosity: The Case of a Specified Number of Alleles. *Theoretical Population Biology* 97: 20–34.

Edge, M.D., and Rosenberg, N.A. 2015. Implications of the Apportionment of Human Genetic Diversity for the Apportionment of Human Phenotypic Diversity. *Studies in History and Philosophy of Science, Part C: Studies in History and Philosophy of Biological and Biomedical Sciences* 52: 32–45.

Edwards, A.W.F. 1972. *Likelihood.* Cambridge: Cambridge University Press.

Edwards, A.W.F. 1994. The Fundamental Theorem of Natural Selection. *Biological Reviews* 69(4): 443–474. Reprinted as Paper 140 in Winther 2018a, pp. 191–224.

Edwards, A.W.F. 2003. Human Genetic Diversity: Lewontin's Fallacy. *BioEssays* 25(8): 798–801. Reprinted as Paper 192 in Winther 2018a, pp. 249–253.

Edwards, A.W.F. 2014. R.A. Fisher's Gene-Centred View of Evolution and the Fundamental Theorem of Natural Selection. *Biological Reviews* 89(1): 135–147. Reprinted as Paper 238 in Winther 2018a, pp. 295–308.

Edwards, A.W.F. 2021. Population Genetics and the 1951 UNESCO Statement on the Nature of Race and Race Differences by Physical Anthropologists and Geneticists. Circular distributed to the Fellows of Gonville and Caius College, University of Cambridge, UK.

Edwards, A.W.F., and Cavalli-Sforza, L.L. 1964. Reconstruction of Evolutionary Trees. In *Phenetic and Phylogenetic Classification*, ed. V.H. Heywood and J. McNeill. London: The Systematics Association, pp. 67–76. Reprinted as Paper 27 in Winther 2018a, pp. 17–27.

Eernisse, D.J., and Kluge, A.G. 1993. Taxonomic Congruence versus Total Evidence, and Amniote Phylogeny Inferred from Fossils, Molecules, and Morphology. *Molecular Biology and Evolution* 10(6): 1170–1195.

Efron, B. 1998. R.A. Fisher in the 21st Century. *Statistical Science* 13(2): 95–122.

Eisenstadt, L. 2017. After a Decade of Genome-Wide Association Studies, a New Phase of Discovery Pushes On. Broad Institute (MIT, Harvard). Available: www.broadinstitute.org/news/after-decade-genome-wide-association-stud ies-new-phase-discovery-pushes Accessed April 2020.

Elwick, J. 2012. Layered History: Styles of Reasoning as Stratified Conditions of Possibility. *Studies in History and Philosophy of Science, Part A* 43(4): 619–627.

Emmeche, C. 1990. *Det biologiske informationsbegreb.* Doctoral Dissertation, Institute of Biological Chemistry, Faculty of Science, University of Copenhagen, Copenhagen.

Enard, D., and Petrov, D.A. 2018. Evidence that RNA Viruses Drove Adaptive Introgression Between Neanderthals and Modern Humans. *Cell* 175(2): 360–371.

Enard, D., Messer, P.W., and Petrov, D.A. 2014. Genome-Wide Signals of Positive Selection in Human Evolution. *Genome Research* 24(6): 885–895.

Enard, D., Cai, L., Gwennap, C., and Petrov, D.A. 2016. Viruses Are a Dominant Driver of Protein Adaptation in Mammals. *eLife* 5: art. e12469.

Enard, W., Przeworski, M., Fisher, S.E., Lai, C.S., Wiebe, V., et al. 2002. Molecular Evolution of *FOXP2*, a Gene Involved in Speech and Language. *Nature* 418(6900): 869–872.

Endler, J.A. 1986. *Natural Selection in the Wild*. Princeton, NJ: Princeton University Press.

Eriksson, A., and Manica, A. 2011. Detecting and Removing Ascertainment Bias in Microsatellites from the HGDP-CEPH Panel. *G3 (Bethesda, Md.)* 1(6): 479–488.

Esposito, M. 2018. A.W.F. Edwards, R.A. Fisher, and *The Genetical Theory of Natural Selection*. In *Phylogenetic Inference, Selection Theory, and History of Science: Selected Papers of A.W.F. Edwards with Commentaries*, ed. R.G. Winther. Cambridge: Cambridge University Press, pp. 376–385.

Ewens, W.J. 2009. Mathematical Population Genetics: Introduction to the Stochastic Theory. Lecture Notes. Available: http://www.rgwinther.com/Ewens2009MathematicalPopulationGeneticsTheGuanajuatoLectures.pdf Accessed August 2022.

Ewens, W.J. 2011. What Is the Gene Trying to Do? *The British Journal for the Philosophy of Science* 62(1): 155–176.

Ewens, W.J. 2018. A Conversation about Fisher (1930, 1958). In *Phylogenetic Inference, Selection Theory, and History of Science: Selected Papers of A.W.F. Edwards with Commentaries*, ed. R.G. Winther. Cambridge: Cambridge University Press, pp. 363–370.

Excoffier, L., Smouse, P.E., and Quattro, J.M. 1992. Analysis of Molecular Variance Inferred from Metric Distances Among DNA Haplotypes: Application to Human Mitochondrial DNA Restriction Data. *Genetics* 131(2): 479–491.

Eyre-Walker, A. 2006. The Genomic Rate of Adaptive Evolution. *Trends in Ecology & Evolution* 21(10): 569–575.

Fairley, S., Lowy-Gallego, E., Perry, E., and Flicek, P. 2020. The International Genome Sample Resource (IGSR) Collection of Open Human Genomic Variation Resources. *Nucleic Acids Research* 48(D1): D941–D947.

Falconer, D.S., and Mackay, T.F.C. 1996. *Introduction to Quantitative Genetics*. 4th ed. Harlow: Longman.

Falush, D., Stephens, M., and Pritchard, J.K. 2003. Inference of Population Structure Using Multilocus Genotype Data: Linked Loci and Correlated Allele Frequencies. *Genetics* 164(4): 1567–1587.

Falush, D., Stephens, M., and Pritchard, J.K. 2007. Inference of Population Structure Using Multilocus Genotype Data: Dominant Markers and Null Alleles. *Molecular Ecology Notes* 7(4): 574–578.

Fanon, F. 2008 [1952]. *Black Skin, White Masks* (original: *Peau noire, masques blancs*). New York: Grove Press.

Fanon, F. 2004 [1961]. *The Wretched of the Earth* (original: *Les damnés de la terre*). New York: Grove Press.

Fausto-Sterling, A. 2020. *Sexing the Body: Gender Politics and the Construction of Sexuality*. New York: Basic Books.

Feldman, M.W. 2010. The Biology of Ancestry: DNA, Genomic Variation, and Race. In *Doing Race: 21 Essays for the 21st Century*, ed. H.R. Markus and P.M.L. Moya. New York: W.W. Norton, pp. 136–159.

Feldman, M.W., and Lewontin, R.C. 1975. The Heritability Hang-up. *Science* 190(4220): 1163–1168.

Feldman, M.W., and Lewontin, R.C. 2008. Race, Ancestry, and Medicine. In *Revisiting Race in a Genomic Age*, ed. B.A. Koenig, S.S.-J. Lee, and S.S. Richardson. New Brunswick, NJ: Rutgers University Press, pp. 89–101.

Felsenstein, J. 2004. *Inferring Phylogenies*. Sunderland, MA: Sinauer Associates.

Felsenstein, J. 2018. Anthony Edwards, Luca Cavalli-Sforza, and Phylogenies. In *Phylogenetic Inference, Selection Theory, and History of Science: Selected Papers of A.W.F. Edwards with Commentaries*, ed. R.G. Winther. Cambridge: Cambridge University Press, pp. 325–333.

Field, I.C., Meekan, M.G., Buckworth, R.C., and Bradshaw, C.J.A. 2009. Protein Mining the World's Oceans: Australasia as an Example of Illegal Expansion-and-Displacement Fishing. *Fish and Fisheries* 10(3): 323–328.

Fischer, M.C., Rellstab, C., Leuzinger, M., Roumet, M., Gugerli, F., et al. 2017. Estimating Genomic Diversity and Population Differentiation: An Empirical Comparison of Microsatellite and SNP Variation in *Arabidopsis halleri*. *BMC Genomics* 18(1): art. 69.

Fisher, R.A. 1912. On an Absolute Criterion for Fitting Frequency Curves. *Messenger of Mathematics* 41: 155–160.

Fisher, R.A. 1918. The Correlation Between Relatives on the Supposition of Mendelian Inheritance. *Transactions of the Royal Society of Edinburgh* 52: 399–433.

Fisher, R.A. 1921. On the "Probable Error" of a Coefficient of Correlation Deduced from a Small Sample. *Metron* 1: 3–32.

Fisher, R.A. 1922. On the Mathematical Foundations of Theoretical Statistics. *Philosophical Transactions of the Royal Society of London. Series A, Containing Papers of a Mathematical or Physical Character* 222(602): 309–368.

Fisher, R.A. 1925. *Statistical Methods for Research Workers*. Edinburgh: Oliver and Boyd.

Fisher, R.A. 1926. The Arrangement of Field Experiments. *Journal of the Ministry of Agriculture* 33: 503–513.

Fisher, R.A. 1958 [1930]. *The Genetical Theory of Natural Selection*. 2nd rev. ed. New York: Dover Publications.

Fisher, R.A. 1934. Foreword. *Annals of Eugenics* 6(1): i.

Fisher, R.A. 1935. *The Design of Experiments*. Edinburgh: Oliver and Boyd.

Fisher, R.A. 1959. *Smoking: The Cancer Controversy – Some Attempts to Assess the Evidence*. Edinburgh: Oliver and Boyd.

Fisher, R.A., and Stock, C.S. 1915. Cuénot on Pre-Adaptation: A Criticism. *The Eugenics Review* 7(1): 46–61.

Foldès, P., and Buisson, O. 2009. The Clitoral Complex: A Dynamic Sonographic Study. *The Journal of Sexual Medicine* 6(5): 1223–1231.

Forgacs, G., and Newman, S.A. 2005. *Biological Physics of the Developing Embryo*. Cambridge: Cambridge University Press.

Frank, S., and Slatkin, M. 1992. Fisher's Fundamental Theorem of Natural Selection. *Trends in Ecology & Evolution* 7(3): 92–95.

Frankel, O.H. 1974. Genetic Conservation: Our Evolutionary Responsibility. *Genetics* 78(1): 53–65.

Franklin, A., Edwards, A.W.F., Fairbanks, D.J., Hartl, D.L., and Seidenfeld, T. 2008. *Ending the Mendel–Fisher Controversy*. Pittsburgh, PA: University of Pittsburgh Press.

Frase, P. 2016. *Four Futures: Visions of the World after Capitalism*. London: Verso.

Freedman, D.A. 1991. Statistical Models and Shoe Leather. *Sociological Methodology* 21: 291–313.

Freedman, D.A. 1994. From Association to Causation Via Regression. Talk presented at the 1993 Notre Dame conference "Causality in Crisis." Available: http://fitelson.org/woodward/freedman.pdf Accessed August 2021.

Freedman, D.A. 2009. *Statistical Models: Theory and Practice*. New York: Cambridge University Press.

Frenkel, E. 2014. Ad Infinitum: Review of *Our Mathematical Universe* by Max Tegmark. *The New York Times Book Review*, February 14. Available: www .nytimes.com/2014/02/16/books/review/our-mathematical-universe-by-max-tegmark.html Accessed May 2014.

Friedlaender, J.S., Friedlaender, F.R., Reed, F.A., Kidd, K.K., Kidd, J.R., et al. 2008. The Genetic Structure of Pacific Islanders. *PLoS Genetics* 4(1): art. e19.

Frodeman, R. 1995. Geological Reasoning: Geology as an Interpretative and Historical Science. *GSA Bulletin* 107(8): 960–968.

Fuentes, A., Ackermann, R.R., Athreya, S., Bolnick, D., Lasisi, T., et al. 2019. AAPA Statement on Race and Racism. *American Journal of Physical Anthropology* 169(3): 400–402.

Fujimura, J.H., Bolnick, D.A., Rajagopalan, R., Kaufman, J.S., Lewontin, R.C., et al. 2014. Clines without Classes: How to Make Sense of Human Variation. *Sociological Theory* 32(3): 208–227.

Fumagalli, M., Moltke, I., Grarup, N., Racimo, F., Bjerregaard, P., et al. 2015. Greenlandic Inuit Show Genetic Signatures of Diet and Climate Adaptation. *Science* 349(6254): 1343–1347.

Fung, T., and Keenan, K. 2014. Confidence Intervals for Population Allele Frequencies: The General Case of Sampling from a Finite Diploid Population of Any Size. *PLoS ONE* 9(1): art. e85925.

Funk, W.C., McKay, J.K., Hohenlohe, P.A., and Allendorf, F.W. 2012. Harnessing Genomics for Delineating Conservation Units. *Trends in Ecology & Evolution* 27(9): 489–496.

Futuyma, D.J. 2017. Evolutionary Biology Today and the Call for an Extended Synthesis. *Interface Focus* 7(5): art. 20160145.

Gaddis, J.L. 2004. *The Landscape of History: How Historians Map the Past*. New York: Oxford University Press.

Galeano, E. 1973 [1971]. *Open Veins of Latin America: Five Centuries of the Pillage of a Continent* (original: *Las venas abiertas de América Latina*). New York: Monthly Review Press.

Galison, P. 2003. *Einstein's Clocks and Poincaré's Maps: Empires of Time*. New York: W.W. Norton.

Galton, F. 1889. *Natural Inheritance*. London: Macmillan.

Gamble, C. 2013. *Settling the Earth: The Archaeology of Deep Human History*. New York: Cambridge University Press.

Gannett, L. 2004. The Biological Reification of Race. *The British Journal for the Philosophy of Science* 55(2): 323–345.

Gannett, L. 2008. The Human Genome Project. In *The Stanford Encyclopedia of Philosophy* (Winter 2014 Edition), ed. E.N. Zalta. Available: http://plato .stanford.edu/archives/win2014/entries/human-genome Accessed May 2015.

Gannett, L. 2022. Human Genetic Diversity: Fact and Fallacy. In *Remapping Race in a Global Context*. L. Lorusso and R.G. Winther. New York: Routledge, pp. 51–73.

Gao, H., Williamson, S., and Bustamante, C.D. 2007. A Markov Chain Monte Carlo Approach for Joint Inference of Population Structure and Inbreeding Rates from Multilocus Genotype Data. *Genetics* 176(3): 1635–1651.

Gardner, H. 1983. *Frames of Mind: The Theory of Multiple Intelligences*. New York: Basic Books.

Gaston, K.J., and Fuller, R.A. 2008. Commonness, Population Depletion and Conservation Biology. *Trends in Ecology & Evolution* 23(1): 14–19.

Gelman, A. 2008. Variance, analysis of. In *The New Palgrave Dictionary of Economics*, ed. S.N. Durlauf and L.E. Blume. Basingstoke: Palgrave Macmillan. Available: https://doi:10.1057/978-1-349-95121-5_2402-1 Accessed January 2014.

Gelman, A., and Vehtari, A. 2021. What Are the Most Important Statistical Ideas of the Past 50 Years? *Journal of the American Statistical Association* 116(536): 2087–2097.

Gelman, A., Carlin, J.B., Stern, H.S., Dunson, D.B., Vehtari, A., et al. 2013. *Bayesian Data Analysis*. 3rd ed. Boca Raton, FL: Chapman and Hall/CRC.

Gerbault, P., Liebert, A., Itan, Y., Powell, A., Currat, M., et al. 2011. Evolution of Lactase Persistence: An Example of Human Niche Construction. *Philosophical Transactions of the Royal Society B: Biological Sciences* 366(1566): 863–877.

Gerlach, G., Jueterbock, A., Kraemer, P., Deppermann, J., and Harmand, P. 2010. Calculations of Population Differentiation Based on G_{ST} and D: Forget G_{ST} But Not All of Statistics! *Molecular Ecology* 19(18): 3845–3852.

Gerstein, M.B., Bruce, C., Rozowsky, J.S., Zheng, D., Du, J., et al. 2007. What Is a Gene, Post-ENCODE? History and Updated Definition. *Genome Research* 17(6): 669–681.

Giblett, E.R. 1969. *Genetic Markers in Human Blood*. Oxford: Blackwell.

Gibson, G. 2012. Rare and Common Variants: Twenty Arguments. *Nature Reviews Genetics* 13(2): 135–140.

Gilbert, K.J., Andrew, R.K., Bock, D.G., Franklin, M.T., Kane, N.C., et al. 2012. Recommendations for Utilizing and Reporting Population Genetic Analyses: The Reproducibility of Genetic Clustering Using the Program STRUCTURE. *Molecular Ecology* 21(20): 4925–4930.

Gilbert, S.F. 2000. *Developmental Biology*. 6th ed. Sunderland, MA: Sinauer Associates. Available: www.ncbi.nlm.nih.gov/books/NBK9980/ Accessed July 2021.

Gillespie, J.H. 2004. *Population Genetics: A Concise Guide*. 2nd ed. Baltimore, MD: Johns Hopkins University Press.

Gini, C.W. 1912. *Variabilità e mutabilità: contributo allo studio delle distribuzioni e delle relazioni statistiche*. Bologna: Tipografia di Paolo Cuppini.

Glasgow, J. 2009. *A Theory of Race*. New York: Routledge.

Glennan, S.S. 1997. Probable Causes and the Distinction Between Subjective and Objective Chance. *Noûs* 31(4): 496–519.

Godfrey-Smith, P. 2001. Three Kinds of Adaptationism. In *Adaptationism and Optimality*, ed. S.H. Orzack and E. Sober. New York: Cambridge University Press, pp. 335–357.

Godfrey-Smith, P. 2009. *Darwinian Populations and Natural Selection*. Oxford: Oxford University Press.

Gómez-Robles, A. 2019. Dental Evolutionary Rates and Its Implications for the Neanderthal–Modern Human Divergence. *Science Advances* 5(5): art. eaaw1268.

Goodall, J. 1964. Tool-Using and Aimed Throwing in a Community of Free-Living Chimpanzees. *Nature* 201(4926): 1264–1266.

Goodall, J. 2010 [1971]. *In the Shadow of Man*. New York: Mariner.

Goodwin, B. 2001. *How the Leopard Changed Its Spots: The Evolution of Complexity*. Princeton, NJ: Princeton University Press.

Goswami, A., Smaers, J.B., Soligo, C., and Polly, P.D. 2014. The Macroevolutionary Consequences of Phenotypic Integration: From Development to Deep Time. *Philosophical Transactions of the Royal Society B: Biological Sciences* 369(1649): art. 20130254.

Goudet, J. 2005. HIERFSTAT, a Package for R to Compute and Test Hierarchical *F*-Statistics. *Molecular Ecology Notes* 5(1): 184–186.

Gould, S.J. 1996 [1981]. *The Mismeasure of Man*. New York: W.W. Norton.

Gould, S.J. 1993. Male Nipples and Clitoral Ripples. *Columbia: A Journal of Literature and Art* 20(Summer): 80–96.

Gould, S.J., and Lewontin, R.C. 1979. The Spandrels of San Marco and the Panglossian Paradigm: A Critique of the Adaptationist Programme. *Philosophical Transactions of the Royal Society of London. Series B, Biological Sciences* 205(1161): 581–598.

Grant, B.R., and Grant, P.R. 1989. *Evolutionary Dynamics of a Natural Population: The Large Cactus Finch of the Galapagos*. Chicago, IL: University of Chicago Press.

Grant, P.R. 1977. Review of *Island Biology: Illustrated by the Land Birds of Jamaica* by D. Lack. *Bird-Banding* 48(3): 296–300.

Gravlee, C.C. 2009. How Race Becomes Biology: Embodiment of Social Inequality. *American Journal of Physical Anthropology* 139(1): 47–57.

Greely, H.T. 2016. *The End of Sex and the Future of Human Reproduction*. Cambridge, MA: Harvard University Press.

Greenbaum, G., Rubin, A., Templeton, A., and Rosenberg, N.A. 2019. Network-Based Hierarchical Population Structure Analysis for Large Genomic Data Sets. *Genome Research* 29(12): 2020–2033.

Gregorius, H.-R. 1978. The Concept of Genetic Diversity and Its Formal Relationship to Heterozygosity and Genetic Distance. *Mathematical Biosciences* 41(3–4): 253–271.

Gregorius, H.-R. 1987. The Relationship Between the Concepts of Genetic Diversity and Differentiation. *Theoretical and Applied Genetics* 74(3): 397–401.

Griesemer, J.R. 1990. Modeling in the Museum: On the Role of Remnant Models in the Work of Joseph Grinnell. *Biology and Philosophy* 5(1): 3–36.

Griesemer, J.R. 1991. Material Models in Biology. *PSA: Proceedings of the Biennial Meeting of the Philosophy of Science Association* 1990(2): 79–94.

Griesemer, J. 2020. A Data Journey Through Dataset-Centric Population Genomics. In *Data Journeys in the Sciences*, ed. S. Leonelli and N. Tempini. Cham: Springer, pp. 145–167.

Griffiths, P.E. 2016. Proximate and Ultimate Information in Biology. In *The Philosophy of Philip Kitcher*, ed. M. Couch and J. Pfeifer. New York: Oxford University Press, pp. 74–97.

Griffiths, P.E., and Stoltz, K. 2013. *Genetics and Philosophy: An Introduction*. Cambridge: Cambridge University Press.

Grodwohl, J.-B. 2018. Reading and Misreading Fisher: Anthony Edwards and the History of Population Genetics. In *Phylogenetic Inference, Selection Theory, and History of Science: Selected Papers of A.W.F. Edwards with Commentaries*, ed. R.G. Winther. Cambridge: Cambridge University Press, pp. 386–398.

Gunnarsdóttir, E.D., Nandineni, M.R., Li, M., Myles, S., Gil, D., et al. 2011. Larger Mitochondrial DNA than Y-Chromosome Differences Between Matrilocal and Patrilocal Groups from Sumatra. *Nature Communications* 2: art. 228.

Günther, T., Valdiosera, C., Malmström, H., Ureña, I., Rodriguez-Varela, R., et al. 2015. Ancient Genomes Link Early Farmers from Atapuerca in Spain to Modern-Day Basques. *Proceedings of the National Academy of Sciences of the United States of America* 112(38): 11917–11922.

Gurdasani, D., Carstensen, T., Tekola-Ayele, F., Pagani, L., Tachmazidou, I., et al. 2015. The African Genome Variation Project Shapes Medical Genetics in Africa. *Nature* 517(7534): 327–332.

Gusella, J.F., Wexler, N.S., Conneally, P.M., Naylor, S.L., Anderson, M.A., et al. 1983. A Polymorphic DNA Marker Genetically Linked to Huntington's Disease. *Nature* 306(5940): 234–238.

Haak, W., Lazaridis, I., Patterson, N., Rohland, N., Mallick, S., et al. 2015. Massive Migration from the Steppe Is a Source for Indo-European Languages in Europe. *Nature* 522(7555): 207–211.

Haavelmo, T. 1943. The Statistical Implications of a System of Simultaneous Equations. *Econometrica* 11(1): 1–12.

Hacking, I. 1983a. *Representing and Intervening: Introductory Topics in the Philosophy of Natural Science*. Cambridge: Cambridge University Press.

Hacking, I. 1983b. Nineteenth Century Cracks in the Concept of Determinism. *Journal of the History of Ideas* 44(3): 455–475.

Hacking, I. 1990. *The Taming of Chance*. New York: Cambridge University Press.

Hacking, I. 1995. The Looping Effect of Human Kinds. In *Causal Cognition: An Interdisciplinary Approach*, ed. D. Sperber, D. Premack, and A.J. Premack. Oxford: Oxford University Press, pp. 351–383.

Hacking, I. 1999. *The Social Construction of What?* Cambridge, MA: Harvard University Press.

Hacking, I. 2002. *Historical Ontology*. Cambridge, MA: Harvard University Press.

Hacking, I. 2005. Why Race Still Matters. *Dædalus* (Winter): 102–116.

Hacking, I. 2007a. Kinds of People: Moving Targets. *Proceedings of the British Academy* 151: 285–318.

Hacking, I. 2007b. Natural Kinds: Rosy Dawn, Scholastic Twilight. *Royal Institute of Philosophy Supplements* 61: 203–239.

Hacking, I. 2007c. On Not Being a Pragmatist: Eight Reasons and a Cause. In *New Pragmatists*, ed. C. Misak. New York: Oxford University Press, pp. 32–49.

Hacking, I. 2014. *Why Is There Philosophy of Mathematics at All?* New York: Cambridge University Press.

Hagmann, D., Loewenstein, G., and Ubel, P.A. 2020. Antibody Tests Might Be Deceptively Dangerous: Blame the Math. *The Washington Post*. Available: www.washingtonpost.com/outlook/2020/04/30/antibody-tests-might-be-deceptively-dangerous-blame-math/ Accessed May 2020.

Haldane, J.B.S. 1964. A Defense of Beanbag Genetics. *Perspectives in Biology and Medicine* 7(3): 343–359.

Hale, M.L., Burg, T.M., and Steeves, T.E. 2012. Sampling for Microsatellite-Based Population Genetic Studies: 25 to 30 Individuals per Population Is Enough to Accurately Estimate Allele Frequencies. *PLoS ONE* 7(9): art. e45170.

Hammer, M.F., Karafet, T.M., Redd, A.J., Jarjanazi, H., Santachiara-Benerecetti, S., et al. 2001. Hierarchical Patterns of Global Human Y-Chromosome Diversity. *Molecular Biology and Evolution* 18(7): 1189–1203.

Hammond, A.S., Royer, D.F., and Fleagle, J.G. 2017. The Omo-Kibish I Pelvis. *Journal of Human Evolution* 108: 199–219.

Hand, D.J. 2008. *Statistics: A Very Short Introduction*. Oxford: Oxford University Press.

Handley, L.J.L., Manica, A., Goudet, J., and Balloux, F. 2007. Going the Distance: Human Population Genetics in a Clinal World. *Trends in Genetics* 23(9): 432–439.

Happe, K.E. 2013. *The Material Gene: Gender, Race, and Heredity after the Human Genome Project*. New York: New York University Press.

Harawa, N.T., and Ford, C.L. 2009. The Foundation of Modern Racial Categories and Implications for Research on Black/White Disparities in Health. *Ethnicity & Disease* 19(2): 209–217.

Hardimon, M.O. 2003. The Ordinary Concept of Race. *The Journal of Philosophy* 100(9): 437–455.

Hartl, D.L., and Clark, A.G. 1989. *Principles of Population Genetics*. 2nd ed. Sunderland, MA: Sinauer Associates.

Harvati, K., Röding, C., Bosman, A.M., Karakostis, F.A., Grün, R., et al. 2019. Apidima Cave Fossils Provide Earliest Evidence of *Homo sapiens* in Eurasia. *Nature* 571(7766): 500–504.

Harville, D.A. 2018. *Linear Models and the Relevant Distributions and Matrix Algebra*. Boca Raton, FL: Taylor and Francis.

Haslanger, S. 2000. Gender and Race: (What) Are They? (What) Do We Want Them to Be? *Noûs* 34(1): 31–55.

Haslanger, S. 2003. Social Construction: The "Debunking" Project. In *Socializing Metaphysics: The Nature of Social Reality*, ed. F.F. Schmitt. Lanham, MD: Rowman and Littlefield, pp. 301–325.

Haslanger, S. 2008. A Social Constructionist Analysis of Race. In *Revisiting Race in a Genomic Age*, ed. B.A. Koenig, S.S.-J. Lee, and S.S. Richardson. New Brunswick, NJ: Rutgers University Press, pp. 56–69.

Hausman, C.R. 1993. *Charles S. Peirce's Evolutionary Philosophy*. Cambridge: Cambridge University Press.

Hawks, J., Hunley, K., Lee, S.-H., and Wolpoff, M. 2000. Population Bottlenecks and Pleistocene Human Evolution. *Molecular Biology and Evolution* 17(1): 2–22.

Hedrick, P.W. 2005. *The Genetics of Populations*. 3rd ed. Sudbury, MA: Jones and Bartlett Publishers.

Henig, R.M. 2001. *The Monk in the Garden: The Lost and Found Genius of Gregor Mendel, the Father of Genetics*. Boston, MA: Houghton Mifflin.

Henn, B.M., Gignoux, C.R., Jobin, M., Granka, J.M., Macpherson, J.M., et al. 2011. Hunter-Gatherer Genomic Diversity Suggests a Southern African Origin for Modern Humans. *Proceedings of the National Academy of Sciences of the United States of America* 108(13): 5154–5162.

Henn, B.M., Botigué, L.R., Peischl, S., Dupanloup, I., Lipatov, M., et al. 2016. Distance from Sub-Saharan Africa Predicts Mutational Load in Diverse Human Genomes. *Proceedings of the National Academy of Sciences of the United States of America* 113(4): E440–E449.

Herr, D.G. 1980. On the History of the Use of Geometry in the General Linear Model. *The American Statistician* 34(1): 43–47.

Herrnstein, R.J., and Murray, C. 1995. *The Bell Curve: Intelligence and Class Structure in American Life*. New York: The Free Press.

Hesse, M. 1966. *Models and Analogies in Science*. South Bend, IN: University of Notre Dame Press.

Hesse, M. 1974. *The Structure of Scientific Inference*. Berkeley, CA: University of California Press.

Heyer, E., Chaix, R., Pavard, S., and Austerlitz, F. 2012. Sex-Specific Demographic Behaviours that Shape Human Genomic Variation. *Molecular Ecology* 21(3): 597–612.

Hill, M.O. 1973. Diversity and Evenness: A Unifying Notation and Its Consequences. *Ecology* 54(2): 427–432.

Hitchcock, C. 2012. Probabilistic Causation. In *The Stanford Encyclopedia of Philosophy* (Winter 2012 Edition), ed. E.N. Zalta. Available: http://plato.stanford.edu/archives/win2012/entries/causation-probabilistic/ Accessed May 2019.

Hite, S. 1976. *The Hite Report: A Nationwide Study of Female Sexuality*. New York: Seven Stories Press.

Ho, S.S., Urban, A.E., and Mills, R.E. 2020. Structural Variation in the Sequencing Era. *Nature Reviews Genetics* 21(3): 171–189.

Hoban, S.M., Hauffe, H.C., Pérez-Espona, S., Arntzen, J.W., Bertorelle, G., et al. 2013. Bringing Genetic Diversity to the Forefront of Conservation Policy and Management. *Conservation Genetics Resources* 5(2): 593–598.

Hochman, A. 2013. Against the New Racial Naturalism. *The Journal of Philosophy* 110(6): 331–351.

Hochman, A. 2016. Race: Deflate or Pop? *Studies in History and Philosophy of Science, Part C: Studies in History and Philosophy of Biological and Biomedical Sciences* 57: 60–68.

Hocking, R.R. 1996. *Methods and Applications of Linear Models: Regression and the Analysis of Variance*. New York: John Wiley & Sons.

Hoffmann, S., and Hoffmann, A. 2008. Is There a "True" Diversity? *Ecological Economics* 65(2): 213–215.

Hofrichter, J., Jost, J., and Tran, T.D. 2017. *Information Geometry and Population Genetics: The Mathematical Structure of the Wright–Fisher Model*. Cham: Springer.

Holland, P.W. 1986. Statistics and Causal Inference. *Journal of the American Statistical Association* 81(396): 945–960.

Holowka, N.B., and Lieberman, D.E. 2018. Rethinking the Evolution of the Human Foot: Insights from Experimental Research. *Journal of Experimental Biology* 221(17): art. jeb174425.

Holsinger, K.E., and Weir, B.S. 2009. Genetics in Geographically Structured Populations: Defining, Estimating and Interpreting F_{ST}. *Nature Reviews Genetics* 10(9): 639–650.

Houle, D., Pélabon, C., Wagner, G.P., and Hansen, T.F. 2011. Measurement and Meaning in Biology. *The Quarterly Review of Biology* 86(1): 3–34.

Hubby, J.L., and Lewontin, R.C. 1966. A Molecular Approach to the Study of Genic Heterozygosity in Natural Populations: I. The Number of Alleles at Different Loci in *Drosophila pseudoobscura*. *Genetics* 54(2): 577–594.

Hubisz, M.J., Falush, D., Stephens, M., and Pritchard, J.K. 2009. Inferring Weak Population Structure with the Assistance of Sample Group Information. *Molecular Ecology Resources* 9(5): 1322–1332.

Hublin, J.-J. 2013. The Middle Pleistocene Record: On the Origin of Neandertals, Modern Humans and Others. In *A Companion to Paleoanthropology*, ed. D. Begun. Oxford: Wiley-Blackwell, pp. 517–537.

Hublin, J.-J., Ben-Ncer, A., Bailey, S.E., Freidline, S.E., Neubauer, S., et al. 2017. New Fossils from Jebel Irhoud, Morocco and the Pan-African Origin of *Homo sapiens*. *Nature* 546(7657): 289–292.

Hudson, N. 1996. From "Nation" to "Race": The Origin of Racial Classification in Eighteenth-Century Thought. *Eighteenth-Century Studies* 29(3): 247–264.

Huerta-Sánchez, E., Jin, X., Asan, Bianba, Z., Peter, B.M., et al. 2014. Altitude Adaptation in Tibetans Caused by Introgression of Denisovan-Like DNA. *Nature* 512(7513): 194–197.

Huson, D.H., Rupp, R., and Scornavacca, C. 2010. *Phylogenetic Networks: Concepts, Algorithms and Applications*. Cambridge: Cambridge University Press.

Ilardo, M.A. 2018. *Man and the Sea: Genetics in Maritime Populations*. Doctoral Dissertation, Center for GeoGenetics, Natural History Museum of Denmark, Faculty of Science, University of Copenhagen, Copenhagen.

Ilardo, M.A., and Nielsen, R. 2018. Human Adaptation to Extreme Environmental Conditions. *Current Opinion in Genetics & Development* 53: 77–82.

Ilardo, M.A., Moltke, I., Korneliussen, T.S., Cheng, J., Stern, A.J., et al. 2018. Physiological and Genetic Adaptations to Diving in Sea Nomads. *Cell* 173(3): 569–580.

Jacobs, G.S., Hudjashov, G., Saag, L., Kusuma, P., Darusallam, C.C., et al. 2019. Multiple Deeply Divergent Denisovan Ancestries in Papuans. *Cell* 177(4): 1010–1021.

Jagoda, E., Lawson, D.J., Wall, J.D., Lambert, D., Muller, C., et al. 2018. Disentangling Immediate Adaptive Introgression from Selection on Standing Introgressed Variation in Humans. *Molecular Biology and Evolution* 35(3): 623–630.

Jakobsson, M., Edge, M.D., and Rosenberg, N.A. 2013. The Relationship Between F_{ST} and the Frequency of the Most Frequent Allele. *Genetics* 193(2): 515–528.

Jannini, E.A., Burri, A.V., Jern, P., and Novelli, G. 2015. Genetics of Human Sexual Behavior: Where We Are, Where We Are Going. *Sexual Medicine Reviews* 3(2): 65–77.

Janssens, A.C.J.W. 2019. Proprietary Algorithms for Polygenic Risk: Protecting Scientific Innovation or Hiding the Lack of It? *Genes* 10(6): art. 448.

Jeffreys, H. 1955. *Scientific Inference*. 3rd ed. Cambridge: Cambridge University Press.

Jeffers, C. 2013. The Cultural Theory of Race: Yet Another Look at Du Bois's "The Conservation of Races." *Ethics* 123(3): 403–426.

Jeffers, C. 2015. Finding Our Way with Taylor's *Race: A Philosophical Introduction*. Talk presented at the 2015 American Philosophical Association Central Division meeting. Available: https://www.academia.edu/27875884/Finding_Our_Way_with_Taylors_Race_A_Philosophical_Introduction Accessed February 2022.

Jeffers, C. 2019. Cultural Constructionism. In *What Is Race? Four Philosophical Views*, ed. J. Glasgow, S. Haslanger, C. Jeffers, and Q. Spencer. New York: Oxford University Press, pp. 38–72.

Jensen, A. 1969. How Much Can We Boost IQ and Scholastic Achievement? *Harvard Educational Review* 39(1): 1–123.

Jensen, J.D., Payseur, B.A., Stephan, W., Aquadro, C.F., Lynch, M., et al. 2018. The Importance of the Neutral Theory in 1968 and 50 years on: A Response to Kern and Hahn 2018. *Evolution* 73(1): 111–114.

Jeong, C., Alkorta-Aranburu, G., Basnyat, B., Neupane, M., Witonsky, D.B., et al. 2014. Admixture Facilitates Genetic Adaptations to High Altitude in Tibet. *Nature Communications* 5: art. 3281.

Jobling, M.A., and Tyler-Smith, C. 2003. The Human Y Chromosome: An Evolutionary Marker Comes of Age. *Nature Reviews Genetics* 4: 598–612.

Jolliffe, I.T., and Cadima, J. 2016. Principal Component Analysis: A Review and Recent Developments. *Philosophical Transactions of the Royal Society A: Mathematical, Physical and Engineering Sciences* 374(2065): art. 20150202.

Jorde, L.B., Watkins, W.S., Bamshad, M.J., Dixon, M.E., Ricker, C.E., et al. 2000. The Distribution of Human Genetic Diversity: A Comparison of Mitochondrial, Autosomal, and Y-Chromosome Data. *The American Journal of Human Genetics* 66(3): 979–988.

Jost, L. 2006. Entropy and Diversity. *Oikos* 113(2): 363–375.

Jost, L. 2008. G_{ST} and Its Relatives Do Not Measure Differentiation. *Molecular Ecology* 17(18): 4015–4026.

Judson, H.F. 1996. *The Eighth Day of Creation: The Makers of the Revolution in Biology*. Woodbury, NY: Cold Spring Harbor Laboratory Press.

Judson, O. 2005. Anticlimax: Review of *The Case of the Female Orgasm: Bias in the Science of Evolution* by E.A. Lloyd. *Nature* 436: 916–917.

Kadrow, S. 2018. South-Eastern Group of Funnel Beaker culture. In *Papers and Materials of the Archaeological and Ethnographic Museum in Łódź*. Łódź, Poland: Muzeum Archeologiczne i Etnograficzne w Łodzi, pp. 255–256.

Kahneman, D. 2002. Maps of Bounded Rationality: A Perspective on Intuitive Judgment and Choice. In *Les Prix Nobel: The Nobel Prizes 2002*, ed. T. Frängsmyr. Stockholm: Almquist & Wiksell International, pp. 449–489.

Kahneman, D. 2011. *Thinking, Fast and Slow*. New York: Farrar, Straus, and Giroux.

Kalinowski, S.T. 2011. The Computer Program STRUCTURE Does Not Reliably Identify the Main Genetic Clusters Within Species: Simulations and Implications for Human Population Structure. *Heredity* 106(4): 625–632.

Kanton, S., Boyle, M.J., He, Z., Santel, M., Weigert, A., et al. 2019. Organoid Single-Cell Genomic Atlas Uncovers Human-Specific Features of Brain Development. *Nature* 574(7778): 418–422.

Kaplan, J.M. 2000. *The Limits and Lies of Human Genetic Research*. New York: Routledge.

Kaplan, J.M. 2010. When Socially Determined Categories Make Biological Realities: Understanding Black/White Health Disparities in the U.S. *The Monist* 93(2): 283–299.

Kaplan, J.M. 2011. "Race": What Biology Can Tell Us about a Social Construct. In *Encyclopedia of Life Sciences*. Chichester: Wiley. Available: https://onlinelibrary .wiley.com/doi/abs/10.1002/9780470015902.a0005857 Accessed May 2012.

Karczewski, K.J., Francioli, L.C., Tiao, G., Cummings, B.B., Alföldi, J., et al. 2020. The Mutational Constraint Spectrum Quantified from Variation in 141,456 Humans. *Nature* 581(7809): 434–443.

Karki, R., Pandya, D., Elston, R.C., and Ferlini, C. 2015. Defining "Mutation" and "Polymorphism" in the Era of Personal Genomics. *BMC Medical Genomics* 8: art. 37.

Karmin, M., Saag, L., Vicente, M., Sayres, M.A.W., Jäyre, M., et al. 2015. A Recent Bottleneck of Y Chromosome Diversity Coincides with a Global Change in Culture. *Genome Research* 25(4): 459–466.

Kauffman, S.A. 1993. *The Origins of Order: Self-Organization and Selection in Evolution.* New York: Oxford University Press.

Keinan, A., and Clark, A.G. 2012. Recent Explosive Human Population Growth Has Resulted in an Excess of Rare Genetic Variants. *Science* 336(6082): 740–743.

Keller, E.F. 2002. *The Century of the Gene.* Cambridge, MA: Harvard University Press.

Kelly, T.E., Chase, G.A., Kaback, M.M., Kumor, K., and McKusick, V.A. 1975. Tay-Sachs Disease: High Gene Frequency in a Non-Jewish Population. *The American Journal of Human Genetics* 27(3): 287–291.

Kendi, I.X. 2019. *How to Be an Antiracist.* New York: One World.

Kern, A.D., and Hahn, M.H. 2018. The Neutral Theory in Light of Natural Selection. *Molecular Biology and Evolution* 35(6): 1366–1371.

Kim, S.Y., Lohmueller, K.E., Albrechtsen, A., Li, Y., Korneliussen, T., et al. 2011. Estimation of Allele Frequency and Association Mapping Using Next-Generation Sequencing Data. *BMC Bioinformatics* 12: art. 231.

Kimbel, W.H., and Villmoare, B. 2016. From *Australopithecus* to *Homo*: The Transition that Wasn't. *Philosophical Transactions of the Royal Society B: Biological Sciences* 371(1698): art. 20150248.

Kingman, J.F.C. 1982a. On the Genealogy of Large Populations. *Journal of Applied Probability* 19(A): 27–43.

Kingman, J.F.C. 1982b. The Coalescent. *Stochastic Processes and their Applications* 13(3): 235–248.

Kingman, J.F.C. 2000. Origins of the Coalescent. 1974–1982. *Genetics* 156(4): 1461–1463.

Kingsland, S. 1995. *Modeling Nature: Episodes in the History of Population Ecology.* Chicago, IL: University of Chicago Press.

Kitcher, P. 2007. Does "Race" Have a Future? *Philosophy and Public Affairs* 35(4): 293–317.

Klein, R.G. 2009. *The Human Career: Human Biological and Cultural Origins.* 3rd ed. Chicago, IL: University of Chicago Press.

Kliman, R., Sheehy, B., and Schultz, J. 2008. Genetic Drift and Effective Population Size. *Nature Education* 1(3): art. 3.

Kline, R.B. 2016. *Principles and Practice of Structural Equation Modeling*. 4th ed. New York: Guilford Press.

Kluge, A.G. 1989. A Concern for Evidence and a Phylogenetic Hypothesis of Relationships Among *Epicrates* (Boidae, Serpentes). *Systematic Zoology* 38(1): 7–25.

Kohler, R.E. 2002. *Landscapes and Labscapes: Exploring the Lab–Field Border in Biology*. Chicago, IL: University of Chicago Press.

Koskela, J., Lefèvre, F., Schueler, S., Kraigher, H., Olrik, D.C., et al. 2013. Translating Conservation Genetics into Management: Pan-European Minimum Requirements for Dynamic Conservation Units of Forest Tree Genetic Diversity. *Biological Conservation* 157: 39–49.

Krantz, D.H., Luce, R.D., Suppes, P., and Tversky, A. 1971. *Foundations of Measurement. Volume 1: Additive and Polynomial Representations*. San Diego, CA: Academic Press.

Kronenberg, Z.N., Fiddes, I.T., Gordon, D., Murali, S., Cantsilieris, S., et al. 2018. High-Resolution Comparative Analysis of Great Ape Genomes. *Science* 360(6393): art. eaar6343.

Krüger, T.H.C., Hartmann, U., and Schedlowski, M. 2005. Prolactinergic and Dopaminergic Mechanisms Underlying Sexual Arousal and Orgasm in Humans. *World Journal of Urology* 23(2): 130–138.

Kruskal, W., and Mosteller, F. 1980. Representative Sampling, IV: The History of the Concept in Statistics, 1895–1939. *International Statistical Review* 48(2): 169–195.

Kuhn, T.S. 1970. *The Structure of Scientific Revolutions*. 2nd ed. Chicago, IL: University of Chicago Press.

Kulynych, J., and Greely, H.T. 2017. Clinical Genomics, Big Data, and Electronic Medical Records: Reconciling Patient Rights with Research When Privacy and Science Collide. *Journal of Law and the Biosciences* 4(1): 94–132.

Kumar, R., Seibold, M.A., Aldrich, M.C., Williams, L.K., Reiner, A.P., et al. 2010. Genetic Ancestry in Lung-Function Predictions. *New England Journal of Medicine* 363(4): 321–330.

Kunimatsu, Y., Nakatsukasa, M., Sawada, Y., Sakai, T., Hyodo, M., et al. 2007. A New Late Miocene Great Ape from Kenya and Its Implications for the Origins of African Great Apes and Humans. *Proceedings of the National Academy of Sciences of the United States of America* 104(49): 19220–19225.

Kusuma, P., Brucato, N., Cox, M.P., Letellier, T., Manan, A., et al. 2017. The Last Sea Nomads of the Indonesian Archipelago: Genomic Origins and Dispersal. *European Journal of Human Genetics* 25(8): 1004–1010.

Lachance, J., and Tishkoff, S.A. 2013. SNP Ascertainment Bias in Population Genetic Analyses: Why It Is Important, and How to Correct It. *BioEssays* 35(9): 780–786.

Lack, D. 1945. The Galapagos Finches (Geospizinae): A Study in Variation. *Occasional Papers of the California Academy of Sciences* 21: 1–159.

Lack, D. 1983 [1947]. *Darwin's Finches*, ed. L.M. Ratcliffe and P.T. Boag. Cambridge: Cambridge University Press.

Laland, K.N., Uller, T., Feldman, M.W., Sterelny, K., Müller, G.B., et al. 2015. The Extended Evolutionary Synthesis: Its Structure, Assumptions, and Predictions. *Proceedings of the Royal Society B: Biological Sciences* 282(1813): art. 20151019.

Lamason, R.L., Mohideen, M.-A.P.K., Mest, J.R., Wong, A.C., Norton, H.L., et al. 2005. SLC24A5, a Putative Cation Exchanger, Affects Pigmentation in Zebrafish and Humans. *Science* 310(5755): 1782–1786.

Lamotte, M. 1959. Polymorphism of Natural Populations of *Cepaea nemoralis*. *Cold Spring Harbor Symposium on Quantitative Biology* 24: 65–86.

Lande, R., and Arnold, S.J. 1983. The Measurement of Selection on Correlated Characters. *Evolution* 37(6): 1210–1226.

Langergraber, K.E., Prüfer, K., Rowney, C., Boesch, C., Crockford, C., et al. 2012. Generation Times in Wild Chimpanzees and Gorillas Suggest Earlier Divergence Times in Great Ape and Human Evolution. *Proceedings of the National Academy of Sciences of the United States of America* 109(39): 15716–15721.

Laplace, P.-S. 1902 [1814]. *A Philosophical Essay on Probabilities* (original: *Essai philosophique sur les probabilités*), transl. F.W. Truscott and F.L. Emory. New York: John Wiley & Sons. Available: https://en.wikisource.org/wiki/A_Philosophical_Essay_on_Probabilities/Chapter_2 Accessed January 2014.

Latter, B.D.H. 1980. Genetic Differences Within and Between Populations of the Major Human Subgroups. *The American Naturalist* 116(2): 220–237.

Lawson, D.J., van Dorp, L., and Falush, D. 2018. A Tutorial on How Not to Over-Interpret STRUCTURE and ADMIXTURE Bar Plots. *Nature Communications* 9: art. 3258.

Layzer, D. 1972. Science or Superstition? (A Physical Scientist Looks at the IQ Controversy). *Cognition* 1(2–3): 265–299.

Leasca, S. 2017. It's Time to Redefine the Clitoris, According to Sex Education Experts. *Mic*. Available: www.mic.com/articles/180753/its-time-to-redefine-the-clitoris-according-to-sex-education-experts Accessed June 2020.

Leeners, B., Tilmann, T.H.C., Brody, S., Schmidlin, S., Naegeli, E., et al. 2013. The Quality of Sexual Experience in Women Correlates with Post-Orgasmic Prolactin Surges: Results from an Experimental Prototype Study. *The Journal of Sexual Medicine* 10(5): 1313–1319.

Leonelli, S. 2007. What Is in a Model? Using Theoretical and Material Models to Develop Intelligible Theories. In *Modeling Biology: Structures, Behaviors, Evolution*, ed. M. Laubichler and G.B. Muller. Cambridge, MA: MIT Press, pp. 15–36.

Levins, R. 1966. The Strategy of Model Building in Population Biology. *American Scientist* 54(4): 421–431.

Levins, R. 1968. *Evolution in Changing Environments: Some Theoretical Explorations*. Princeton, NJ: Princeton University Press.

Levins, R. 2006. Strategies of Abstraction. *Biology and Philosophy* 21(5): 741–755.

Levins, R., and Lewontin, R.C. 1985. *The Dialectical Biologist*. Cambridge, MA: Harvard University Press.

Lewontin, R.C. 1967. Population Genetics. *Annual Review of Genetics* 1: 37–70.

Lewontin, R.C. 1970. Race and Intelligence. *Bulletin of the Atomic Scientists* 26(March): 2–8.

Lewontin, R.C. 1972. The Apportionment of Human Diversity. In *Evolutionary Biology*, vol. 6, ed. T. Dobzhansky, M.K. Hecht, and W.C. Steere. New York: Springer, pp. 381–398.

Lewontin, R.C. 1974. *The Genetic Basis of Evolutionary Change*. New York: Columbia University Press.

Lewontin, R.C. 1978. Single- and Multiple-Locus Measures of Genetic Distance Between Groups. *The American Naturalist* 112(988): 1138–1139.

Lewontin, R.C. 1995 [1982]. *Human Diversity*. New York: Scientific American Books.

Lewontin, R.C. 1993. *Biology as Ideology: The Doctrine of DNA*. New York: Harper Perennial.

Lewontin, R.C. 1998. The Evolution of Cognition: Questions We Will Never Answer. In *An Invitation to Cognitive Science. Methods, Models, and Conceptual Issues*, vol. 4, ed. D. Scarborough, and S. Sternberg. Cambridge, MA: MIT Press, pp. 106–132.

Lewontin, R.C., and Hubby, J.L. 1966. A Molecular Approach to the Study of Genic Heterozygosity in Natural Populations: II. Amount of Variation and Degree of Heterozygosity in Natural Populations of *Drosophila pseudoobscura*. *Genetics* 54(2): 595–609.

Lewontin, R.C., Rose, S., and Kamin, L.J. 1984. *Not in Our Genes: Biology, Ideology and Human Nature*. New York: Pantheon.

Li, C.C. 1955. *Population Genetics*. Chicago, IL: University of Chicago Press.

Li, H., and Ralph, P. 2019. Local PCA Shows How the Effect of Population Structure Differs Along the Genome. *Genetics* 211(1): 289–304.

Li, J.Z., Absher, D.M., Tang, H., Southwick, A.M., Casto, A.M., et al. 2008. Worldwide Human Relationships Inferred from Genome-Wide Patterns of Variation. *Science* 319(5866): 1100–1104.

Li, W.-H., and Sadler, L.A. 1991. Low Nucleotide Diversity in Man. *Genetics* 129(2): 513–523.

Li, W.-H., Wu, C.-I., and Luo, C.-C. 1985. A New Method for Estimating Synonymous and Nonsynonymous Rates of Nucleotide Substitution Considering the Relative Likelihood of Nucleotide and Codon Changes. *Molecular Biology and Evolution* 2(2): 150–174.

Lieberson, S. 1987. *Making It Count: The Improvement of Social Research and Theory*. Berkeley, CA: University of California Press.

Liebert, A., López, S., Jones, B.L., Montalva, N., Gerbault, P., et al. 2017. World-Wide Distributions of Lactase Persistence Alleles and the Complex Effects of Recombination and Selection. *Human Genetics* 136(11–12): 1445–1453.

Lightman, A. 2013. *The Accidental Universe: The World You Thought You Knew*. New York: Pantheon Books.

Liou, S. 2010. Population Genetics and Huntington's Disease. HOPES. Available: https://hopes.stanford.edu/population-genetics-and-hd/ Accessed April 2020.

Lippold, S., Xu, H., Ko, A., Li, M., Renaud, G., et al. 2014. Human Paternal and Maternal Demographic Histories: Insights from High-Resolution Y Chromosome and mtDNA Sequences. *Investigative Genetics* 5: art. 13.

Livingstone, F.B. 1962. On the Non-Existence of Human Races (with Reply). *Current Anthropology* 3(3): 279–281.

Lloyd, E.A. 1994. *The Structure and Confirmation of Evolutionary Theory*. Princeton, NJ: Princeton University Press.

Lloyd, E.A. 2005a. *The Case of the Female Orgasm: Bias in the Science of Evolution*. Cambridge, MA: Harvard University Press.

Lloyd, E.A. 2005b. Elisabeth Lloyd's Views on the New Heritability Study of Female Orgasm. Philosophy of Biology Weblog [now defunct], June 9,

2005. Mirrored at www.arlindo-correia.com/female_orgasm.html Accessed February 2018.

Lodé, T. 2020. A Brief Natural History of the Orgasm. *All Life* 13(1): 34–44.

Long, J.C. 1986. The Allelic Correlation Structure of Gainj- and Kalam-Speaking People. I. The Estimation and Interpretation of Wright's *F*-Statistics. *Genetics* 112(3): 629–647.

Long, J.C. 2009. Update to Long and Kittles's "Human Genetic Diversity and the Nonexistence of Biological Races" (2003): Fixation on an Index. *Human Biology* 81(5–6): 799–803.

Long, J.C., and Kittles, R.A. 2003. Human Genetic Diversity and the Nonexistence of Biological Races. *Human Biology* 75(4): 449–471.

Long, J.C., Smouse, P.E., and Wood, J.W. 1987. The Allelic Correlation Structure of Gainj- and Kalam-Speaking People. II. The Genetic Distance Between Population Subdivisions. *Genetics* 117(2): 273–283.

Long, J.C., Li, J., and Healy, M.E. 2009. Human DNA Sequences: More Variation and Less Race. *American Journal of Physical Anthropology* 139(1): 23–34.

Longino, H.E. 2002. *The Fate of Knowledge*. Princeton, NJ: Princeton University Press.

Longino, H.E. 2013. *Studying Human Behavior: How Scientists Investigate Aggression and Sexuality*. Chicago, IL: University of Chicago Press.

Longino, H.E. 2021. Scaling Up; Scaling Down: What's Missing? *Synthese* 198(4): 2849–2863.

López Beltrán, C. (ed.) 2011. *Genes & mestizos: genómica y raza en la biomedicina Mexicana*. Mexico City: UNAM.

Lorusso, L., and Winther, R.G. (ed.) 2022. *Remapping Race in a Global Context*. London: Routledge.

Lovelock, J. 1979. *Gaia: A New Look at Life on Earth*. Oxford: Oxford University Press.

Lu, Y.-F., Goldstein, D.B., Angrist, M., and Cavalleri, G. 2014. Personalized Medicine and Human Genetic Diversity. *Cold Spring Harbor Perspectives in Medicine* 4(9): art. a008581.

Ludwig, D. 2015. Against the New Metaphysics of Race. *Philosophy of Science* 82(2): 244–265.

Lynch, K.E. 2017. Heritability and Causal Reasoning. *Biology and Philosophy* 32(1): 25–49.

Lynch, M., and Ho, W.-C. 2020. The Limits to Estimating Population-Genetic Parameters with Temporal Data. *Genome Biology and Evolution* 12(4): 443–455.

Lynch, M., and Walsh, J.B. 1998. *Genetics and Analysis of Quantitative Traits*. Sunderland, MA: Sinauer Press.

Lynn, R., and Vanhanen, T. 2002. *IQ and the Wealth of Nations*. Westport, CT: Praeger.

MacKenzie, D.A. 1981. *Statistics in Britain, 1865–1930: The Social Construction of Scientific Knowledge*. Edinburgh: Edinburgh University Press.

Maddison, W.P. 1997. Gene Trees in Species Trees. *Systematic Biology* 46(3): 523–536.

Malécot, G. 1969 [1948]. *The Mathematics of Heredity* (original: *Les mathématiques de l'hérédité*), transl., ed., and rev. D.M. Yermanos. San Francisco, CA: W.H. Freeman.

Manco, J. 2016. *Ancestral Journeys: The Peopling of Europe from the First Venturers to the Vikings*. London: Thames and Hudson.

Maples, B.K., Gravel, S., Kenny, E.E., and Bustamante, C.D. 2013. RFMix: A Discriminative Modeling Approach for Rapid and Robust Local-Ancestry Inference. *The American Journal of Human Genetics* 93(2): 278–288.

Margulis, L. 1975. Symbiotic Theory of the Origin of Eukaryotic Organelles: Criteria for Proof. *Symposia of the Society for Experimental Biology* 29: 21–38.

Margulis, L. 2004. Serial Endosymbiotic Theory (SET) and Composite Individuality: Transition from Bacterial to Eukaryotic Genomes. *Microbiology Today* 31(November): 172–174.

Marks, J. 1995. *Human Biodiversity*. New York: Routledge.

Marks, S.J., Levy, H., Martínez-Cadenas, C., Montinaro, F., and Capelli, C. 2012. Migration Distance Rather than Migration Rate Explains Genetic Diversity in Human Patrilocal Groups. *Molecular Ecology* 21(20): 4958–4969

Maslin, M. 2017. *The Cradle of Humanity: How the Changing Landscape of Africa Made Us So Smart*. Oxford: Oxford University Press.

Masters, W., and Johnson, V.E. 1966. *Human Sexual Response*. Boston, MA: Little, Brown, & Company.

Mathieson, I., and Scally, A. 2020. What Is Ancestry? *PLoS Genetics* 16(3): art. e1008624.

Matthen, M., and Ariew, A. 2002. Two Ways of Thinking about Fitness and Natural Selection. *Journal of Philosophy* 99(2): 55–83.

Matthews, L.J., and Turkheimer, E. 2021. Across the Great Divide: Pluralism and the Hunt for Missing Heritability. *Synthese* 198: 2297–2311.

Maturana, H., and Varela, F. 1980. *Autopoiesis and Cognition: The Realization of the Living*. Dordrecht: D. Reidel.

Maynard Smith, J., and Szathmáry, E. 1995. *The Major Transitions in Evolution*. New York: W.H. Freeman.

McDonald, D. 2008. Distances Summary. Available: www.uwyo.edu/dbmcd/mol mark/GenDistEqns.pdf Accessed April 2020.

McDonald, J.H., and Kreitman, M. 1991. Adaptive Protein Evolution at the *Adh* Locus in *Drosophila*. *Nature* 351(6328): 652–654.

McGary, H. 2012. *The Post-Racial Ideal*. Milwaukee, WI: Marquette University Press.

McWhorter, J. 2021. *Woke Racism: How a New Religion Has Betrayed Black America*. New York: Portfolio/Penguin.

M'charek, A. 2005. *The Human Genome Diversity Project: An Ethnography of Scientific Practice*. Cambridge: Cambridge University Press.

Meehl, P.E. 1978. Theoretical Risks and Tabular Asterisks: Sir Karl, Sir Ronald, and the Slow Progress of Soft Psychology. *Journal of Consulting and Clinical Psychology* 46(4): 806–834.

Mendel, G. 1996 [1866]. Experiments in Plant Hybridization (1865), transl. W. Bateson (1901), with edits by R. Blumberg (1996). Electronic Scholarly Publishing Project. Available: www.esp.org/foundations/genetics/classical/gm-65.pdf (original: 1866. Versuche über Plflanzen-Hybriden. *Verhandlungen des naturforschenden Vereines in Brünn*, Bd. IV für das Jahr 1865, *Abhandlungen*, S. 3–47).

Mendez, F.L., Krahn, T., Schrack, B., Krahn, A.-M., Veeramah, K.R. et al. 2013. An African American Paternal Lineage Adds an Extremely Ancient Root to the

Human Y Chromosome Phylogenetic Tree. *The American Journal of Human Genetics* 92(3): 454–459.

Mendez, F.L., Poznik, G.D., Castellano, S., and Bustamante, C.D. 2016. The Divergence of Neandertal and Modern Human Y Chromosomes. *The American Journal of Human Genetics* 98(4): 728–734.

Mertz, D.B., Cawthon, D.A., and Park, T. 1976. An Experimental Analysis of Competitive Indeterminacy in *Tribolium. Proceedings of the National Academy of Sciences of the United States of America* 73(4): 1368–1372.

Messer, P.W., and Petrov, D.A. 2013a. Frequent Adaptation and the McDonald–Kreitman Test. *Proceedings of the National Academy of Sciences of the United States of America* 110(21): 8615–8620.

Messer, P.W., and Petrov, D.A. 2013b. Population Genomics of Rapid Adaptation by Soft Selective Sweeps. *Trends in Ecology & Evolution* 28(11): 659–669.

Mills, C.W. 1997. *The Racial Contract.* Ithaca, NY: Cornell University Press.

Mills, C.W. 1998. "But What Are You *Really*?" The Metaphysics of Race. In *Blackness Visible: Essays on Philosophy and Race*, ed. C.W. Mills. Ithaca, NY: Cornell University Press, pp. 41–66.

Millstein, R.L. 2009. Populations as Individuals. *Biological Theory* 4(3): 267–273.

Millstein, R.L. 2010. The Concepts of Population and Metapopulation in Evolutionary Biology and Ecology. In *Evolution Since Darwin: The First 150 Years*, ed. M.A. Bell, D.J. Futuyma, W.F. Eanes, and J.S. Levinton. Sunderland, MA: Sinauer, pp. 61–86.

Millstein, R.L. 2014. How the Concept of Population Resolves Concepts of Environment. *Philosophy of Science* 81(4): 741–755.

Millstein, R.L. 2015. Thinking about Populations and Races in Time. *Studies in History and Philosophy of Science, Part C: Studies in History and Philosophy of Biological and Biomedical Sciences* 52: 5–11.

Mithen, S. 2015. On Ancestor Apes in Europe: Review of *The Real Planet of the Apes: A New Story of Human Origins* by D.R. Begun. *The New York Review of Books* 62(18). Available: www.nybooks.com/articles/2015/11/19/ancestor-apes-europe/ Accessed December 2015.

Mitman, G. 1992. *The State of Nature: Ecology, Community, and American Social Thought, 1900–1950.* Chicago, IL: University of Chicago Press.

Mitton, J.B. 1977. Genetic Differentiation of Races of Man as Judged by Single-Locus and Multilocus Analyses. *The American Naturalist* 111(978): 203–212.

Moorjani, P., Amorim, C.E.G., Arndt, P.F., and Przeworski, M. 2016. Variation in the Molecular Clock of Primates. *Proceedings of the National Academy of Sciences of the United States of America* 113(38): 10607–10612.

Morell, V. 1995. *Ancestral Passions: The Leakey Family and the Quest for Humankind's Beginnings.* New York: Simon and Schuster.

Morgan, M. 2012. *The World in the Model: How Economists Work and Think.* New York: Cambridge University Press.

Morning, A. 2011. *The Nature of Race: How Scientists Think and Teach about Human Differences.* Berkeley, CA: University of California Press.

Morrison, D.A. 2014. Is the Tree of Life the Best Metaphor, Model, or Heuristic for Phylogenetics? *Systematic Biology* 63(4): 628–638.

Morrison, M. 2000. *Unifying Scientific Theories: Physical Concepts and Mathematical Structures.* New York: Cambridge University Press.

Morrison, M. 2002. Modelling Populations: Pearson and Fisher on Mendelism and Biometry. *The British Journal for the Philosophy of Science* 53(1): 39–68.

Mounier, A., and Mirazón Lahr, M. 2019. Deciphering African Late Middle Pleistocene Hominin Diversity and the Origin of Our Species. *Nature Communications* 10(1): art. 3406.

Mountain, J.L., and Cavalli-Sforza, L.L. 1997. Multilocus Genotypes, a Tree of Individuals, and Human Evolutionary History. *The American Journal of Human Genetics* 61(3): 705–718.

Mourant, A.E. 1954. *The Distribution of the Human Blood Groups*. Oxford: Blackwell.

Mukherjee, S. 2016. *The Gene: An Intimate History*. New York: Scribner.

Mukhopadhyay, C.C., and Moses, Y.T. 1997. Reestablishing "Race" in Anthropological Discourse. *American Anthropologist* 99(3): 517–533.

Muller, K.E., and Fetterman, B.A. 2002. *Regression and ANOVA: An Integrated Approach Using SAS Software*. Cary, NC: SAS Institute.

Murray, C. 2020. *Human Diversity: The Biology of Gender, Race, and Class*. New York: Twelve.

Nagatsu, K. 2017. Maritime Diaspora and Creolization: Genealogy of the Sama-Bajau in Insular Southeast Asia. *Senri Ethnological Studies* 95: 35–64.

Najmi, A. 2018. A Geometric Interpretation of "Human Genetic Diversity: Lewontin's Fallacy." In *Phylogenetic Inference, Selection Theory, and History of Science: Selected Papers of A.W.F. Edwards with Commentaries*, ed. R.G. Winther. Cambridge: Cambridge University Press, pp. 483–487.

Nakhleh, L., Ruths, D., and Innan, H. 2009. Gene Trees, Species Trees, and Species Networks. In *Meta-analysis and Combining Information in Genetics and Genomics*, ed. R. Guerra and D.R. Goldstein. Boca Raton, FL: Chapman & Hall/CRC Press, pp. 275–293.

Nei, M. 1973. Analysis of Gene Diversity in Subdivided Populations. *Proceedings of the National Academy of Sciences of the United States of America* 70(12): 3321–3323.

Nei, M. 1978. Estimation of Average Heterozygosity and Genetic Distance from a Small Number of Individuals. *Genetics* 89(3): 583–590.

Nei, M. 1987. *Molecular Evolutionary Genetics*. New York: Columbia University Press.

Neigel, J.E. 1997. A Comparison of Alternative Strategies for Estimating Gene Flow from Genetic Markers. *Annual Review of Ecology and Systematics* 28: 105–128.

Nengo, I., Tafforeau, P., Gilbert, C.C, Fleagle, J.G., Miller, E.R., et al. 2017. New Infant Cranium from the African Miocene Sheds Light on Ape Evolution. *Nature* 548(7666): 169–174.

Neubauer, S., Hublin, J.-J., and Gunz, P. 2018. The Evolution of Modern Human Brain Shape. *Science Advances* 4(1): art. eaao5961.

Newman, M.E.J. 2010. *Networks: An Introduction*. Oxford: Oxford University Press.

Neyman, J. 1923 [1990]. On the Application of Probability Theory to Agricultural Experiments. Essay on Principles. Section 9, transl. and ed. D.M. Dabrowska and T.P. Speed. *Statistical Science* 5(4): 465–472.

Neyman, J., and Pearson, E.S. 1928. On the Use and Interpretation of Certain Test Criteria for Purposes of Statistical Inference: Part I. *Biometrika* 20A(1–2): 175–240.

Neyman, J., and Pearson, E.S. 1933. On the Problem of the Most Efficient Tests of Statistical Hypotheses. *Philosophical Transactions of the Royal Society of London.*

Series A, Containing Papers of a Mathematical or Physical Character 231(702): 289–337.

Neyman, J., Park, T., and Scott, E.L. 1956. Struggle for Existence: The *Tribolium* Model – Biological and Statistical Aspects. *Proceedings of the Third Berkeley Symposium on Mathematical Statistics and Probability* 4: 41–79.

Nielsen, R. 2005. Molecular Signatures of Natural Selection. *Annual Review of Genetics* 39: 197–218.

Nielsen, R. 2009. Adaptionism: 30 Years after Gould and Lewontin. *Evolution* 63(10): 2487–2490.

Nielsen, R. 2018. The Historic Split and Merger of Gene Trees and Species Trees. In *Phylogenetic Inference, Selection Theory, and History of Science: Selected Papers of A.W.F. Edwards with Commentaries*, ed. R.G. Winther. Cambridge: Cambridge University Press, pp. 334–342.

Nielsen, R. 2022. Modern Population Genetics and Race. In *Remapping Race in a Global Context*, ed. L. Lorusso and R.G. Winther. London: Routledge, pp. 157–163.

Nielsen, R., and Slatkin, M. 2013. *An Introduction to Population Genetics: Theory and Applications*. Sunderland, MA: Sinauer Associates.

Nielsen, R., Akey, J.M., Jakobsson, M., Pritchard, J.K., Tishkoff, S., and Willerslev, E. 2017. Tracing the Peopling of the World Through Genomics. *Nature* 541(7637): 302–310.

Niklas, K.J. 2016. *Plant Evolution: An Introduction to the History of Life*. Chicago, IL: University of Chicago Press.

Noble, D. 2006. *The Music of Life: Biology Beyond Genes*. Oxford: Oxford Univesity Press.

Novembre, J., and Peter, B.M. 2016. Recent Advances in the Study of Fine-Scale Population Structure in Humans. *Current Opinion in Genetics & Development* 41: 98–105.

Novembre, J., Johnson, T., Bryc, K., Kutalik, Z., Boyko, A.R., et al. 2008. Genes Mirror Geography Within Europe. *Nature* 456(7218): 98–101.

O'Brien, S.J., Roelke, M.E., Marker, L., Newman, A., Winkler, C.A., et al. 1985. Genetic Basis for Species Vulnerability in the Cheetah. *Science* 227(4693): 1428–1434.

O'Connell, H.E., Sanjeevan, K.V., and Hutson, J.M. 2005. Anatomy of the Clitoris. *The Journal of Urology* 174(4, Part 1): 1189–1195.

Odling-Smee, F.J., Laland, K.N., and Feldman, M.W. 2003. *Niche Construction: The Neglected Process in Evolution*. Princeton, NJ: Princeton University Press.

Offner, S. 2011. Mendel's Peas & the Nature of the Gene: Genes Code for Proteins & Proteins Determine Phenotype. *The American Biology Teacher* 73(7): 382–387.

Ohlsson, S., and Lehtinen, E. 1997. Abstraction and the Acquisition of Complex Ideas. *International Journal of Educational Research* 27(1): 37–48.

Okasha, S. 2004. The "Averaging Fallacy" and the Levels of Selection. *Biology and Philosophy* 19(2): 167–184.

Okasha, S. 2006. *Evolution and the Levels of Selection*. Oxford: Oxford University Press.

Okasha, S. 2008. Fisher's Fundamental Theorem of Natural Selection: A Philosophical Analysis. *The British Journal for the Philosophy of Science* 59(3): 319–351.

Olalde, I., Brace, S., Allentoft, M.E., Armit, I., Kristiansen, K., et al. 2018. The Beaker Phenomenon and the Genomic Transformation of Northwest Europe. *Nature* 555(7695): 190–196.

Olby, R.C. 1979. Mendel No Mendelian? *History of Science* 17(1): 53–72.

Olby, R.C. 1994. *The Path to the Double Helix: The Discovery of DNA.* New York: Dover.

Olson, M.V. 1999. When Less Is More: Gene Loss as an Engine of Evolutionary Change. *The American Journal of Human Genetics* 64(1): 18–23.

Omi, M., and Winant, H. 1986. *Racial Formation in the United States: From the 1960s to the 1980s.* New York: Routledge.

Oota, H., Settheetham-Ishida, W., Tiwawech, D., Ishida, T., and Stoneking, M. 2001. Human mtDNA and Y-Chromosome Variation Is Correlated with Matrilocal versus Patrilocal Residence. *Nature Genetics* 29(1): 20–21.

Oreskes, N., and Conway, E. 2011. *Merchants of Doubt: How a Handful of Scientists Obscured the Truth on Issues from Tobacco Smoke to Global Warming.* New York: Bloomsbury Press.

Orr, H.A. 2014. Stretch Genes: Review of *A Troublesome Inheritance: Genes, Race, and Human History* by N. Wade. *The New York Review of Books* 61(10). Available: https://www.nybooks.com/articles/2014/06/05/stretch-genes/ Accessed August 2014.

Ostrer, H., and Skorecki, K. 2013. The Population Genetics of the Jewish People. *Human Genetics* 132(2): 119–127.

Ousley, S., Jantz, R., and Freid, D. 2009. Understanding Race and Human Variation: Why Forensic Anthropologists Are Good at Identifying Race. *American Journal of Physical Anthropology* 139(1): 68–76.

Oyama, S. 2000a. *The Ontogeny of Information: Developmental Systems and Evolution.* 2nd ed. Durham, NC: Duke University Press.

Oyama, S. 2000b. *Evolution's Eye: A Systems View of the Biology–Culture Divide.* Durham, NC: Duke University Press.

Pagani, L., Kivisild, T., Tarekegn, A., Ekong, R., Plaster, C., et al. 2012. Ethiopian Genetic Diversity Reveals Linguistic Stratification and Complex Influences on the Ethiopian Gene Pool. *The American Journal of Human Genetics* 91(1): 83–96.

Pagani, L., Schiffels, S., Gurdasani, D., Danecek, P., Scally, A., et al. 2015. Tracing the Route of Modern Humans Out of Africa by Using 225 Human Genome Sequences from Ethiopians and Egyptians. *The American Journal of Human Genetics* 96(6): 986–991.

Palmer, C., and Pe'er, I. 2017. Statistical Correction of the Winner's Curse Explains Replication Variability in Quantitative Trait Genome-Wide Association Studies. *PLoS Genetics* 13(7): art. e1006916.

Panizzon, M.S., Vuoksimaa, E., Spoon, K.M., Jacobson, K.C., Lyons, M.J., et al. 2014. Genetic and Environmental Influences on General Cognitive Ability: Is *g* a Valid Latent Construct? *Intelligence* 43: 65–76.

Panofsky, A., and Bliss, C. 2017. Ambiguity and Scientific Authority: Population Classification in Genomic Science. *American Sociological Review* 82(1): 59–87.

Park, T. 1955. Experimental Competition in Beetles, with Some General Implications. In *The Numbers of Man and Animals*, ed. J.B. Cragg and N.W. Pirie. London: Oliver and Boyd, pp. 69–82.

Parker, A., and Preston, J. 2013. Paper on Immigrant I.Q. Dogs Critic of Overhaul. *The New York Times*. Available: www.nytimes.com/2013/05/09/us/heritage-analysts-dissertation-on-immigrant-iq-causes-furor.html Accessed May 2017.

Partridge, C. 2018. *Introduction to World Religions*. Minneapolis, MN: Fortress Press.

Parvez, M.K., and Parveen, S. 2017. Evolution and Emergence of Pathogenic Viruses: Past, Present, and Future. *Intervirology* 60(1–2): 1–7.

Paul, D. 1984. Eugenics and the Left. *Journal of the History of Ideas* 45(4): 567–590.

Pavličev, M., and Wagner, G. 2016. The Evolutionary Origin of Female Orgasm. *Journal of Experimental Zoology Part B: Molecular and Developmental Evolution* 326(6): 326–337.

Pavličev, M., Zupan, A.M., Barry, A., Walters, S., Milano, K.M., et al. 2019. An Experimental Test of the Ovulatory Homolog Model of Female Orgasm. *Proceedings of the National Academy of Sciences of the United States of America* 116(41): 20267–20273.

Pearl, J. 2000. *Causality: Models, Reasoning, and Inference*. New York: Cambridge University Press.

Pearl, J., and Mackenzie, D. 2018. *The Book of Why: The New Science of Cause and Effect*. New York: Basic Books.

Pedersen, H.D., and Mikkelsen, L.F. 2019. Göttingen Minipigs as Large Animal Model in Toxicology. In *Biomarkers in Toxicology*, ed. R.C. Gupta. 2nd ed. Amsterdam: Elsevier, pp. 75–89.

Peirce, C.S. 1893a. Evolutionary Love. *The Monist* 3(2): 176–200.

Peirce, C.S. 1893b. Reply to the Necessitarians: Rejoinder to Dr. Carus. *The Monist* 3(4): 526–570.

Peter, B.M. 2016. Admixture, Population Structure, and *F*-Statistics. *Genetics* 202(4): 1485–1501.

Pigliucci, M., and Kaplan, J.M. 2003. On the Concept of Biological Race and Its Applicability to Humans. *Philosophy of Science* 70(5): 1161–1172.

Pigliucci, M., and Müller, G.B. 2010. *Evolution: The Extended Synthesis*. Cambridge, MA: MIT Press.

Plomin, R. 2018. *Blueprint: How DNA Makes Us Who We Are*. Cambridge, MA: MIT Press.

Plomin, R., and Daniels, D. 1987. Why Are Children in the Same Family so Different from One Another? *Behavioral and Brain Sciences* 10(1): 1–16.

Plomin, R., and von Stumm, S. 2018. The New Genetics of Intelligence. *Nature Reviews Genetics* 19(3): 148–159.

Plomin, R., DeFries, J.C., Knopik, V.S., and Neiderhiser, J.M. 2016. Top 10 Replicated Findings from Behavioral Genetics. *Perspectives on Psychological Science* 11(1): 3–23.

Plutynski, A. 2006. What was Fisher's Fundamental Theorem of Natural Selection and What Was It for? *Studies in History and Philosophy of Science, Part C: Studies in History and Philosophy of Biological and Biomedical Sciences* 37(1): 59–82.

Polderman, T.J.C., Benyamin, B., de Leeuw, C.A., Sullivan, P.F., van Bochoven, A., et al. 2015. Meta-analysis of the Heritability of Human Traits Based on Fifty Years of Twin Studies. *Nature Genetics* 47(7): 702–709.

Poldrack, R.A. 2018. *Statistical Thinking for the 21st Century*. Available: http://web.stanford.edu/group/poldracklab/statsthinking21/index.html#an-open-source-book Accessed May 2019.

Pollex, R.L., and Hegele, R.A. 2007. Copy Number Variation in the Human Genome and Its Implications for Cardiovascular Disease. *Circulation* 115(24): 3130–3138.

Poznick, G.D., Henn, B.M., Yee, M.-C., Sliwerska, E., Eukirchen, G.M., et al. 2013. Sequencing Y Chromosomes Resolves Discrepancy in Time to Common Ancestor of Males versus Females. *Science* 341(6145): 562–565.

Pritchard, J.K., Stephens, M., and Donnelly, P. 2000. Inference of Population Structure Using Multilocus Genotype Data. *Genetics* 155(2): 945–959.

Provine, W.B. 1971. *The Origins of Theoretical Population Genetics*. Chicago, IL: University of Chicago Press.

Provine, W.B. 1986. *Sewall Wright and Evolutionary Biology*. Chicago, IL: University of Chicago Press.

Prüfer, K., Racimo, F., Patterson, N., Jay, F., Sankararaman, S., et al. 2014. The Complete Genome Sequence of a Neanderthal from the Altai Mountains. *Nature* 505(7481): 43–49.

Prüfer, K., de Filippo, C., Grote, S., Mafessoni, F., Korlević, P., et al. 2017. A High-Coverage Neandertal Genome from Vindija Cave in Croatia. *Science* 358(6363): 655–658.

Prugnolle, F., Manica, A., and Balloux, F. 2005. Geography Predicts Neutral Genetic Diversity of Human Populations. *Current Biology* 15(5): R159–R160.

Putnam, A.I., and Carbone, I. 2014. Challenges in Analysis and Interpretation of Microsatellite Data for Population Genetic Studies. *Ecology and Evolution* 4(22): 4399–4428.

Race, Ethnicity, and Genetics Working Group. 2005. The Use of Racial, Ethnic, and Ancestral Categories in Human Genetics Research. *The American Journal of Human Genetics* 77(4): 519–532.

Racimo, F., Sankararaman, S., Nielsen, R., and Huerta-Sánchez, E. 2015. Evidence for Archaic Adaptive Introgression in Humans. *Nature Reviews Genetics* 16(6): 359–371.

Racimo, F., Berg, J.J., and Pickrell, J.K. 2018. Detecting Polygenic Adaptation in Admixture Graphs. *Genetics* 208(4): 1565–1584.

Raff, J. 2017. Did Human Women Contribute to Neanderthal Genomes over 200,000 years ago? *The Guardian*. Available: www.theguardian.com/science/2017/jul/18/did-human-women-contribute-to-neanderthal-genomes-over-200000-years-ago Accessed June 2020.

Ramachandran, S., Deshpande, O., Roseman, C.C., Rosenberg, N.A., Feldman, M.W., et al. 2005. Support from the Relationship of Genetic and Geographic Distance in Human Populations for a Serial Founder Effect Originating in Africa. *Proceedings of the National Academy of Sciences of the United States of America* 102(44): 15942–15947.

Rampino, M.R., and Shen, S-Z. 2021. The End-Guadalupian (259.8 Ma) Biodiversity Crisis: The Sixth Major Mass Extinction? *Historical Biology* 33(5): 716–722.

Ranciaro, A., Campbell, M.C., Hirbo, J.B., Ko, W.-Y., Froment, A., et al. 2014. Genetic Origins of Lactase Persistence and the Spread of Pastoralism in Africa. *The American Journal of Human Genetics* 94(4): 496–510.

Rao, C.R. 1992. R.A. Fisher: The Founder of Modern Statistics. *Statistical Science* 7(1): 34–48.

Ray, L. 2018. The Menstrual Cycle: More Than Just Your Period. *Clue*. Available: https://helloclue.com/articles/cycle-a-z/the-menstrual-cycle-more-than-just-the-period Accessed June 2020.

Reardon, J. 2005. *Race to the Finish: Identity and Governance in an Age of Genomics.* Princeton, NJ: Princeton University Press.

Regalado, A. 2019. More Than 26 Million People Have Taken an At-Home Ancestry Test. *MIT Technology Review*. Available: www.technologyreview .com/2019/02/11/103446/more-than-26-million-people-have-taken-an-at-home-ancestry-test/ Accessed May 2020.

Reich, D. 2018. *Who We Are and How We Got Here: Ancient DNA and the New Science of the Human Past.* New York: Pantheon.

Reich, D., Thangaraj, K., Patterson, N., Price, A.L., and Singh, L. 2009. Reconstructing Indian Population History. *Nature* 461(7263): 489–494.

Reich, D., Green, R.E., Kircher, M., Krause, J., Patterson, N., et al. 2010. Genetic History of an Archaic Hominin Group from Denisova Cave in Siberia. *Nature* 468(7327): 1053–1060.

Reid, J.B., and Ross, J.J. 2011. Mendel's Genes: Toward a Full Molecular Characterization. *Genetics* 189(1): 3–10.

Relethford, J.H. 1998. Genetics of Modern Human Origins and Diversity. *Annual Review of Anthropology* 27: 1–23.

Rembert, N., He, K., Judd, S.E., and McClure, L.A. 2017. The Geographic Distribution of Trace Elements in the Environment: The REGARDS Study. *Environmental Monitoring and Assessment* 189(2): art. 84.

Renfrew, C. 1990. *Archaeology and Language: The Puzzle of Indo-European Origins.* Cambridge: Cambridge University Press.

Richardson, S.S. 2013. *Sex Itself: The Search for Male and Female in the Human Genome.* Chicago, IL: Chicago University Press.

Rieppel, O. 2009. "Total evidence" in Phylogenetic Systematics. *Biology and Philosophy* 24(5): 607–622.

Risch, N., Buchard, E., Ziv, E., and Hua, T. 2002. Categorization of Humans in Biomedical Research: Genes, Race and Disease. *Genome Biology* 3(7): art. comment2007.

Robinson, A.H. 1985. Arno Peters and His New Cartography. *American Cartographer* 12(2): 103–111.

Robinson, G. 1979. *A Prelude to Genetics: Theories of a Material Substance of Heredity, Darwin to Weismann.* Lawrence, KS: Coronado Press.

Rochmyaningsih, D. 2018. Did a Study of Indonesian People Who Spend Most of Their Days Under Water Violate Ethical Rules? *Science*. Available: www .sciencemag.org/news/2018/07/did-study-indonesian-people-who-spend-their-days-under-water-violate-ethical-rules Accessed May 2019.

Rohde, D.L., Olson, S., and Chang, J.T. 2004. Modelling the Recent Common Ancestry of All Living Humans. *Nature* 431(7008): 562–566.

Rohlfs, R.V., Fullerton, S.M., and Weir, B.S. 2012. Familial Identification: Population Structure and Relationship Distinguishability. *PLoS Genetics* 8(2): art. e1002469.

Roseman, C.C. 2021. Lewontin Did Not Commit Lewontin's Fallacy, His Critics Do: Why Racial Taxonomy Is Not Useful for the Scientific Study of Human Variation. *BioEssays* 43(12): art. 2100204.

Rosenberg, N.A. 2011. A Population-Genetic Perspective on the Similarities and Differences Among Worldwide Human Populations. *Human Biology* 83(6): 659–684.

Rosenberg, N.A. 2018. Variance-Partitioning and Classification in Human Population Genetics. In *Phylogenetic Inference, Selection Theory, and History of Science: Selected Papers of A.W.F. Edwards with Commentaries*, ed. R.G. Winther. Cambridge: Cambridge University Press, pp. 399–403.

Rosenberg, N.A., and Nordborg, M. 2002. Genealogical Trees, Coalescent Theory and the Analysis of Genetic Polymorphisms. *Nature Reviews Genetics* 3(5): 380–390.

Rosenberg, N.A., Pritchard, J.K., Weber, J.L., Cann, H.M., Kidd, K.K., et al. 2002. Genetic Structure of Human Populations. *Science* 298(5602): 2381–2385.

Rosenberg, N.A., Mahajan, S., Ramachandran, S., Zhao, C., Pritchard, J.K., et al. 2005. Clines, Clusters, and the Effect of Study Design on the Inference of Human Population Structure. *PLoS Genetics* 1(6): art. e70.

Rosenberg, N.A., Huang, L., Jewett, E.M., Szpiech, Z.A., Jankovic, I., et al. 2010. Genome-Wide Association Studies in Diverse Populations. *Nature Reviews Genetics* 11(5): 356–366.

Roughgarden, J. 2009. *The Genial Gene: Deconstructing Darwinian Selfishness.* Berkeley, CA: University of California Press.

Routledge, R.D. 1979. Diversity Indices: Which Ones Are Admissible? *Journal of Theoretical Biology* 76(4): 503–515.

Rushton, J.P. 1995. *Race, Evolution, and Behavior: A Life History Perspective.* New Brunswick, NJ: Transaction Publishers.

Russell, B. 1997 [1912]. The Value of Philosophy. In *The Problems of Philosophy,* with an Introduction by J. Perry. Oxford: Oxford University Press, pp. 153–161.

Rutherford, A.C. 2020. *How to Argue with a Racist: What Our Genes Do (and Don't) Say about Human Difference.* New York: The Experiment.

Ryckman, T. 2017. *Einstein.* New York: Routledge.

Ryman, N., Chakraborty, R., and Nei, M. 1983. Differences in the Relative Distribution of Human Gene Diversity Between Electrophoretic and Red and White Cell Antigen Loci. *Human Heredity* 33(2): 93–102.

Sabeti, P.C., Schaffner, S.F., Fry, B., Lohmueller, J., Varilly, P., et al. 2006. Positive Natural Selection in the Human Lineage. *Science* 312(5780): 1614–1620.

Sabeti, P.C., Varilly, P., Fry, B., Lohmueller, J., Hostetter, E., et al., and the International HapMap Consortium. 2007. Genome-Wide Detection and Characterization of Positive Selection in Human Populations. *Nature* 449(7164): 913–918.

Sagan, L. 1967. On the Origin of Mitosing Cells. *Journal of Theoretical Biology* 14(3): 225–274.

Säll, T., and Bengtsson, B.O. 2017. *Understanding Population Genetics: Through the Derivation of Ten Major Results.* Chichester: Wiley.

Sandel, M.J. 2020. *The Tyranny of Merit: What's Become of the Common Good?* New York: Farrar, Straus and Giroux.

Sankararaman, S., Mallick, S., Dannemann, M., Prüfer, K., Kelso, J, et al. 2014. The Genomic Landscape of Neanderthal Ancestry in Present-Day Humans. *Nature* 507(7492): 354–357.

Sapp, J. 1994. *Evolution by Association: A History of Symbiosis.* New York: Oxford University Press.

Sarich, V.M., and Wilson, A.C. 1967. Immunological Time Scale for Hominid Evolution. *Science* 158(3805): 1200–1203.

Sarkar, S. 1998. *Genetics and Reductionism*. Cambridge: Cambridge University Press.

Sarkar, T. 2020. False Positives/Negatives and Bayes Rule for COVID-19 Testing. *Medium*. Available: https://towardsdatascience.com/false-positives-negatives-and-bayes-rule-for-covid-19-testing-750eaba84acd Accessed May 2020.

Sather, C. 1997. *The Bajau Laut: Adaptation, History, and Fate in a Maritime Fishing Society of South-Eastern Sabah*. Oxford: Oxford University Press.

Savage, J.E., Jansen, P.R., Stringer, S., Watanabe, K., Bryois, J., et al. 2018. Genome-Wide Association Meta-analysis in 269,867 Individuals Identifies New Genetic and Functional Links to Intelligence. *Nature Genetics* 50(7): 912–919.

Sayres, M.A.W. 2018. Genetic Diversity on the Sex Chromosomes. *Genome Biology and Evolution* 10(4): 1064–1078.

Scerri, E.M.L., Thomas, M.G., Manica, A., Gunz, P., Stock, J.T., et al. 2018. Did Our Species Evolve in Subdivided Populations across Africa, and Why Does It Matter? *Trends in Ecology & Evolution* 33(8): 582–594.

Schabath, M.B., Cress, W.D., and Muñoz-Antonia, T. 2016. Racial and Ethnic Differences in the Epidemiology of Lung Cancer and the Lung Cancer Genome. *Cancer Control* 23(4): 338–346.

Schaffer, J. 2016. The Metaphysics of Causation. In *The Stanford Encyclopedia of Philosophy* (Fall 2016 Edition), ed. E.N. Zalta. Available: https://plato.stanford.edu/archives/fall2016/entries/causation-metaphysics/ Accessed March 2017.

Schagatay, E., Richardson, M.X., and Lodin-Sundström, A. 2012. Size Matters: Spleen and Lung Volumes Predict Performance in Human Apneic Divers. *Frontiers in Physiology* 3: art. 173.

Schlebusch, C.M., Malmström, H., Günther, T., Sjödin, P., Coutinho, A., et al. 2017. Southern African Ancient Genomes Estimate Modern Human Divergence to 350,000 to 260,000 Years Ago. *Science* 358(6363): 652–655.

Ségurel, L., Thompson, E.E., Flutre, T., Lovstad, J., Venkat, A., et al. 2012. The ABO Blood Group Is a Trans-Species Polymorphism in Primates. *Proceedings of the National Academy of Sciences of the United States of America* 109(45): 18493–18498.

Seielstad, M.T., Minch, E., and Cavalli-Sforza, L.L. 1998. Genetic Evidence for a Higher Female Migration Rate in Humans. *Nature Genetics* 20(3): 278–280.

Serre, D., and Pääbo, S. 2004. Evidence for Gradients of Human Genetic Diversity Within and Among Continents. *Genome Research* 14(9): 1679–1685.

Servedio, M.R., Brandvain, Y., Dhole, S., Fitzpatrick, C.L., Goldberg, E.E., et al. 2014. Not Just a Theory: The Utility of Mathematical Models in Evolutionary Biology. *PLoS Biology* 12(12): art. e1002017.

Sesardić, N. 2005. *Making Sense of Heritability*. Cambridge: Cambridge University Press.

Sesardić, N. 2010. Race: A Social Destruction of a Biological Concept. *Biology and Philosophy* 25(2): 143–162.

Sesardić, N. 2013. Confusions about Race: A New Installment. *Studies in History and Philosophy of Science, Part C: Studies in History and Philosophy of Biological and Biomedical Sciences* 44(3): 287–293.

Shapin, S. 1996. *The Scientific Revolution*. Chicago, IL: Chicago University Press.

Shapin, S., and Schaffer, S. 1985. *Leviathan and the Air-Pump: Hobbes, Boyle, and the Experimental Life*. Princeton, NJ: Princeton University Press.

Shea, N. 2013. Inherited Representations Are Read in Development. *The British Journal for the Philosophy of Science* 64(1): 1–31.

Shiao, J.L., Bode, T., Beyer, A., and Selvig, D. 2012. The Genomic Challenge to the Social Construction of Race. *Sociological Theory* 30(2): 67–88.

Shih, C., Cold, C.J., and Yang, C.C. 2013. Cutaneous Corpuscular Receptors of the Human Glans Clitoris: Descriptive Characteristics and Comparison with the Glans Penis. *The Journal of Sexual Medicine* 10(7): 1783–1789.

Shipley, B. 2000. *Cause and Correlation in Biology: A User's Guide to Path Analysis, Structural Equations and Causal Inference*. Cambridge: Cambridge University Press.

Simon, H.A. 1953. Causal Ordering and Identifiability. In *Studies in Econometric Method*, ed. W.C. Hood and J.C. Koopmans. New York: Wiley, pp. 49–74.

Simon, S. 2019. Gap in Cancer Death Rates. American Cancer Society. Available: www.cancer.org/latest-news/gap-in-cancer-death-rates-between-blacks-and-whites-narrows.html Accessed May 2020.

Simpson, E.H. 1949. Measurement of Diversity. *Nature* 163(4148): 688.

Sinervo, B., and Basolo, A.L. 1996. Testing Adaptation Using Phenotypic Manipulations. In *Adaptation*, ed. M.R. Rose and G.V. Lauder. San Diego, CA: Academic Press, pp. 149–185.

Skoglund, P., Thompson, J.C., Prendergast, M.E., Mittnik, A., Sirak, K., et al. 2017. Reconstructing Prehistoric African Population Structure. *Cell* 171(1): 59–71.

Skov, L., Coll Macià, M., Sveinbjörnsson, G., Mafessoni, F, Lucotte, E.A., et al. 2020. The Nature of Neanderthal Introgression Revealed by 27,566 Icelandic Genomes. *Nature* 582(7810): 78–83.

Slatkin, M. 1985. Gene Flow in Natural Populations. *Annual Review of Ecology and Systematics* 16: 393–430.

Slatkin, M. 1991. Inbreeding Coefficients and Coalescence Times. *Genetics Research* 58(2): 167–175.

Slatkin, M. 2008. Linkage Disequilibrium: Understanding the Evolutionary Past and Mapping the Medical Future. *Nature Reviews Genetics* 9(6): 477–485.

Smith, B.C. 1996. *On the Origin of Objects*. Cambridge, MA: MIT Press.

Smouse, P.E., Spielman, R.S., and Park, M.H. 1982. Multiple-Locus Allocation of Individuals to Groups as a Function of the Genetic Variation Within and Differences Among Human Populations. *The American Naturalist* 119(4): 445–463.

Sniekers, S., Stringer, S., Watanabe, K., Jansen, P.R., Coleman, J.R.I., et al. 2017. Genome-Wide Association Meta-analysis of 78,308 Individuals Identifies New Loci and Genes Influencing Human Intelligence. *Nature Genetics* 49(7): 1107–1112.

Sober, E. 2008. *Evidence and Evolution: The Logic Behind the Science*. Cambridge: Cambridge University Press.

Sober, E. 2020. A.W.F. Edwards on Phylogenetic Inference, Fisher's Theorem, and Race: Review of *Phylogenetic Inference, Selection Theory, and History of Science: Selected Papers of A.W.F. Edwards with Commentaries* edited by R.G. Winther. *The Quarterly Review of Biology* 95(2): 125–129.

Sober, E., and Lewontin, R.C. 1982. Artifact, Cause and Genic Selection. *Philosophy of Science* 49(2): 157–180.

Sober, E., and Wilson, D.S. 1998. *Unto Others: The Evolution and Psychology of Unselfish Behavior*. Cambridge, MA: Harvard University Press.

Sokal, R.R., and Rohlf, F.J. 1995. *Biometry*. 3rd ed. New York: W.H. Freeman.

Solomon, P., Cavanaugh, M.M., and Draine, J. 2009. *Randomized Controlled Trials: Design and Implementation for Community-Based Psychosocial Interventions*. New York: Oxford University Press.

Sommer, M. 2015. Population-Genetic Trees, Maps, and Narratives of the Great Human Diasporas. *History of the Human Sciences* 28(5): 108–145.

Spencer, Q. 2013. Biological Theory and the Metaphysics of Race: A Reply to Kaplan and Winther. *Biological Theory* 8(1): 114–120.

Spencer, Q. 2014. A Radical Solution to the Race Problem. *Philosophy of Science* 81(5): 1025–1038.

Spencer, Q. 2015. Philosophy of Race Meets Population Genetics. *Studies in History and Philosophy of Science, Part C: Studies in History and Philosophy of Biological and Biomedical Sciences* 52: 46–55.

Spencer, Q. 2019. How to Be a Biological Racial Realist. In *What Is Race? Four Philosophical Views*, ed. J. Glasgow, S. Haslanger, C. Jeffers, and Q. Spencer. New York: Oxford University Press, pp. 73–110.

Spirtes, P., Glymour, C., and Scheines, R. 1993. *Causation, Prediction, and Search*. Dordrecht: Springer.

Stegenga, J. 2010. Population Is Not a Natural Kind of Kinds. *Biological Theory* 5(2): 154–160.

Sterelny, K., and Kitcher, P. 1988. The Return of the Gene. *The Journal of Philosophy* 85(7): 339–361.

Stern, A.J., and Nielsen, R. 2019. Detecting Natural Selection. In *Handbook of Statistical Genomics*, ed. D.J. Balding, I. Moltke, and J. Marioni. 4th ed. Hoboken, NJ: Wiley, pp. 397–420.

Stern, A.J., Speidel, L., Zaitlen, N.A., and Nielsen, R. 2021. Disentangling Selection on Genetically Correlated Polygenic Traits Using Whole-Genome Genealogies. *The American Journal of Human Genetics* 108(2): 219–239.

Sternberg, R.J. 1985. *Beyond IQ: A Triarchic Theory of Intelligence*. Cambridge: Cambridge University Press.

Stift, M., Kolář, F., and Meirmans, P.G. 2019. *Structure* Is More Robust than Other Clustering Methods in Simulated Mixed-Ploidy Populations. *Heredity* 123(4): 429–441.

Stigler, S.M. 2016. *The Seven Pillars of Statistical Wisdom*. Cambridge, MA: Harvard University Press.

Stolley, P.D. 1991. When Genius Errs: R.A. Fisher and the Lung Cancer Controversy. *American Journal of Epidemiology* 133(5): 416–425.

Stoltzfus, A. 2017. Why We Don't Want Another "Synthesis." *Biology Direct* 12: art. 23.

Storz, J.F., Ramakrishnan, U., and Alberts, S.C. 2001. Determinants of Effective Population Size for Loci with Different Modes of Inheritance. *Journal of Heredity* 92(6): 497–502.

Strait, D., Grine, F.E., and Fleagle, J.G. 2015. Analyzing Hominin Phylogeny: Cladistic Approach. In *Handbook of Paleoanthropology*, ed. W. Henke and I. Tattersall. Berlin: Springer, pp. 1989–2014.

Stringer, C. 2012. *Lone Survivors: How We Came to Be the Only Humans on Earth*. New York: Henry Holt.

Stringer, C. 2016. The Origin and Evolution of *Homo sapiens*. *Philosophical Transactions of the Royal Society of London B: Biological Sciences* 371(1698): art. 20150237.

Sudmant, P.H., Rausch, T., Gardner, E.J., Handsaker, R.E., Abyzov, A., et al. 2015. An Integrated Map of Structural Variation in 2,504 Human Genomes. *Nature* 526(7571): 75–81.

Suppes, P. 1984. *Probabilistic Metaphysics*. Oxford: Blackwell.

Suzuki, K., Harada, A., Suzuki, H., Miyamoto, M., and Kimura, H. 2016. TAK-063, a PDE10A Inhibitor with Balanced Activation of Direct and Indirect Pathways, Provides Potent Antipsychotic-Like Effects in Multiple Paradigms. *Neuropsychopharmacology* 41(9): 2252–2262.

Swallow, D. 2015. Lactase Persistence: A Case of Evolution in Modern Humans. *The Biochemist* 37(5): 20–23.

Swindle, M.M., Makin, A., Herron, A.J., Clubb, Jr., F.J., and Frazier, K.S. 2012. Swine as Models in Biomedical Research and Toxicology Testing. *Veterinary Pathology* 49(2): 344–356.

Symons, D. 1979. *The Evolution of Human Sexuality*. New York: Oxford University Press.

Szilagyi, A. 2015. Adult Lactose Digestion Status and Effects on Disease. *Canadian Journal of Gastroenterology and Hepatology* 29(3): 149–156.

Szpak, M., Xue, Y., Ayub, Q., and Tyler-Smith, C. 2019. How Well Do We Understand the Basis of Classic Selective Sweeps in Humans? *FEBS Letters* 593(13): 1431–1448.

Szpiech, Z.A., Jakobsson, M., and Rosenberg, N.A. 2008. ADZE: A Rarefaction Approach for Counting Alleles Private to Combinations of Populations. *Bioinformatics* 24(21): 2498–2504.

Tabery, J. 2014. *Beyond Versus: The Struggle to Understand the Interaction of Nature and Nurture*. Cambridge, MA: MIT Press.

Takahata, N. 1993. Allelic Genealogy and Human Evolution. *Molecular Biology and Evolution* 10(1): 2–22.

Tal, O. 2012. The Cumulative Effect of Genetic Markers on Classification Performance: Insights from Simple Models. *Journal of Theoretical Biology* 293: 206–218.

TallBear, K. 2013. *Native American DNA: Tribal Belonging and the False Promise of Genetic Science*. Minneapolis, MN: University of Minnesota Press.

Tang, H., Coram, M., Wang, P., Zhu, X., and Risch, N. 2006. Reconstructing Genetic Ancestry Blocks in Admixed Individuals. *The American Journal of Human Genetics* 79(1): 1–12.

Taylor, P.C. 2014. Taking Postracialism Seriously: From Movement Mythology to Racial Formation. *Du Bois Review* 11(1): 9–25.

Taylor, P.C. 2022. *Race: A Philosophical Introduction*. 3rd ed. Cambridge: Polity.

Taylor, W.R. 1997. Residual Colours: A Proposal for Aminochromography. *Protein Engineering* 10(7): 743–746.

Tegmark, M. 2014. *Our Mathematical Universe: My Quest for the Ultimate Nature of Reality*. New York: Knopf.

Teller, E. 1985. A Few Memories. In *Niels Bohr: A Centenary Volume*, ed. A.P. French and P.J. Kennedy. Cambridge, MA: Harvard University Press, pp. 181–182.

Templeton, A.R. 1997. Out of Africa? What Do Genes Tell Us? *Current Opinion in Genetics & Development* 7(6): 841–847.

Templeton, A.R. 1999. Human Races: A Genetic and Evolutionary Perspective. *American Anthropologist* 100(3): 632–650.

Templeton, A.R. 2002. Out of Africa Again and Again. *Nature* 416(6876): 45–51.

Thompson, E.A. 2013. Identity by Descent: Variation in Meiosis, Across Genomes, and in Populations. *Genetics* 194(2): 301–326.

Thompson, E.A. 2018. Likelihood Inference in Models for the Genetic Diversity of Populations. In *Phylogenetic Inference, Selection Theory, and History of Science: Selected Papers of A.W.F. Edwards with Commentaries*, ed. R.G. Winther. Cambridge: Cambridge University Press, pp. 347–351.

Tishkoff, S.A., Reed, F.A., Friedlaender, F.R., Ehret, C., Ranciaro, A., et al. 2009. The Genetic Structure and History of Africans and African Americans. *Science* 324(5930): 1035–1044.

Tremellen, K.P., Valbuena, D., Landeras, J., Ballesteros, A., Martinez, J., et al. 2000. The Effect of Intercourse on Pregnancy Rates During Assisted Human Reproduction. *Human Reproduction* 15(12): 2653–2658.

Turkheimer, E. 2000. Three Laws of Behavior Genetics and What They Mean. *Current Directions in Psychological Science* 9(5): 160–164.

Turkheimer, E., and Waldron, M. 2000. Nonshared Environment: A Theoretical, Methodological, and Quantitative Review. *Psychological Bulletin* 126(1): 78–108.

Underhill, P.A., and Kivisild, T. 2007. Use of Y Chromosome and Mitochondrial DNA Population Structure in Tracing Human Migrations. *Annual Review of Genetics* 41: 539–564.

UNESCO. 1952. *The Race Concept: Results of an Inquiry*. Paris: UNESCO.

Uricchio, L.H., Petrov, D.A., and Enard, D. 2019. Exploiting Selection at Linked Sites to Infer the Rate and Strength of Adaptation. *Nature Ecology & Evolution* 3(6): 977–984.

Vaesen, K., Scherjon, F., Hemerik, L., and Verpoorte, A. 2019. Inbreeding, Allee Effects and Stochasticity Might Be Sufficient to Account for Neanderthal Extinction. *PLoS ONE* 14(11): art. e0225117.

Valles, S.A. 2018. *Philosophy of Population Health: Philosophy for a New Public Health Era*. London: Routledge.

van Fraassen, B. 1980. *The Scientific Image*. Oxford: Oxford University Press.

van Oven, M., and Kayser, M. 2009. Updated Comprehensive Phylogenetic Tree of Global Human Mitochondrial DNA Variation. *Human Mutation* 30(2): E386–E394.

Van Valen, L. 1973. Festschrift: Review of *Evolutionary Biology*, vol. 6, edited by Dobzhansky, T., Hecht, M.K., and Steere, W.C. New York: Springer. *Science* 180(4085): 488.

Venter, J.C., Adams, M.D., Myers, E.W., Li, P.W., Mural, R.J., et al. 2001. The Sequence of the Human Genome. *Science* 291(5507): 1304–1351.

Vergara-Silva, F. 2009. Pattern Cladistics and the "Realism–Antirealism Debate" in the Philosophy of Biology. *Acta Biotheoretica* 57(1–2): 269–294.

Visscher, P.M., and Goddard, M.E. 2019. From R.A. Fisher's 1918 Paper to GWAS a Century Later. *Genetics* 211(4): 1125–1130.

Visscher, P.M., Hill, W.G., and Wray, N.R. 2008. Heritability in the Genomics Era: Concepts and Misconceptions. *Nature Reviews Genetics* 9(4): 255–266.

Visscher, P.M., Brown, M.A., McCarthy, M.I., and Yang, J. 2012. Five Years of GWAS Discovery. *The American Journal of Human Genetics* 90(1): 7–24.

Visscher, P.M., Wray, N.R., Zhang, Q., Sklar, P., McCarthy, M.I., et al. 2017. 10 Years of GWAS Discovery: Biology, Function, and Translation. *The American Journal of Human Genetics* 101(1): 5–22.

Vitti, J.J., Cho, M.K., Tishkoff, S.A., and Sabeti, P.C. 2012. Human Evolutionary Genomics: Ethical and Interpretive Issues. *Trends in Genetics* 28(3): 137–145.

Vitti, J.J., Grossman, S.R., and Sabeti, P.C. 2013. Detecting Natural Selection in Genomic Data. *Annual Review of Genetics* 47: 97–120.

Wade, M.J. 1992. Sewall Wright: Gene Interaction and the Shifting Balance Theory. In *Oxford Surveys of Evolutionary Biology*, vol. 6, ed. J. Antonovics and D. Futuyma. New York: Oxford University Press, pp. 35–62.

Wade, M.J. 2002. A Gene's Eye View of Epistasis, Selection, and Speciation. *Journal of Evolutionary Biology* 15(3): 337–346.

Wade, M.J. 2016. *Adaptation in Metapopulations: How Interaction Changes Evolution*. Chicago, IL: University of Chicago Press.

Wade, M.J., Winther, R.G., Agrawal, A.F., and Goodnight, C.J. 2001. Alternative Definitions of Epistasis: Dependence and Interaction. *Trends in Ecology & Evolution* 16(9): 498–504.

Wade, N. 2014. *A Troublesome Inheritance: Genes, Race, and Human History*. New York: Penguin.

Wahlsten, D. 1990. Insensitivity of the Analysis of Variance to Heredity–Environment Interaction. *Behavioral and Brain Sciences* 13(1): 109–120.

Wakeley, J. 2008. *Coalescent Theory: An Introduction*. New York: W.H. Freeman.

Wall, J.D., Cox, M.P., Mendez, F.L., Woerner, A., Severson, T., et al. 2008. A Novel DNA Sequence Database for Analyzing Human Demographic History. *Genome Research* 18(8): 1354–1361.

Wall, J.D., Yang, M.A., Jay, F., Kim, S.K., Durand, E.Y., et al. 2013. Higher Levels of Neanderthal Ancestry in East Asians than in Europeans. *Genetics* 194(1): 199–209.

Wallen, K., and Lloyd, E.A. 2008. Clitoral Variability Compared with Penile Variability Supports Nonadaptation of Female Orgasm. *Evolution and Development* 10(1): 1–2.

Wallen, K., and Lloyd, E.A. 2011. Female Sexual Arousal: Genital Anatomy and Orgasm in Intercourse. *Hormones and Behavior* 59(5): 780–792.

Walsh, D. 2012. Mechanism and Purpose: A Case for Natural Teleology. *Studies in History and Philosophy of Science, Part C: Studies in History and Philosophy of Biological and Biomedical Sciences* 43(1): 173–181.

Walsh, D. 2013. Mechanism, Emergence, and Miscibility: The Autonomy of Evo-Devo. In *Functions: Selection and Mechanisms*, ed. P. Huneman. Dordrecht: Springer, pp. 43–65.

Walsh, D. 2015. *Organisms, Agency, and Evolution.* Cambridge: Cambridge University Press.

Ward, P. 2004. The Father of All Mass Extinctions. *Conservation in Practice* 5(3): 12–17.

Warren, M. 2018. The Power of Many. *Nature* 562(7726): 181–183.

Webster, G., and Goodwin, B. 1996. *Form and Transformation: Generative and Relational Principles in Biology.* Cambridge: Cambridge University Press.

Weiner, J. 2014 [1994]. *The Beak of the Finch.* New York: Random House.

Weir, B.S. 1996. *Genetic Data Analysis II.* Sunderland, MA: Sinauer Associates.

Weir, B.S. 2012. Estimating *F*-Statistics: A Historical View. *Philosophy of Science* 79(5): 637–643.

Weir, B.S., and Cockerham, C.C. 1984. Estimating *F*-Statistics for the Analysis of Population Structure. *Evolution* 38(6): 1358–1370.

Weir, B.S., and Hill, W.G. 2002. Estimating *F*-Statistics. *Annual Review of Genetics* 36: 721–750.

Weisberg, M. 2013. *Simulation and Similarity: Using Models to Understand the World.* New York: Oxford University Press.

Weiss, K.M., and Fullerton, S.M. 2005. Racing Around, Getting Nowhere. *Evolutionary Anthropology* 14(5): 165–169.

Weiss, K.M., and Long, J.C. 2009. Non-Darwinian Estimation: My Ancestors, My Genes' Ancestors. *Genome Research* 19(5): 703–710.

Weiss, M.C., Sousa, F.L., Mrnjavac, N., Neukirchen, S., Roettger, M., et al. 2016. The Physiology and Habitat of the Last Universal Common Ancestor. *Nature Microbiology* 1(9): art. 16116.

Weitzman, S. 2017. *The Origin of the Jews: The Quest for Roots in a Rootless Age.* Princeton, NJ: Princeton University Press.

Wells, S. 2003. *The Journey of Man: A Genetic Odyssey.* New York: Random House.

Welsh, A.H., Peterson, A.T., and Altmann, S.A. 1988. The Fallacy of Averages. *The American Naturalist* 132(2): 277–288.

West-Eberhard, M.J. 2003. *Developmental Plasticity and Evolution.* New York: Oxford University Press.

Wexler, A. 2010. Stigma, History, and Huntington's Disease. *The Lancet* 376(9734): 18–19.

Whiteley, A.R., Bhat, A., Martins, E.P., Mayden, R.L., Arunachalam, M., et al. 2011. Population Genomics of Wild and Laboratory Zebrafish (*Danio rerio*). *Molecular Ecology* 20(20): 4259–4276.

Whitlock, M.C. 2011. G'_{ST} and D Do Not Replace F_{ST}. *Molecular Ecology* 20(6): 1083–1091.

Wilde, S., Timpson, A., Kirsanow, K., Kaiser, E., Kayser, M., et al. 2014. Direct Evidence for Positive Selection of Skin, Hair, and Eye Pigmentation in Europeans During the Last 5,000 y. *Proceedings of the National Academy of Sciences of the United States of America* 111(13): 4832–4837.

Wilder, J.A., Kingan, S.B., Mobasher, Z., Pilkington, M.M., and Hammer, M.F. 2004. Global Patterns of Human Mitochondrial DNA and Y-Chromosome Structure Are Not Influenced by Higher Migration Rates of Females versus Males. *Nature Genetics* 36(10): 1122–1125.

Wilkins, A. 2012. The Scientists Behind Mitochondrial Eve Tell Us about the "Lucky Mother" Who Changed Human Evolution Forever. *Gizmodo.* Available: https://gizmodo.com/the-scientists-behind-mitochondrial-eve-tell-us-about-t-5879991 Accessed May 2020.

Wilkins, J.F., and Marlowe, F.W. 2006. Sex-Biased Migration in Humans: What Should We Expect from Genetic Data? *BioEssays* 28(3): 290–300.

Williams, M. 1982. The Importance of Prediction Testing in Evolutionary Biology. *Erkenntnis* 17(3): 291–306.

Willing, E.-M., Dreyer, C., and van Oosterhout, C. 2012. Estimates of Genetic Differentiation Measured by F_{ST} Do Not Necessarily Require Large Sample Sizes When Using Many SNP markers. *PLoS ONE* 7(8): art. e42649.

Wimsatt, W.C. 2007. *Re-Engineering Philosophy for Limited Beings: Piecewise Approximations to Reality.* Cambridge, MA: Harvard University Press.

Winther, R.G. 1996. *Constructions of Form in Biological Systems.* Masters Thesis, Philosophy Department, Stanford University.

Winther, R.G. 2000. Darwin on Variation and Heredity. *Journal of the History of Biology* 33(3): 425–455.

Winther, R.G. 2001. August Weismann on Germ-Plasm Variation. *Journal of the History of Biology* 34(3): 517–555.

Winther, R.G. 2005. An Obstacle to Unification in Biological Social Science: Formal and Compositional Styles of Science. *Graduate Journal of Social Science* 2(2): 40–100.

Winther, R.G. 2004. A Molecular and Evolutionary Study of the Obligate Endosymbiont *Wolbachia* in *Tribolium confusum.* Masters Thesis, Ecology and Evolutionary Biology, Department of Biology, Indiana University. Available: www.rgwinther.com/Publications/ WintherRGWolbachiaTriboliumMastersThesis.pdf Accessed March 2010.

Winther, R.G. 2006a. Fisherian and Wrightian Perspectives in Evolutionary Genetics and Model-Mediated Imposition of Theoretical Assumptions. *Journal of Theoretical Biology* 240(2): 218–232.

Winther, R.G. 2006b. Parts and Theories in Compositional Biology. *Biology and Philosophy* 21: 471–499.

Winther, R.G. 2006c. On the Dangers of Making Scientific Models Ontologically Independent: Taking Richard Levins' Warnings Seriously. *Biology and Philosophy* 21: 703–724.

Winther, R.G. 2008. Systemic Darwinism. *Proceedings of the National Academy of Sciences of the United States of America* 105(33): 11833–11838.

Winther, R.G. 2009a. Character Analysis in Cladistics: Abstraction, Reification, and the Search for Objectivity. *Acta Biotheoretica* 57(1–2): 129–162.

Winther, R.G. 2009b. Prediction in Selectionist Evolutionary Theory. *Philosophy of Science* 76(5): 889–901.

Winther, R.G. 2011. Part-Whole Science. *Synthese* 178(3): 397–427.

Winther, R.G. 2012a. Interweaving Categories: Styles, Paradigms, and Models. *Studies in History and Philosophy of Science, Part A* 43(4): 628–639.

Winther, R.G. 2012b. Mathematical Modeling in Biology: Philosophy and Pragmatics. *Frontiers in Plant Science* 3: art. 102.

Winther, R.G. 2014. James and Dewey on Abstraction. *The Pluralist* 9(2): 1–28.

Winther, R.G. (ed.). 2018a. *Phylogenetic Inference, Selection Theory, and History of Science: Selected Papers of A.W.F. Edwards with Commentaries.* Cambridge: Cambridge University Press.

Winther, R.G. 2018b. Introduction. In *Phylogenetic Inference, Selection Theory, and History of Science: Selected Papers of A.W.F. Edwards with Commentaries*, ed. R.G. Winther. Cambridge: Cambridge University Press, pp. 1–12.

Winther, R.G. 2018c. Race and Biology. In *The Routledge Companion to the Philosophy of Race*, ed. P. Taylor, L. Alcoff, and L. Anderson. New York: Routledge, pp. 305–320.

Winther, R.G. 2019. Mapping the Deep Blue Oceans. In *The Philosophy of GIS*, ed. T. Tambassi. New York: Springer, pp. 99–123.

Winther, R.G. 2020a. *When Maps Become the World.* Chicago, IL: University of Chicago Press.

Winther, R.G. 2020b. Cutting the Cord: A Corrective for World Navels in Cartography and Science. *The Cartographic Journal (British Cartographic Society)* 57(2): 147–159.

Winther, R.G. 2021a. The Structure of Scientific Theories. In *The Stanford Encyclopedia of Philosophy* (Spring 2021 Edition), ed. E.N. Zalta. Available: https://plato.stanford.edu/archives/spr2021/entries/structure-scientific-theories Accessed May 2022.

Winther, R.G. 2021b. Lewontin 1972 The Apportionment of Human Diversity Apportionment Check and Recalculation Winther 2022. Harvard Dataverse. https://doi.org/10.7910/DVN/64CKYN.

Winther, R.G. 2021c. Lewontin as Master Dialectician: Rest in Power, Dick. In *Science for the People,* ed. The Editorial Collective. Available: https://magazine.scienceforthepeople.org/lewontin-special-issue/lewontin-as-master-dialectician/ Accessed November 2021.

Winther, R.G. 2022a. Lewontin (1972). In *Remapping Race in a Global Context*, ed. L. Lorusso and R.G. Winther. London: Routledge, pp. 9–47.

Winther, R.G. 2022b. Map Thinking Across the Life Sciences. In *The Routledge Handbook of Geospatial Technologies and Society*, ed. A.J. Kent and D. Specht. London: Routledge.

Winther, R.G., Wade, M.J., and Dimond, C.C. 2013. Pluralism in Evolutionary Controversies: Styles and Averaging Strategies in Hierarchical Selection Theories. *Biology and Philosophy* 28(6): 957–979.

Winther, R.G., Millstein, R.L., and Nielsen, R. 2015. Introduction: Genomics and Philosophy of Race. *Studies in History and Philosophy of Science, Part C: Studies in History and Philosophy of Biological and Biomedical Sciences* 52: 1–4.

Witherspoon, D.J., Wooding, S., Rogers, A.R., Marchani, E.E., Watkins, W.S., et al. 2007. Genetic Similarities Within and Between Human Populations. *Genetics* 176(1): 351–359.

Withrock, I.C., Anderson, S.J., Jefferson, M.A., McCormack, G.R., Mlynarczyk, G.S.A., et al. 2015. Genetic Diseases Conferring Resistance to Infectious Diseases. *Genes & Diseases* 2(3): 247–254.

Woese, C.R., Kandler, O., and Wheelis, M.L. 1990. Towards a Natural System of Organisms: Proposal for the Domains Archaea, Bacteria, and Eucarya. *Proceedings of the National Academy of Sciences of the United States of America* 87(12): 4576–4579.

Wolfe, N., Dunavan, C.P., and Diamond, J. 2007. Origins of Major Human Infectious Diseases. *Nature* 447(7142): 279–283.

Wolfram, S. 2002. *A New Kind of Science*. Champaign, IL: Wolfram Media.

Woodward, J. 2011. Scientific Explanation. In *The Stanford Encyclopedia of Philosophy* (Winter 2011 Edition), ed. E.N. Zalta. Available: http://plato.stanford.edu/archives/win2011/entries/scientific-explanation/ Accessed September 2012.

Wrangham, R. 2019. *The Goodness Paradox: The Strange Relationship Between Virtue and Violence in Human Evolution*. New York: Pantheon.

Wright, S. 1920. The Relative Importance of Heredity and Environment in Determining the Piebald Pattern of Guinea-Pigs. *Proceedings of the National Academy of Sciences of the United States of America* 6(6): 320–332.

Wright, S. 1921. Correlation and Causation. *Journal of Agricultural Research* 20(7): 557–585.

Wright, S. 1922. Coefficients of Inbreeding and Relationship. *The American Naturalist* 56(645): 330–338.

Wright, S. 1931a. Evolution in Mendelian Populations. *Genetics* 16(2): 97–159.

Wright, S. 1931b. Statistical Methods in Biology. *Journal of the American Statistical Association* 26(173A): 155–163.

Wright, S. 1934. The Method of Path Coefficients. *The Annals of Mathematical Statistics* 5(3): 161–215.

Wright, S. 1938. Size of Population and Breeding Structure in Relation to Evolution. *Science* 87(2263): 430–431.

Wright, S. 1943. Isolation by Distance. *Genetics* 28(2): 114–138.

Wright, S. 1949. The Genetical Structure of Populations. *Annals of Eugenics* 15(1): 323–354.

Wright, S. 1965. The Interpretation of Population Structure by *F*-Statistics with Special Regards to Systems of Mating. *Evolution* 19(3): 395–420.

Wright, S. 1968. *Evolution and the Genetics of Population. Vol. 1: Genetic and Biometric Foundations*. Chicago, IL: University of Chicago Press.

Wright, S. 1969. *Evolution and the Genetics of Populations. Vol. 2.: The Theory of Gene Frequencies*. Chicago, IL: University of Chicago Press.

Yang, R.-C. 1998. Estimating Hierarchical *F*-Statistics. *Evolution* 52(4): 950–956.

Yang, Z., and Bielawski, J.P. 2000. Statistical Methods for Detecting Molecular Adaptation. *Trends in Ecology & Evolution* 15(12): 496–503.

Yang, Z., and Nielsen, R. 2000. Estimating Synonymous and Nonsynonymous Substitution Rates Under Realistic Evolutionary Models. *Molecular Biology and Evolution* 17(1): 32–43.

Yi, X., Liang, Y., Huerta-Sanchez, E., Jin, X., Cuo, Z.X.P., et al. 2010. Sequencing of 50 Human Exomes Reveals Adaptation to High Altitude. *Science* 329(5987): 75–78.

Yong, E. 2018. The New Story of Humanity's Origins in Africa. *The Atlantic*. Available: www.theatlantic.com/science/archive/2018/07/the-new-story-of-humanitys-origins/564779/ Accessed June 2020.

Yu, N., Chen, F.-C., Ota, S., Jorde, L.B., Pamilo, P., et al. 2002. Larger Genetic Differences Within Africans Than Between Africans and Eurasians. *Genetics* 161(1): 269–274.

Yu, N., Jensen-Seaman, M.I., Chemnick, L., Ryder, O., and Li, W.-H. 2004. Nucleotide Diversity in Gorillas. *Genetics* 166(3): 1375–1383.

Zack, N. 1999. White Ideas. In *Whiteness: Feminist Philosophical Reflections*, ed. C.J. Cuomo and K.Q. Hall. Lanham, MD: Rowman & Littlefield, pp. 77–84.

Zerubavel, E. 1991. *The Fine Line: Making Distinctions in Everyday Life.* Chicago, IL: University of Chicago Press.

Zerubavel, E. 1996. Lumping and Splitting: Notes on Social Classification. *Sociological Forum* 11(3): 421–433.

Zietsch, B.P., and Santtila, P. 2013. No Direct Relationship Between Human Female Orgasm Rate and Number of Offspring. *Animal Behaviour* 86(2): 253–255.

Zuckerkandl, E., and Pauling, L. 1965. Molecules as Documents of Evolutionary History. *Journal of Theoretical Biology* 8(2): 357–366.

Zuk, M., and Travisano, M. 2018. Models on the Runway: How Do We Make Replicas of the World? *The American Naturalist* 192(1): 1–9.

Index

Page numbers in **bold (e.g., 123)** refer to figures; in ***bold/italics (e.g., 123)*** refer to tables; in *italics (e.g., 123)* refer to footnotes